Signal Transduction Protocols

METHODS IN MOLECULAR BIOLOGY™

John M. Walker, SERIES EDITOR

METHODS IN MOLECULAR BIOLOGY™

Signal Transduction Protocols

Second Edition

Edited by

Robert C. Dickson

and

Michael D. Mendenhall

Department of Molecular and Cellular Biochemistry, College of Medicine, University of Kentucky, Lexington, KY

HUMANA PRESS ✳ TOTOWA, NEW JERSEY

Production Editor: C. Tirpak
Cover design by Patricia F. Cleary.

For additional copies, pricing for bulk purchases, and/or information about other Humana titles, contact Humana at the above address or at any of the following numbers: Tel.: 973-256-1699; Fax: 973-256-8341; E-mail: humana@humanapr.com; or visit our Website: www.humanapress.com

Printed in the United States of America. 10 9 8 7 6 5 4 3 2 1

ISSN 1064-3745

E-ISBN 1-59259-816-1

Library of Congress Cataloging-in-Publication Data

Signal transduction protocols / edited by Robert C. Dickson and Michael D. Mendenhall.-- 2nd ed.
    p. ; cm. -- (Methods in molecular biology ; 284)
  Includes bibliographical references and index.
  ISBN 1-58829-245-2 (alk. paper)
  1. Cellular signal transduction--Laboratory manuals. 2. G proteins--Laboratory manuals.
3. Cell receptors--Laboratory manuals.
  [DNLM: 1. GTP-Binding Proteins--analysis--Laboratory Manuals. 2. Signal Transduction--physiology--Laboratory Manuals. QU 25 S578 2004] I. Dickson, Robert C. II. Mendenhall, Michael Dean. III. Series: Methods in molecular biology (Clifton, N.J.) ; 284.

QP517.C45S555 2004

2004005164

Preface

In 1995, *Signal Transduction Protocols*, edited by David A. Kendall and Stephen J. Hill, was published in the *Methods in Molecular Biology* series. This second edition represents an update to that previous work with an emphasis on new methodologies that have developed in the last few years. The goal, then and now, is to provide procedures written by experts with first-hand experience in a detail that goes far beyond what is generally encountered in the "methods" section of most journals and thus actually permits a particular procedure to be replicated. In addition, we have had as a secondary goal the identification of protocols for the assay of general classes of signal transduction components that, ideally, can be adapted to the assay of any member of that class. The ability to do this has resulted in large part from the use of affinity-based assays, the ease with which specific proteins can be specifically tagged, and an explosion in the availability of highly specific antibodies from commercial sources, especially antibodies raised against signaling proteins of human origin. The number of available approaches is, fortunately for those working in signaling research, far too great to fit within the confines of this volume, so hard choices as to what to include had to be made. We have attempted to avoid those procedures for which detailed instructions are readily available (such as those associated with commercial kits), or which require highly specialized, expensive, or generally unavailable apparati. This has still left us with a rich variety of approaches from which to choose and, in many cases, these choices necessarily reflect the editors' interests.

We hope that we have met our goals with the protocols in this second edition of *Signal Transduction Protocol* and that postdocs, graduate students, technicians, and researchers will find them useful in their work.

Robert C. Dickson
Michael D. Mendenhall

Contents

Contributors

NATALIE G. AHN • *Howard Hughes Medical Institute, Department of Chemistry and Biochemistry, University of Colorado at Boulder, Boulder, CO*

DOUGLAS A. ANDRES • *Department of Molecular and Cellular Biochemistry, University of Kentucky College of Medicine, Lexington, KY*

ROLAND S. ANNAN • *Proteomics and Biological Mass Spectrometry Laboratory, Department of Computational, Analytical and Structural Sciences, GlaxoSmithKline, King of Prussia, PA*

MARK T. BEDFORD • *Department of Carcinogenesis, Science Park-Research Division, The University of Texas M.D. Anderson Cancer Center, Smithville, TX*

MEGHAN E. BOYER • *Molecular and Cellular Pharmacology Program, Department of Pharmacology, University of Wisconsin Medical School, Madison, WI*

JAN BRÁBEK • *Department of Cell and Developmental Biology, Vanderbilt University School of Medicine, Nashville, TN*

PAUL C. BRANDT • *Department of Medical Pharmacology and Toxicology, Texas A&M University System Health Science Center, College Station, TX*

EMERY H. BRESNICK • *Molecular and Cellular Pharmacology Program, Department of Pharmacology, University of Wisconsin Medical School, Madison, WI*

GEORGE M. CARMAN • *Department of Food Science, Rutgers University, New Brunswick, NJ*

DONGHANG CHENG • *Department of Carcinogenesis, Science Park-Research Division, The University of Texas M.D. Anderson Cancer Center, Smithville, TX*

WIN DEN CHEUNG • *Department of Biological Chemistry, The Johns Hopkins University School of Medicine, Baltimore, MD*

DARYLL B. DEWALD • *Department of Biology, Utah State University, Logan, UT and Center for Cell Signaling, Salt Lake City, UT*

ROBERT C. DICKSON • *Department of Molecular and Cellular Biochemistry, University of Kentucky College of Medicine, Lexington, KY*

JACK E. DIXON • *Department of Pharmacology and Department of Cellular and Molecular Medicine, Chemistry and Biochemistry, University of California San Diego, La Jolla, CA*

ELAINE A. ELION • *Department of Biological Chemistry and Molecular Pharmacology, Harvard Medical School, Boston, MA*

SANDRINE EVELLIN • *Venetian Institute for Molecular Medicine, Dulbecco Telethon Institute, Padova, Italy*

MAOFU FU • *Lombardi Cancer Center, Department of Oncology, Georgetown University, Washington, DC*

MICHAEL H. GELB • *Departments of Chemistry and Biochemistry, University of Washington, Seattle, WA*

JEFFREY A. GRASS • *Department of Pharmacology, Molecular and Cellular Pharmacology Program, University of Wisconsin Medical School, Madison, WI*

HIROKO HAMA • *Department of Biochemistry and Molecular Biology, The Medical University of South Carolina, Charleston, SC*

GIL-SOO HAN • *Department of Food Science, Rutgers University, New Brunswick, NJ*

STEVEN K. HANKS • *Department of Cell and Developmental Biology, Vanderbilt University School of Medicine, Nashville, TN*

GERALD WARREN HART • *Department of Biological Chemistry, The Johns Hopkins University School of Medicine, Baltimore, MD*

KENNETH HENRY • *Assay Development Department, Sequenom Inc., San Diego, CA*

MICHAEL J. HUDDLESTON • *Proteomics and Biological Mass Spectrometry Laboratory, Department of Computational, Analytical and Structural Sciences, GlaxoSmithKline, King of Prussia, PA*

HOGUNE IM • *Department of Pharmacology, Molecular and Cellular Pharmacology Program, University of Wisconsin Medical School, Madison, WI*

KIRBY D. JOHNSON • *Molecular and Cellular Pharmacology Program, Department of Pharmacology, University of Wisconsin Medical School, Madison, WI*

CARL HIRSCHIE JOHNSON • *Department of Biological Sciences, Vanderbilt University, Nashville TN*

YUKIHITO KABUYAMA • *Department of Chemistry and Biochemistry, Howard Hughes Medical Institute, University of Colorado at Boulder, Boulder, CO*

RUTH KLUCK • *Walter and Eliza Hall Institute, Parkville, Australia*

LAURENT KURAS • *Centre de Genetique Molecularie, Centre National de la Recherche Scientifique, Gif-sur-Yvette, France*

JAEHO LEE • *Department of Carcinogenesis, Science Park-Research Division, The University of Texas M.D. Anderson Cancer Center, Smithville, TX*

CHRISTINA C. LESLIE • *Program in Cell Biology, Department of Pediatrics, National Jewish Medical and Research Center, Denver, CO, and Departments of Pathology and Pharmacology, University of Colorado School of Medicine, Denver, CO*

VALENTINA LISSANDRON • *Venetian Institute for Molecular Medicine, Dulbecco Telethon Institute, Padova, Italy*

MICHAEL D. MENDENHALL • *Department of Molecular and Cellular Biochemistry, University of Kentucky College of Medicine, Lexington, KY*

MARCO MONGILLO • *Dulbecco Telethon Institute and Venetian Institute for Molecular Medicine, Padova, Italy*

DANIEL J. NOONAN • *Department of Molecular and Cellular Biochemistry, University of Kentucky, Lexington, KY*

RICHARD G. PESTELL • *Lombardi Cancer Center, Department of Oncology, Georgetown University Washington, DC*

MATTHIAS PETER • *Swiss Federal Institute of Technology Zurich, Institute of Biochemistry, Zurich, Switzerland*

KIRSI K. POLVINEN • *Howard Hughes Medical Institute, Department of Chemistry and Biochemistry, University of Colorado at Boulder, Boulder, CO*

GLENN D. PRESTWICH • *Department of Medicinal Chemistry, The University of Utah, Salt Lake City, UT and Center for Cell Signaling, Salt Lake City, UT*

KATHERYN A. RESING • *Department of Chemistry and Biochemistry, Howard Hughes Medical Institute, University of Colorado at Boulder, Boulder, CO*

ROHAN STEEL • *Cell Biology, Research Division, Peter MacCallum Cancer Center, Melbourne, Australia*

CHITRA SUBRAMANIAN • *Department of Botany, The University of Tennessee, Knoxville TN*

GREGORY S. TAYLOR • *Department of Pharmacology, University of California San Diego, La Jolla, CA*

ANNA TERRIN • *Venetian Institute for Molecular Medicine, Dulbecco Telethon Institute, Padova, Italy*

JAVAD TORABINEJAD • *Department of Biology, Utah State University, Logan, UT*

JOSEPH A. TRAPANI • *Cancer Immunology, Research Division, Peter MacCallum Cancer Center, Melbourne, Australia*

MICHELLE L. TWAROSKI • *US Food and Drug Administration, Office of Food Additive Safety, Washington, DC*

FRANK VAN DROGEN • *Department of Molecular Biology, The Scripps Research Institute, La Jolla, CA*

THOMAS C. VANAMAN • *Department of Molecular and Cellular Biochemistry, University of Kentucky College of Medicine, Lexington, KY*

ALBRECHT G. VON ARNIM • *Department of Botany, The University of Tennessee, Knoxville TN*

CHENGUANG WANG • *Department of Oncology, Lombardi Cancer Center, Georgetown University Washington, DC*

YUNMEI WANG • *Department of Biological Chemistry and Molecular Pharmacology, Harvard Medical School, Boston, MA*

NIGEL J. WATERHOUSE • *Cancer Immunology, Research Division, Peter MacCallum Cancer Center, Melbourne, Australia*

JING WU • *Molecular and Cellular, Pharmacology Program, Department of Pharmacology, University of Wisconsin Medical School, Madison, WI*

YAO XU • *Department of Biological Sciences, Vanderbilt University, Nashville TN*

MANUELA ZACCOLO • *Venetian Institute for Molecular Medicine, Dulbecco Telethon Institute, Padova, Italy*

NATASHA ELIZABETH ZACHARA • *Department of Biological Chemistry, The Johns Hopkins University School of Medicine, Baltimore, MD*

FRANCESCA ZAPPACOSTA • *Proteomics and Biological Mass Spectrometry Laboratory; Department of Computational, Analytical and Structural Sciences; GlaxoSmithKline, King of Prussia, PA*

XUEPING ZHANG • *Department of Oncology, Lombardi Cancer Center, Georgetown University Washington, DC*

PENGBO ZHOU • *Department of Pathology and Laboratory Medicine, Weill Medical College of Cornell University, New York, NY*

1

Making Protein Immunoprecipitates

Elaine A. Elion and Yunmei Wang

Summary

A wide variety of methods used in the study of signal transduction in eukaryotes rely on the ability to precipitate proteins from whole cell extracts. Immunoprecipitation and related methods of affinity purification are routinely used to assess binding partner interactions and enzyme activity in addition to the size of a protein, rates of protein synthesis and turnover, and protein abundance, thus making it a mainstay of a wide variety of protocols. This chapter will provide starting-point methods for immunoprecipitation of proteins under denaturing and nondenaturing conditions and the detection of protein-protein interactions by co-precipitation. The Notes section gives recommendations on how to troubleshoot potential problems that can arise while doing these methodologies.

Key Words: Immunoprecipitation; precipitation; co-immunoprecipitation; co-precipitation; immune complex; affinity purification; affinity matrix; whole-cell extracts; *Saccharomyces cerevisiae*.

1. Introduction

Protein precipitation involves the formation of protein aggregates out of solution followed by their recovery by centrifugation. A variety of methods can be used to make proteins aggregate out of aqueous solution, including nonspecific methods such as salt and trichloroacetic acid, and specific methods directed against a particular protein, such as antibodies or other affinity matrixes. When antibodies are used, the method is termed immunoprecipitation; when other affinity-based methods are used, the method is termed precipitation. This

From: *Methods in Molecular Biology, vol. 284:*
Signal Transduction Protocols
Edited by: R. C. Dickson © Humana Press Inc., Totowa, NJ

methodology is often used to detect or confirm physical interactions between two proteins. When the method is for detecting physical associations, it is referred to as co-immunoprecipitation.

There are three major reasons to incorporate immunoprecipitation of proteins into analysis. First, it is a simple and rapid method of affinity purification. Second, it is adaptable and can be done on either small or large scales. Third, it is amenable for the detection of both strong and weak physical interactions between proteins that may or may not withstand the rigors of purification methods involving substantial dilution of the initial cell extract.

Many protocols are available for immunoprecipitation and co-precipitation. All immunoprecipitation protocols follow a common series of ordered steps: (1) lysing cells and preparing cell extracts, (2) binding the antibody to protein through a specific antigen on the protein that is recognized by the antibody, (3) precipitating the antibody-antigen complex, and (4) washing the precipitate to remove nonspecific proteins. An immunoprecipitation can be done under native conditions that preserve enzyme activity and associations with other proteins or it can be done under more stringent conditions that are likely to reduce nonspecific interactions with the protein in question, but may abolish enzyme activity and protein complexes. This chapter will provide a basic methodology that can be adjusted to be more or less stringent depending on the experimental considerations.

Once an immunoprecipitate has been isolated, it can be used directly in an enzyme assay or after the protein has been dissociated from the antibody through a solution-based method or by gel electrophoresis. After gel electrophoresis, the protein is detected typically either by virtue of its being radiolabeled prior to cell lysis or by immunoblot analysis with either the same or different antibody. A candidate associated-protein is typically detected by immunoblot analysis in analytical studies. When immunoprecipitation is done on a large enough scale, it is possible to detect the immunoprecipitated protein and potential binding partners by silver staining or Coommassie Blue staining of polyacrylamide gels. The resolution of detection of an average-sized protein is approx 1–10 ng/band for silver stain detection and approx 0.1–1 μg/band for Coommassie Blue detection.

Of the many strategies possible, this section will describe: (1) options available for detecting proteins, (2) basic protocol for making whole-cell extracts, (3) basic protocol for immunoprecipitation under native conditions, (4) immunoprecipitation under more stringent conditions, (5) co-immunoprecipitation, (6) controls to test specificity of interaction, and (7) notes for troubleshooting. For an in-depth discussion on the generation and use of antibodies, see Harlow and Lane (*1,2*). For an in-depth review of co-precipitation and other approaches to detect protein-protein interactions, see Phizicky and Fields *(3)*.

1.1. Detecting the Protein(s) in Question

How well an immunoprecipitation will work depends on a variety of factors, the most important being the affinity of the antibody to the antigenic site on the protein. Antibody affinity can vary over a wide range, but for an immunoprecipitation to work efficiently the affinity of the antibody to the antigen should be at least 10^7 mol^{-1} to 10^9 mol^{-1} *(1)*. The simplest way to improve the detection of an antibody-antigen complex is to increase the concentration of the antibody and the antigen. This will only be effective under conditions in which the antibody is not saturating, which must be empirically determined by doing a titration of the amount of antibody for a given amount of antigen in a given reaction volume. When the quantity of antibody is limited, it is easiest to reduce the reaction volume. When an epitope-tag is used, it is also possible to improve detection by inserting multiple copies of the tag onto the protein to allow for multivalent binding by the antibody. To determine the amount of antibody in your preparation, run some of it on a sodium dodecyl sulfate (SDS)-polyacrylamide gel and compare the intensity of the heavy and light chains to standard controls. If the antibody is pure, then one can determine its concentration by its absorbance at 280 nm using the relationship 1 OD = approx 0.75 mg/mL purified antibody.

The first step is to generate an antibody to the protein in question. Information for generating antibodies can be found in Harlow and Lane *(1,2)*. Alternatively, a protein can be tagged in a variety of ways to allow their detection with commercially available antibodies against the epitope tag or other affinity reagents. The tagged proteins are then introduced into the host organism using expression vectors. All tagged proteins must be assessed for function in vivo. A frequently-used option is to add a short peptide or eptiope that is recognized by a commercially available high-affinity monoclonal antibody (MAb). The epitope is added typically at the amino or carboxyl terminus, although internal positions that do not disrupt function can also be used. Two frequently-utilized epitopes are derived from the influenza hemagglutinin protein (HA) and human c-Myc; both are recognized by high-affinity MAbs 12CA5 and 9E10, respectively *(4)*. However, others such as the leader peptide of gene 10, product of bacteriophage T7 (FLAG 5,6) are also available (BioSupplyNet Source Book). The choice of the epitope may be dictated by its amino-acid composition. It is often useful to insert tandem copies of the epitope in order to increase sensitivity. The number of tandem copies can range widely from one *(7)* to several (e.g., *3,8*) to many (e.g., *9*).

Proteins can also be fused to small proteins or peptides that have high affinity to small molecules that can be attached to solid support. This is a particularly valuable approach when the protein to be precipitated co-migrates with immunoglobulin heavy or light chains in a SDS-polyacrylamide gel. Such alternative tagging methods include fusion to glutathione-*S*-transferase (GST) to allow purification by a glutathione affinity matrix or fusion to maltose binding protein (MBP) to allow purification by a maltose affinity matrix. An excellent reference for identifying sources of commercially available antibodies and approaches to tagging proteins can be found in the *BiosupplyNet Source Book. The American Type Culture Collection* and *European Collection of Cell Cultures* can also be resources for hybridoma cell lines.

The second step to a successful co-precipitation is generating whole-cell extracts in which the yield and activity of the proteins you wish to analyze is optimal, using lysis buffer conditions that permit recognition of the proteins by the antibody or affinity matrix. In general, the lysis buffer conditions are not very different from the immunoprecipitation conditions. The yield of total protein in a whole cell extract is not always a reliable indicator of the relative yield and activity of specific proteins, so it is wise to verify both parameters at the onset of an experiment before proceeding on to the immunoprecipitation. Once the extracts are prepared, the co-precipitation can be done within 3–4 h and be ready to load on a gel for immunoblot analysis.

1.2. Basic Protocol for Making Whole-Cell Extracts

Yield and activity can be affected by a number of factors. Small variations in the relative amounts of salt and detergents in the lysis buffer can have large effects on yield and activity, as can the speed and efficiency of cell breakage. Both factors are particularly important for less soluble proteins that associate with macromolecular structures such as membranes or cytoskeleton. In addition, global inhibition of proteolysis through the inclusion of multiple classes of protease inhibitors may be essential. It is recommended that the investigator begin by comparing a series of small-scale extract preparations that vary the amount of salt and nonionic detergent. As a starting point, a basic lysis buffer might contain a buffering agent (such as 25–50 mM Tris-HCl, pH 7.5), a small amount of nonionic detergent (such as 0.1% Triton X-100), some salt (such as 100–250 mM NaCl), a reducing agent (such as 1 mM dithiothreitol [DTT] and 5–10% glycerol as a stabilizer. The lysis buffer should also contain protease inhibitors. Protease inhibitor cocktails are also commercially available. A reasonable starting point would be to include 5 μg/mL each chymostatin, pepstatin A, leupeptin, and antipain as well as 1 mM phenylmethylsulfonylfluoride and 2 mM benzamidine. Ethylene glycol-bis (beta-aminoethyl-ether)-N,N,N',N'-tetraacetic (EGTA) is also commonly included (at approx 15 mM) to chelate divalent metal ions that are essential for metalloproteases. Because EGTA will also inhibit other metal-dependent enzymes, it may be omitted, or combined with the addition of a needed metal ion to the lysis buffer and/or substituted with ethylenediamine tetracetate (EDTA). If the phosphorylation state of the proteins in question is important, a mixture of phosphatase inhibitors should also be included in the lysis buffer. A starting mixture could be 0.5 mM vanadate (0.25 mM each meta- and ortho-vanadate or 0.5 mM sodium vanadate, pH 7.4), 10 mM sodium fluoride (NaF), 10 mM β-glycerol phosphate. Simple modifications of this initial buffer include varying the amount of NaCl (from 0 to 500 mM) and the amount of Triton X-100 (from 0 to 1%).

Total protein concentration in the whole-cell extract generally is assayed using the Bio-Rad protein assay and calculating protein concentration. Extracts should be tested for the amount of each specific protein by immunoblot analysis, analyzing 25–100 μg of total protein. In general, it is best to test for the presence of a second protein (such as a housekeeping enzyme, cytoskeletal, or ribosomal protein or a previously defined component in the pathway being studied) for a positive control of the immunoblot and normalization. The amount of specific protein in the whole-cell extract is then compared to the amount that is recovered by precipitation with an affinity matrix.

A general small-scale glass-bead breakage protocol for preparing a basic whole-cell extract is described below as a starting point, with suggested ranges of salt and nonionic detergent concentrations for initial variations of this protocol. It is recommended that the investigator compare several combinations of salts and nonionic detergent. This method can be scaled up and used with an automated bead beater. The investigator may choose to compare the glass-bead breakage method described below to the liquid nitrogen-grinding method, which keeps the cells colder and may break the cells more efficiently *(10)*.

2. Materials
2.1. Cell-Free Extract Preparation

All solutions used for extract preparation and immunoprecipitation are either filter sterilized or autoclaved.

1. Yeast cells.
2. Ice bath.
3. Dry ice/ethanol bath or liquid nitrogen bath.
4. Autoclaved ice-cold water.
5. Acid-washed glass beads.
6. Ice-cold lysis buffer.
7. 50-mL conical plastic disposable tubes.
8. 15-mL conical plastic disposable tubes.

9. Vortexer.
10. Timer.
11. Microcentrifuge tubes and centrifuge.
12. Acid washed glass beads: soak 0.25–0.600 microns glass beads (Sigma G-8772) for 3 h in concentrated nitric acid. Wash beads thoroughly with large amounts of water, test pH of the wash water, and continue washing until pH is neutral. If necessary, wash several times with 2 *M* Tris-HCl, pH 8.0, or with 50X TE to raise the pH, then wash again several times with water to remove the buffer. Glass-distilled or deionized water should be used in the final washes. Bake beads for 4 h in a baking oven until dry. Acid-washed beads are also available from Sigma.
13. Lysis buffer pre-chilled in an ice bath (*see* **Note 1**): 25 m*M* Tris-HCl, pH 7.5, 15 m*M* EGTA, 1 m*M* EDTA, 150 m*M* NaCl (or in the range of 50–250 m*M* NaCl), 0.1% Triton X-100 (or in the range of 0.1–1.0% Triton X-100), 10% glycerol, 1 m*M* NaN$_3$, 1 m*M* DTT, 1X protease inhibitor mix, 1 m*M* phenylmethylsulfonyl fluoride (PMSF), 2 m*M* benzamidine, phosphatase inhibitors. Add the protease inhibitor mix, benzamidine, PMSF, DTT, and phosphatase inhibitors from concentrated stock solutions prior to use. PMSF is labile in aqueous solution and should be added immediately before use.
14. Phosphatase inhibitors: The phosphorylation state of the proteins in question is frequently important, therefore, a mixture of phosphatase inhibitors should be included in the lysis buffer. A good starting mixture is 1 m*M* sodium vanadate (from equal amounts of meta and ortho forms of vanadate), 10 m*M* NaF, 10 m*M* β-glycerol phosphate. Okadaic acid can also be added if needed.
15. 1000X Protease inhibitor mix: 5 mg/mL chymostatin, 5 mg/mL pepstatin A, 5 mg/mL leupeptin, 5 mg/mL antipain. Dissolve protease inhibitors in dimethyl sulfoxide (DMSO) and store in aliquots at −20°C. Premade mixtures of protease inhibitors are also available commercially.
16. 250X PMSF: 0.25 *M* PMSF in 95% ethanol. Make fresh.
17. Bio-Rad D$_C$ Protein Assay Reagent (Bio-Rad Laboratories, Hercules, CA).

2.2. Immunoprecipitation of Proteins

1. Antibody specific to the protein of interest. Polyclonal antisera, ascites fluid, and culture supernatant of a hybridoma that secretes a MAb can all be used for immunoprecipitation. Use approx 1 µg of antibody per immunoprecipitation. Increase this amount several-fold for antibodies with low affinity to antigen. This has to be determined empirically. Antibody suppliers can be located in the BiosupplyNet Source Book (www.biosupplynet.com) published yearly in collaboration with Cold Spring Harbor

Laboratories. For information on its contents, telephone 516-349-5595, or fax 516-349-5598, or E-mail: info@biosupplynet.com.

2. Protein A Sepharose and/or Protein G Sepharose. Recipe can be scaled up or down. Hydrate 1.5 g of protein A or G Sepharose beads in 30 mL of 50 m*M* Tris-HCl, pH 7.5, for 1–2 h on ice. Pellet beads by gravity or very gentle centrifugation (1 min at 1000 *g*) and then wash four times with immunoprecipitation buffer that lacks (the expensive) protease inhibitor mix and contains 1 m*M* sodium azide. Resuspend the beads in 15 mL of this same buffer to yield a final slurry concentration of approx 100 mg/mL (approx 50% of total volume are the beads). The slurry is stable for months when stored at 4°C.

3. Lysis buffer (*see* **Subheading 2.1.**) with protease and phosphatase inhibitors but without glycerol or NaCl.

4. 5 *M* NaCl.

5. 80% glycerol.

6. Microfuge tubes.

7. 2X loading buffer for SDS-PAGE: 125 m*M* Tris-HCl, pH 6.8, 140 m*M* SDS, 20% (v/v) glycerol, 2% (v/v) β-mercaptoethanol, 10 µg/mL bromphenol blue.

8. Refrigerated centrifuge.

9. Ice bucket.

10. Chilled buffers and centrifuge tubes.

2.3. Immunoprecipitating Proteins Under Stringent Conditions

1. Radio immunoprecipitation assay (RIPA) buffer: 150 m*M* NaCl, 1.0% Triton-X-100, 0.5% sodium deoxycholate, 0.1% SDS, 50 m*M* Tris-HCl, pH 8.0.

3. Methods

3.1. Growth and Harvesting of Cells

1. Cells should be grown under optimal conditions for the proteins under study. In the absence of specific information, it is advisable to harvest logarithmically-growing cells.

2. Grow yeast cells to an A600 of 0.5–1.3 in a volume of medium that will yield 100 A600 Un of cells using appropriate media, temperature, and aeration.

3. Pellet cells by 5 min centrifugation at 5000 *g* in a centrifuge and rotor chilled to 4°C.

4. Wash cells once in 30 mL of ice water. Place pellet cells in a 50-mL sterile plastic conical tube that has been pre-chilled in an ice bucket.

5. Drain liquid rapidly and thoroughly from the pellet and immediately immerse the tube into a dry ice/ethanol, or in liquid nitrogen bath.

6. Store frozen pellets at −80°C or immediately make cell-free extracts.

3.2. Preparation of Cell-Free Extracts

Speed and maintenance of ice-cold conditions are key components of good extract preparation. Extract preparation is done using an ice bucket either at room temperature or in the cold room if the proteins are particularly labile.

1. Label one 15-mL conical tube and four microfuge tubes per sample. Prechill all tubes in an ice bath.
2. Thaw pellets in an ice bath. Begin extract preparation while pellets are still partially frozen (*see* **Note 2**).
3. Add 1 mL lysis buffer and transfer-cell suspension to 15-mL conical plastic tube. Add chilled glass beads to just below the meniscus (*see* **Note 3**).
4. Strongly vortex cells for five 30-s pulses, chilling on ice between pulses. Keep vortexer on the highest setting. When multiple samples are being processed, use two vortexers.
5. Add 0.25 mL more lysis buffer. Vortex sample for 30 s. Check cells under microscope for complete or nearly complete lysis (*see* **Note 4**). Vortex again if necessary.
6. Centrifuge sample 5 min at 5000g in a centrifuge chilled to 4°C.
7. Transfer supernatant to a new microfuge tube and do not be concerned if a few beads are carried along. The supernatant fluid will be turbid if the protein concentration is high.
8. Centrifuge supernatant in a microfuge chilled to 4°C at the fastest speed for 10 min.
9. Transfer supernatant to a new microfuge tube. Cap tube and invert gently to mix contents. Distribute sample into three microfuge tubes. Reserve 5 µL on ice to quantify the protein concentration.
10. Freeze samples at –80°C either directly or after prefreezing in a dry ice/ethanol bath or process immediately for immunoprecipitation.
11. Assay total protein of reserved extract using the Bio-Rad protein assay and calculate protein concentration using bovine serum albumin (BSA) to generate a standard curve. The total protein yield is generally 5–10 mg.
12. Analyze 25–100 µg of protein by immunoblotting to determine if the protein in question is present and readily detected.

3.3. Immunoprecipitation

3.3.1. Immunoprecipitating Proteins Under Native Conditions

In this method, the antibody is added to the whole-cell extracts on ice and then the antibody-antigen complex is precipitated out of solution using protein A or protein G coupled to Sepharose. Protein A and protein G are bacterial cell-wall proteins that bind tightly to a region of the Fc domain in antibodies. The interaction is sensitive to pH and is reduced under conditions of low pH. Protein A and protein G bind with different affinities to antibody subtypes from different animals. Check Harlow and Lane *(1)* for a listing of their relative affinities. If in doubt, use an equal mixture of Protein A and Protein G Sepharose. Protein A Sepharose and Protein G Sepharose can be purchased separately or as mixture.

The amount of antibody and whole-cell extracts in the immunoprecipitation can vary widely, depending on the expression level of the protein and the affinity of the antibody, and needs to be determined empirically. A starting point is to use 0.25–1 mg whole-cell extract with approx 1–3 µg antibody. In general, 1 µg antibody is approx 0.5–1 µL of a polyclonal antiserum, approx 50 µL of a hybridoma culture supernatant or approx 0.5 mL of ascites fluid that contains a single MAb *(1)*.

1. Prepare samples in microfuge tubes on ice: Mix 0.25–1 mg of the whole-cell extract with 1 µg of antibody in a final volume of 0.5 mL, using the lysis buffer without NaCl to bring up the final volume to 0.5 mL. Adjust NaCl concentration to 100 m*M* NaCl and glycerol concentration to 5%. Adjust lysis buffer by adding divalent cation or removing EGTA if necessary for activity of the protein in question (*see* **Note 5**).
2. Invert tube gently several times and incubate on ice for 90 min with occasional tube inversion to allow antibody to bind to antigen.
3. Centrifuge in a chilled microfuge for 10 min to pellet nonspecific aggregates (*see* **Note 6**).
4. Transfer supernatant to a new chilled microfuge tube.
5. Add 30 µL of the protein A or G-Sepharose slurry. Be sure to evenly suspend the slurry before distributing it to the samples.
6. Rotate tube gently at 4°C for 60–90 min to allow the antigen-antibody complex to bind to the protein A or G Sepharose beads (*see* **Note 7**).
7. Gently pellet protein A/G sepharose by spinning 30 s at 2000 *g* in a chilled centrifuge (*see* **Note 8**).
8. Wash pellet twice with 1 mL lysis buffer that has the same amount of NaCl as the immunoprecipitation, then wash twice with lysis buffer that lacks NaCl. For each wash, gently invert tube three times before pelleting. Carefully remove supernatant without disturbing the beads. It is not necessary to have protease inhibitors present in the lysis buffer during the washes (*see* **Note 9**).
9. After the last wash, aspirate away as much liquid as possible without touching the beads. Use the beads directly in an enzyme assay or run the sample on a SDS polyacrylamide gel, or add an equal volume of 2X loading buffer

(approx 25 μL). Samples can be frozen at −80°C or loaded immediately on a SDS-PAGE gel as follows: boil samples for 5 min (or incubate samples for 10 min in a 100°C heating block). Vortex the boiled sample and then centrifuge it briefly in a microfuge to pellet beads before loading onto the SDS-PAGE gel (*see* **Note 10**).

3.3.2. Immunoprecipitating Proteins Under Stringent Conditions

In instances of a great deal of nonspecific background or in cases when the antibody does not recognize a nondenatured epitope, it may be necessary to use more stringent or denaturing conditions. This is accomplished by using RIPA buffer. Follow the previously-outlined protocols for making extracts and immunoprecipitation but use RIPA whenever lysis buffer is needed.

3.3.3. Co-Immunoprecipitating Proteins

Co-immunoprecipitation (Co-IP) is a simple and useful method to detect an interaction between two or more proteins *(1–4)*. Under gentle enough conditions, it is possible to co-immunoprecipitate one or more proteins with the protein that is recognized by an antibody. The standard immunoprecipitation conditions presented can be used along with specific antibodies for each of the proteins in question. Follow the previously-outlined methods and perform the Co-IP in duplicate; run the samples on an SDS-PAGE gel along with whole-cell extracts, and probe by immunoblot analysis for the presence of associated proteins with appropriate antibodies.

3.3.3.1. CO-IP CONTROLS

Controls are essential to verify that the antibody is specific and that a potential protein-protein interaction is specific. For Co-IP experiments, it is important to immunoprecipitate each protein in the absence of the other to determine whether their presence in an immune complex is owing to specific interactions with the protein being bound by antibody or to nonspecific binding to the beads. Additional controls to confirm specificity of the antibody are also needed. These controls are the simplest to set up when proteins are tagged, because one can compare extracts from cells that do not express the tagged protein.

If antibodies to native proteins are being used in a genetically tractable organism, then one can compare extracts made from strains harboring deletions of the proteins in question (obviously only possible if the deletions do not cause inviability) to test for the specificity of the antibody and the interaction. If deletion mutations are not possible, a commonly used approach is to show that the pre-immune serum or an antibody not known to be specific to either of the proteins in question, does not co-precipitate them in a parallel experiment. However, the latter control does not rule out the possibility that the antibody is precipitating the protein in question through an indirect association.

In Co-IP, it is also essential to compare the amount of protein co-precipitated with the amounts of the proteins in question in the whole-cell extract. This allows one to determine whether apparent differences in the ability of two proteins to co-precipitate from sample to sample is owing to differences in abundance of the proteins in the whole-cell extracts. This control is particularly important when an interaction has been established and the investigator wishes to search for regulatory changes in association apart from changes in abundance.

3.3.3.2. TROUBLESHOOTING

In the absence of detecting an interaction, it is worthwhile to try less-stringent conditions or fewer washes after the immunoprecipitation. In addition, it may be necessary to avoid any dilution of the whole-cell extract, by adding the protein A/G-Sepharose directly after an initial clarification centrifugation and using smaller wash volumes. When performing a co-precipitation, it is important to precipitate from both directions (i.e., individually precipitating Protein 1 and Protein 2, and testing for the presence of Protein 2 and Protein 1, respectively), because it is possible the interaction will only be detected in one direction. An inability to detect an interaction in one direction could be owing to a variety of factors, including obstruction of an interaction by the binding of the antibody, or other affinity agent, or differences in pool size representation of each protein. For example, Protein 1 may bind to many proteins besides Protein 2, whereas most of Protein 2 binds to Protein 1. In this scenario, detection of their association may be most efficient when Protein 2 is precipitated.

In addition, it may be necessary to increase the expression levels of the proteins in question to be able to readily detect them in a co-precipitation. A range of expression levels is recommended, because too high a level of expression of proteins can lead to unregulated interactions. Alternatively, one can scale up the Co-IP and use more than 0.5–1 mg of whole-cell extract. Larger-scale extract preparations may be necessary to generate more concentrated extracts. In cases of failure owing to low abundance of the proteins in the host organism, one can overexpress one of the two proteins in another host (such as *Escherichia coli*), concentrate this protein by pre-immobilization on the appropriate affinity matrix, and then incubate the affixed protein with extracts from the host organism (*see* **Note 11**).

4. Notes

1. The EGTA is added to chelate divalent cations that are essential for metalloproteases. The 15 mM concentration of EGTA has been found to be optimal for studies involving mitogen-activated protein kinase (MAPK). However,

because EGTA will also inhibit other metal-dependent enzymes it may be omitted, used at a lower concentration (e.g., 1 mM), or combined with the addition of a needed metal ion to the lysis buffer.

2. The yield can sometimes improve if the thawed sample is refrozen in liquid nitrogen and then rethawed prior to extract preparation.

3. Predetermine the amount of glass beads to add for the given volume, mark that amount on a microfuge tube, and use this tube to scoop equal amounts of beads into each sample.

4. This back-extraction step can be skipped for more concentrated extracts.

5. Greater variability in pipetting occurs with small volumes. When preparing multiple samples, predilute the antibody in immunoprecipitation buffer so that a bigger volume can be added to each sample to ensure equal distribution.

6. This step can decrease the presence of nonspecific background and is particularly important in instances when the immunoprecipitate is being used for an enzyme assay. If background is not a problem, this step can be skipped.

7. Samples can be gently rocked instead of rotated; however, we find this to be a less efficient way to recover the antigen-antibody complex. The 60-min incubation step can be lengthened to several hours to increase the amount of antigen captured. Increasing the length of time can be a tradeoff owing to the potential for sample degradation.

8. Save the supernatant in order to determine the efficiency of immunoprecipitation of the protein in question.

9. This is a variable part of the protocol that depends on the affinity of the antibody-antigen complex. The stringency of the washes can be increased by adding more salt or by having a fixed amount of salt present in a greater number of washes. It is helpful to do a final wash without added NaCl prior to running samples on a SDS-polyacrylamide gel.

10. Include aliquots of the initial whole-cell extract and the supernatant from the immunoprecipitation for comparison to determine what fraction of the total protein is immunoprecipitated and as a positive control for the immunoblot.

11. The most important objective is to generate as great a signal to noise ratio and avoid problems of background. A variety of parameters can be changed to enhance immunoprecipitation. Optimization of the precipitating antibody is one possibility. Protein A-sepharose and Protein G-sepharose should give results comparable to coupling the antibody to sepharose. However, direct coupling of the antibody to sepharose may lead to less background and more quantitative precipitation. In addition, varying the ratio of antibody to whole-cell extract and the total amount of whole-cell extract utilized is suggested to determine the optimal amount of antibody that gives the most precipitation with the least amount of background.

Affinity purification of the antibody may be necessary if the antibody immunoprecipitates additional cross-reacting proteins. Additional approaches can be taken to minimize background. First, the amount of salt and detergent can be increased in both the co-precipitation and the washes to reduce nonspecific binding. Second, increasing the number of washes may also help, although it may reduce the amount of specific protein that remains associated. Third, the whole-cell extract can be preincubated with protein A/G sepharose to remove nonspecific proteins that bind to the solid support. Fourth, both the lysis buffer and the co-precipitation buffer can be supplemented with 1% BSA to reduce the amount of nonspecific binding to the affinity matrix. Finally, one can increase the expression levels of the proteins in question to generate a stronger signal that is above the background binding.

It may be possible to produce a whole-cell extract that is enriched for the proteins in question, such as preparing a nuclear extract if the proteins are known to be in the nucleus. Better clarification of the cell extract can be done by precentrifugation at 100,000 g and the extracts can be directly used for co-precipitation without an intervening freezing step, which can increase the amount of protein precipitation. In instances where one of the proteins binds nonspecifically to sepharose, the substitution of an agarose-based affinity matrix may solve the problem. In this instance, it may be necessary to generate a different set of reagents to precipitate the proteins in question (i.e., different antibodies and/or protein tags).

References

1. Harlow, E. and Lane, D. (1988) *Antibodies: A Laboratory Manual.* Cold Spring Harbor Laboratory Press, Cold Spring Harbor, NY.
2. Harlow, E. and Lane, D. (1998) *Using Antibodies: A Laboratory Manual.* Cold Spring Harbor Laboratory Press, Cold Spring Harbor, NY.
3. Phizicky, E. M. and Fields, S. (1995) Protein-protein interactions: methods for detection and analysis. *Microbiolog. Rev.* **59,** 94–123.
4. Kolodziej, P. A. and Young, R. A. (1991) Epitope Tagging and Protein Surveillance, in *Methods in Enzymology*, vol. 194 Academic Press, San Diego, CA, pp. 508–519.
5. Witzgall, R., O'Leary, E., Bonventure, J.V. (1994) A mammalian expression vector for the expression of GAL4 fusion proteins with an epitope tag and histidine tail. *Anal. Biochem.* **2,** 291–298.
6. Knappik, A. and Pluckthun, A. (1994) An improved affinity tag based on the FLAG peptide for the detection and purification of recombinant antibody fragments. *Biotechniques* **17,** 754–761.
7. Field, J., Nikawa, J., Broek, D., et al. (1988) Purification of a *Ras*-responsive adenylyl cyclase complex from *Saccharomyces cerevisiae* by use of an epitope addition method. *Mol. Cell Biol.* **8,** 2159–2165.

8. Tyers, M., Tokiwa, G., and Futcher, A.B. (1993) Comparison of the *Saccharomyces cerevisiae* G1 cyclins: Cln3 may be an upstream activator of Cln1, Cln2, and other cyclins. *EMBO J.* **11,** 1773–1784.

9. Feng, Y., Song, L.Y., Kincaid, E., et al. (1998) Functional binding between Gbeta and the LIM domain of Ste5 is required to activate the MEKK Ste11. *Curr. Biol.* **8,** 267–278.

10. Sorger, P. K. and Pelham, H. R. (1987) Purification and characterization of a heat-shock element binding protein from yeast. *EMBO J.* **6,** 3035–3041.

2

Signal Transduction Inhibitors in Cellular Function

Maofu Fu, Chenguang Wang, Xueping Zhang, and Richard G. Pestell

Summary

Signal transduction pathways mediate cell–cell interactions and integrate signals from the extracellular environment through specific receptors at the cell membrane. They play a pivotal role in regulating cellular growth and differentiation and in mediating many physiological and pathological processes, such as apoptosis, inflammation, and tumor development. The mitogen-activated protein kinases (MAPKs) constitute a cascade of phosphorylation events that transmit extracellular growth signals through membrane-bound *Ras* to the nucleus of the cell. In this chapter, detailed protocols for analyzing the kinase activities of the key components of the MAPKs pathway—MEK1, ERK1, JNK, and p38 MAPK—are described. A brief introduction to the chemical inhibitors to the MAPKs pathway is provided in the method section of each kinase assay. Inhibitors of other signaling pathways are summarized in **Table 1**. The reporter assay of cyclin D1, a key downstream target gene of MAPKs pathway, is also described in detail.

Key Words: Signal transduction; MAPKs; chemical inhibitors; in vitro kinase assay.

1. Introduction

In multicellular organisms, gene expression is tightly controlled within the cell. Extracellular molecules, such as hormones, growth factors, and cytokines, communicate with the nuclear gene regulatory machinery through the interaction with receptors on the cell membrane and initiate intracellular signaling cascades. Signal transduction can occur between cells and within a single cell. In cancer cells, the integrity of signal transduction cascades is often disrupted by

From: *Methods in Molecular Biology, vol. 284:*
Signal Transduction Protocols
Edited by: R. C. Dickson © Humana Press Inc., Totowa, NJ

Table 1. Signal Transduction Pathways and Inhibitors

Signal Transduction Pathways	Inhibitors	References
Tyrosine Kinase Growth Factor Receptors		
HER2/Neu inhibitor	Herceptin (anti-HER2 antibody)	*(53,54)*
EGF receptors	IMC-C225 (anti-EGFR antidody)	*(55,56)*
EGFR-Tyrosine kinase	ZD1839, pyridopyrimidines	*(54,57,58)*
***Ras* Signaling**		
Inhibitors of *Ras* farnesyltransferase		
FPP analogues		
CAAX peptide analogues	BZA-5B, L-739, Cys-4-ABA-Met FT1-276, SCH44342, SCH66336	*(57–59)*
Bisubstrate inhibitors	BMS-182878, BMS-184467	*(57–59)*
Inhibitors of the Raf protein kinases	ISIS5132, BAY43-9006	*(57–59)*
Mitogen-Activated Protein Kinase Pathways		
Inhibitors of MEK	PD184352, PD098059, U0126, R009-22110	*(17,20,21, 23–27)*
Inhibitors of ERK1 and ERK2	PD098059, E64D, calpeptin	*(5,28)*
Inhibitors of JNKs	SB600125	*(33,34)*
Inhibitors of p38 kinases	SB203580, SB202190, SB242235, SB239063, SB220025, SB202474, SC68376, FR167653	*(36,37, 39–44,46)*
Inhibitors of PI3-Kinase Signaling Pathways	Wortmannin, LY294002	*(60,61)*
Protein Phosphatases Inhibitors	Microcystines, calyculins, cantharidin	*(67)*
Proteasome Inhibitors		
Peptide aldehydes	ALLN, MG132, PSI, MG115	*(63)*
Peptide boronates	MG262	*(63)*
Nonpeptide inhibitors	Lactacystin	*(63)*
DCI	3,4-DCI	*(63,64)*
Peptide vinyl sulfones	NLVS, YL3YS	*(63)*
Epoxyketones	Epoxomicin, eponemycin, Ac-hFLFL-epoxide	*(63,65)*
Bivalent inhibitors	Polyoxyethylene	*(66,67)*
Natural compound inhibitors	TMC-95A, gliotoxin, EGCG	*(63,68)*
Histone Deacetylase inhibitors		
Short-chain fatty acids	Butyrates	*(70,75)*
Hydroxamic acids	Trichostatin A, oxamflatin	*(73)*
Cyclic peptides	Trapoxin A, FR901228, apicidin	
Benzamides	MS-27275	*(78,80)*
SIR2 inhibitors	Nicotinamide, splitomicin	*(78,79)*

gene mutations or altered gene expression. Constitutive activation of signaling cascades contributes to uncontrolled cellular growth *(1,2)*.

The elucidation of signal-transduction pathways in cancer cells, both at the proteomic and the genomic level, has provided the basis of rational screening for chemical inhibitors and targeted drug design. New therapeutics act at specific steps of the signal transduction cascade. The inhibitor may interfere with signaling processes by blocking binding of a ligand to a cell-surface receptor, by inhibiting the receptor tyrosine kinase (RTK) activity of a receptor or by inhibiting downstream components of a signaling pathway *(3)*.

Protein kinases are enzymes that covalently attach phosphate to the side chain of serine, threonine, or tyrosine of specific proteins inside cells. Mitogen-activated protein kinases (MAPKs) are a family of protein kinases whose function and regulation have been conserved during evolution from unicellular organisms to complex organisms, including humans. Multicellular organisms have three subfamilies of MAPKs, namely ERK, JNK, and p38 protein Kinases, which control a vast array of physiological processes *(4)*. The extracellular signal-regulated kinases (ERKs) are involved in the control of cell proliferation and division. The c-Jun amino-terminal kinases (JNKs) are critical regulators of apoptosis and gene transcription. The p38 MAPKs are activated by inflammatory cytokines and environmental stresses *(5–7)*.

Signal transduction inhibitors have been developed to diverse signaling pathways. Limitations of using such inhibitors have been the temporal and spatial control of drug delivery. More recently approaches have been developed to target inhibitors to discrete subcellular compartments, or to activate compounds at a single-cell level using chemical "caging" *(8)*. For example, it has been possible to screen for compounds that are selectively taken up by mitochondria and inhibit growth of tumor cell targets, in part owing to the altered mitochondrial membrane potential of malignant cells *(9)*. Chemical "caging" of small molecules (e.g., ATP, NO, etc.), peptides and proteins, has been useful to define temporal relationships in biochemically mediated processes and to delineate the role of individual proteins in biological phenomena.

The recent application of caging ligands to regulate gene expression will provide important new insights into the mechanisms governing signal transduction in vivo *(8)*. Using light to activate caged molecules at the single-cell level will allow the dissection of intracrine and paracrine signaling at an organismal level. Future development in signal transduction research will integrate microarray technology at a genome-wide level to identify novel signal-transduction inhibitors and, therefore, provide better chemotherapeutic approaches in the treatment of human diseases *(8,10)*.

Here we briefly outline the MAPK signaling pathways and inhibitors that have proven useful for studying such pathways. Stepwise protocols for immunoprecipitating MEK1, ERK1, JNK, and p38 MAP Kinase are described, along with assays for kinase activity. Because cyclin D1 is a key downstream target of the MAPK pathways, the utility of cyclin D1 promoter reporter assays to examine proliferative signaling pathways is also described.

2. Materials

2.1. Measuring MEK1 Kinase Activity (11,12)

1. Cell lysis buffer: 50 mM HEPES, pH 7.5, 150 mM NaCl, 0.5% deoxycholate, 1% Triton X-100, 1% NP-40, 50 mM sodium fluoride (NaF), 1 mM sodium orthovanadate (Na$_3$VO$_4$), 0.01% aprotinin, 4 µg/µL pepstatin A, 10 µg/µL leupeptin, 1 mM phenylmethanesulfonyl fluoride (PMSF), 1 mM dithiothreitol DTT. Add proteinase inhibitors immediately before use and keep solution on ice.
2. Phosphate-Buffered saline (PBS): For preparation of 10 L 1X PBS, dissolve 80 g NaCl, 2 g KCl, 14.2 g Na$_2$HPO$_4$, and 2.4 g KH$_2$PO$_4$ in double-distilled H$_2$O. The pH should be between 7.28 and 7.60.
3. Anti-MEK1 antibody and Protein A-agarose (Santa Cruz Biotechnology, Santa Cruz, CA).
4. MAPK 2/Erk2, (inactive) (Upstate Biotechnology, Lake Placid, NY, cat. no. 14–198.)
5. Nonradioactive adenosine triphosphate (ATP) cocktail: 30 mM β-glycerol phosphate, 60 mM HEPES, pH 7.3, 4 mM EGTA, 1.5 mM DTT, 0.45 mM Na$_3$VO$_4$, 30 mM MgCl$_2$, 0.3 mM ATP, and 0.3 mg/mL BSA.
6. Radioactive ATP cocktail: 2 µCi (γ-^{32}P)-ATP, 10 µg of myelin basic protein (MBP), 30 mM glycerophosphate, 60 mM HEPES, pH 7.3, 4 mM EGTA, 1.5 mM DTT, 0.45 mM Na$_3$VO$_4$, 30 mM MgCl$_2$, and 6 µg of BSA.
7. Myelin basic protein (MBP) (Research Diagnostics, Flanders, NJ, cat. no. RDI-TRK8M79).
8. Cell lines and cell-culture supplies.
9. Disposable cell lifter (Fisher Scientific, Pittsburg, PA).
10. PD-98059 stock (50 mM) in dimethyl sulfoxide (DMSO), store at −20°C (Calbiochem-Novabiochem, La Jolla, CA).
11. Bio-Rad Protein Assay Reagent (Bio-Rad Laboratories).
12. Phosphorimager screen and phosphorimaging scanner (Strom, Amersham Biosciences, Piscataway, NJ).
13. Protein sample loading buffer: 50 mM Tris-HCl, pH 6.8, 10% glycerol, 1% sodium docecyl sulfate (SDS), 1% 2-mercaptoethanol.

2.2. In Vitro ERK1 Kinase Assay (12,13)

1. Cell lysis buffer: 50 mM HEPES, pH 7.5, 0.5% deoxycholate, 1% Triton X-100, 1% NP-40, 150 mM NaCl, 50 mM NaF, 1 mM Na$_3$VO$_4$, 0.01% aprotinin, 4 µg/µL pepstatin A, 10 µg/µL leupeptin, 1 mM PMSF, 1 mM DTT. Add proteinase inhibitors immediately before use and keep solution on ice.

2. Anti-ERK1 antibody (Santa Cruz Biotechnology, cat. no. SC-94).
3. Protein A-agarose and Protein G-agarose (Santa Cruz Biotechnology).
4. Myelin basic protein.
5. Kinase reaction buffer: 10 µCi (γ-^{32}P)-ATP, 50 µM ATP, 20 mM HEPES, pH 8.0, 10 mM MgCl$_2$, 1 mM DTT, 1 mM benzamidine.

2.3. In Vitro JNK Kinase Assay (14)

1. Cell lysis buffer: 20 mM Tris-HCl, pH 7.5, 150 mM NaCl, 1% Triton X-100, 2.5 mM sodium pyrophosphate, 1 mM EDTA, 1 mM EGTA, 1 mM Na$_3$VO$_4$, 1 mM β-glycerol phosphate, 1 mM PMSF, and 1 µg/mL leupeptin.
2. Anti-JNK antibody (Cell Signaling Technology).
3. Protein A-agarose and Protein G-agarose (Santa Cruz Biotechnology).
4. ATF2 fusion protein (Cell Signaling Technology).
5. Kinase buffer: 25 mM Tris-HCl, pH 7.5, 5 mM β-glycerol phosphate, 2 mM DTT, 0.1 mM Na$_3$VO$_4$, 10 mM MgCl$_2$ and 100 µM ATP.
6. 6X SDS sample buffer: For 100 mL, add 35 mL 1 M Tris-HCl (pH 6.8), 10.28 g SDS, 36 mL Glycerin, 9.2 g DTT, 12 mg Bromophenol Blue, adjust volume with dd H$_2$O to 100 mL. Store in aliquots at –20°C.
7. Potter Elvehjem tissue grinder.

2.4. In Vitro p38 MAPK Assay (15)

1. Cell lysis buffer: 50 mM HEPES, pH 7.6, 150 mM NaCl, 10% glycerol (v/v), 1% Triton X-100 (v/v), 30 mM Na$_4$P$_2$O$_7$, 10 mM NaF, 1 mM EDTA, 1 mM PMSF, 1 mM benzamidine, 1 mM Na$_3$VO$_4$, 1 mM DDT, and 100 nM okadaic acid.
2. Anti-p38 MAP kinase antibody (Santa Cruz Biotechnology).
3. Protein A- and Protein G-agarose.
4. ATF2 fusion protein (Cell Signaling Technology) or ATF-2 peptide (New England BioLabs, Beverly, MA).
5. Kinase buffer: 50 mM Tris-HCl, pH 7.5, 10 mM MgCl$_2$, and 1 mM dithiothreitol and 100 µM ATP.
6. Whatman P81 phosphocellulose filter (Whatman, cat. no. 3698-023).
7. 175 mM phosphoric acid.
8. Potter Elvehjem tissue grinder.
9. Polyvinylidene flouride (PVDF) membrane.

2.5. Cyclin D1 Reporter Assay

1. Cell line: Breast cancer cell line MCF-7. Cells are maintained in Dulbecco's Modified Eagle's Medium (DMEM) supplemented with 10% fetal bovine serum (FBS) and 1% Penicillin-streptomycin at 37°C in the presence of 5% CO$_2$.

2. Plasmid DNA: Mammalian expression vector pSV2, pSV2Neu-T, and luciferase reporter constructs 1745D1-LUC, Cyclin A-LUC and *c-fos*-LUC *(16)*.
3. Transfection reagents: SuperFect reagent (Qiagen, cat. no. 301305, 1.2 mL). Store at 4°C.
4. MEK inhibitor: PD098059 (2-amino-3-methoxyflavone) (Calbiochem, cat. no. 513000, 5 mg, M.W. 267.3) Stock Solution: 10 mM in DMSO. (Dissolve 5 mg PD098059 in 1.87 mL DMSO, aliquot into 100 μL/tube and store at -20°C.)
5. 0.5 M Glycylglycine (Glygly) buffer: dissolve 33.05 g in 500 mL distilled water, adjust pH to 7.8 with KOH, store at 4°C.
6. 100 mM Potassium phosphate (K-Phos): Mix 90.8 mL of 1 M K_2HPO_4 with 9.2 mL 1 M KH_2PO_4, adjust volume to 1 L with distilled water and determine if pH is 7.8.
7. 1 M DTT in distilled water, stored at -20°C in 1-mL aliquots.
8. 200 mM ATP in distilled water, store at -20°C in 400 μL aliquots.
9. 1 mM Luciferin substrate (Molecular Probes, cat. no. L-2911, 25 mg). Dissolve in 78.51 mL distilled water, store at -20°C in 1-mL aliquots. Protect from light using aluminum foil.
10. 1 M $MgSO_4$, store at room temperature.
11. GME buffer: 25 mM Glygly, 15 mM $MgSO_4$, 4 mM EGTA, 5 mL 0.5 M glygly, 1.5 mL 1 M Mg $MgSO_4$, 0.8 mL 0.5 M EGTA, adjust volume to 100 mL with distilled water, store at 4°C.
12. Extraction buffer: 1% (w/v) Triton X-100 and 1 mM DTT in GME buffer. To prepare, add 0.5 mL Triton X-100 to 50 mL GME buffer, mix well, and then add 50 μL of 1 M DTT. Prepare freshly before use.
13. ATP assay buffer: For each assay point, mix 300 μL GME buffer with 60 μL 100 mM K Phos buffer, 0.4 μL 1 M DTT, and 4 μL 200 mM ATP. Prepare freshly before use.
14. Luciferin solution: Prepare 100 μL per assay. A 5 mL preparation will be enough for 40 samples (1 mL of 1 mM Luciferin, 4 mL GME buffer, and 50 μL of 1 M DTT). Make fresh and protect from light by wrapping the tubes with aluminum foil. Leave on ice before use.
15. Luciferase assay tubes (Becton Dickinson Labwares, cat. no. 352008).
16. Luminometer (i.e., Autolumat, Model LB953, Berthold).

3. Methods

3.1. Measuring MEK1 Kinase Activity

Ras interacts with and activates Raf1, which in turn phosphorylates and activates the dual-specificity kinase MEK1 (MAP kinase kinase) on two distinct serine residues. Activated MEK1 catalyzes the phosphorylation of p44MAPK (ERK1) and p42MAPK (ERK2) on a tyrosine and a threonine residue (Y183 and T185). These MAP kinases can phosphorylate a variety of substrates, including transcription factors and cell-cycle control genes. The small-molecule inhibitor of MEK, PD184352, directly inhibits MEK1 with a 50% inhibitory concentration (IC_{50}) of 17 nM. PD184352 produces a dose-dependent block in the G1 phase of the cell cycle in colon cancer cells. The cell-culture and in vivo efficacy studies indicate colon tumors are especially sensitive to MEK inhibition *(17)*. When human multiple myeloma or leukemia cell lines are exposed to the MEK/MAPK inhibitor PD184352 and the cell-cycle checkpoint inhibitor UCN-01 the cells show dramatic mitochondrial damage and apoptosis *(18,19)*.

PD098059 is a synthetic inhibitor that selectively blocks the activation of MEK-1 and, to a lesser extent, the activation of MEK-2 *(20)*. The inhibition of MEK-1 activation prevents activation of the MAPKs ERK-1/2 and subsequent phosphorylation of MAPK substrates both in vitro and in intact cells. PD098059 reversed the transformed phenotype of *Ras*-transformed mouse fibroblasts and rat kidney cells and blocked induction of cyclin D1 and cell-cycle progression *(21–24)*. PD098059 does not inhibit JNK/SAPK and the p38 pathways at the concentrations that inhibit ERK activity, demonstrating its specificity for the ERK pathway *(25)*.

U0126 is a newly discovered potent inhibitor of the dual-specificity kinases MEK1 and MEK2 *(26)*. Like PD98059, U0126 is a noncompetitive inhibitor of MEK1/2. U0126 displays significantly higher affinity for all forms of MEK than PD098059. U0126 inhibits phosphorylation of MEK1/2 and ERK1/2, inhibits the invasion of human A375 melanoma cells, and decreases c-Jun expression, a major component of the transcription factor AP-1 *(27)*. U0126 inhibits T-cell proliferation in response to antigenic stimulation and cross-linked anti-CD3 plus anti-CD28 antibodies. U0126 has an inhibitory concentration (IC_{50}) of 50–70 nmol/L, whereas PD098059 has an IC_{50} of 5 μmol/L. Ro 09-2210, another inhibitor of MEK-1 and MEK-2, also inhibits other dual-specificity kinases such as MKK-4, MKK-6, and MKK-7, albeit at 4 to 10-fold higher IC_{50} concentrations compared with its effect on MEK-1*(25)*.

1. After treatment of the cells with proper kinase inhibitors, such as 5 μM PD098059 for 24 h, aspirate the culture medium from the tissue-culture plates.
2. Wash the cells twice with 15 mL ice-cold PBS.
3. Put the culture plates on ice, add 300–500 μL cell lysis buffer. Scrape the cells from the culture plates using a disposable cell lifter. Transfer the cell lysate to an ice-cold 1.5-mL Eppendorf tube.
4. Freeze-thaw twice using liquid nitrogen or a dry ice-ethanol mix.
5. Vortex for 30 s and centrifuge at 14,000 rpm in a microcentrifuge at 4°C for 10 min.

6. Transfer the supernatant to a 1.5-mL Eppendorf tube.
7. Measure the protein concentration by using the Bio-Rad Protein Assay Reagent.
8. Dilute 300–600 μg cell lysate in 400 μL cell lysis buffer; add 10 μL anti-MEK1 antibody. Incubate for 1 h at 4°C, rotating to thoroughly mix the sample.
9. Add 20 μL of Protein A-agarose bead slurry, washed according to the manufacture's instruction, and incubate for 2 h at 4°C with constant rotation to immunoprecipitate the kinase.
10. Pellet the agarose by centrifuging for 15 s at 14,000 rpm in a microcentrifuge.
11. Remove the supernatant fraction and wash the protein A agarose beads twice with 800 μL ice-cold lysis buffer and 800 μL once with ice-cold PBS.
12. Add 5 μL of inactive ERK2 (250 μg/mL) and 10 μL of nonradioactive ATP cocktail and incubate for 10 min at 30°C on a shaking incubator to mix the sample thoroughly.
13. Add 20 μL of the (γ-^{32}P)-ATP mixture and incubate for an additional 10 min (*see* **Note 1**).
14. Stop the reaction by adding 40 μL sample buffer and boil at 95°C for 5 min in heat block, then cool on ice for 2 min (*see* **Note 2** and **3**).
15. Vortex vigorously for 30 s, and centrifuge at 14,000 rpm at room temperature for 5 min.
16. Electrophorese 15 μL of the supernatant fluid on an 15% SDS-polyacrylamide gel (PAGE) gel.
17. Transfer proteins from the SDS-PAGE onto a nitrocellulose membrane and determine the amount of radiolabeled ERK2 by phosphor imager analysis.

3.2. In Vitro ERK1 Kinase Assay (12,13)

ERK1 and ERK2 are widely expressed and are involved in the regulation of meiosis, mitosis, and postmitotic functions in differentiated cells (*5*). ERKs 1 and 2 are both components of a three-kinase phosphorylation module that includes the MKKK c-Raf1, B-Raf, or A-Raf, which can be activated by the proto-oncogene *Ras*. Oncogenic *Ras* persistently activates the ERK1 and ERK2 pathways, which contributes to the increased proliferative rate of tumor cells (*5*). PD098059 specifically inhibits the ERK pathway (*25*). Interestingly, inhibition of cysteine proteinases by either E64D or calpeptin leads to a dramatic inhibition of ERK activity (*28*).

1. Aspirate the culture medium from the tissue culture plates.
2. Wash the cells twice with 15 mL ice-cold PBS.
3. Put the cell-culture plates on ice and add 0.5 mL ice-cold cell lysis buffer. Scrape the cells from the culture plates using a disposable cell lifter.

Transfer the cell lysate to a 1.5-mL Eppendorf tube and incubate on ice for 30 min (*see* **Notes 4** and **5**)

4. Freeze-thaw twice using liquid nitrogen or a dry ice ethanol mix.
5. Vortex 30 s and centrifuge at 14,000 rpm in a microcentrifuge at 4°C for 10 min.
6. Transfer the supernatant to a new 1.5-mL Eppendorf microcentrifuge tube.
7. Measure the protein concentration Bio-Rad Protein Assay Reagent.
8. Dilute 500 µg of cell lysate in 500 µL cell lysis buffer, and incubate with 2 µg ERK1 anti-antibody for 2 h at 4°C, rotating to thoroughly mix the sample.
9. Add 30 µL of washed Protein A Plus-agarose bead slurry and incubate for another 2 h at 4°C to immunoprecipitate the kinase, rotating thoroughly to mix the sample.
10. Pellet the agarose beads by centrifugation in a microcentrifuge at 14,000 rpm for 15 s.
11. Remove the supernatant and wash the pellet twice with 800 µL ice-cold lysis buffer and twice with 100 m*M* NaCl in 50 m*M* HEPES buffer, pH 8.0.
12. Incubate the immunoprecipitated complexes with 0.3 mg/mL MBP at 37°C for 15 min in kinase reaction buffer. Use a shaking incubator to thoroughly mix the sample.
13. Stop the reaction by adding 40 µL sample buffer, boil at 95°C for 5 min, then cool on ice for 2 min (*see* **Note 3**).
14. Vortex vigorously for 30 s, and centrifuge at 14,000 rpm at room temperature for 5 min to pellet the beads.
15. Electrophorese 15 µL of the supernatant fraction on to an 15% SDS-PAGE gel.
16. Transfer proteins from the SDS-PAGE onto a nitrocellulose membrane and determine the amount of radiolabeled MBP by phosphorimager analysis.

3.3. In Vitro JNK Kinase Assay (14)

The JNKs are stress-activated protein kinases *(29,30)*. The JNKs bind and phosphorylate c-Jun, a component of the AP-1 transcription complex, and increase its transcriptional activity *(7,31,32)*. AP-1 is involved in regulation of many cytokine genes and is activated in response to environmental stress, radiation, and growth factors, all stimuli that activate JNKs. The inhibition of JNKs enhances chemotherapy-induced inhibition of tumor cell growth, suggesting that JNKs may provide a molecular target for the treatment of cancer. JNK inhibitors have shown promise inhibiting tumor cell growth and in the treatment of rheumatoid arthritis *(5)*.

SP600125 is a JNK inhibitor that completely blocks IL-1-induced expression of c-Jun and collagenase mRNAs. The inhibitor suppressed IL-1-induced accumulation of phosphorylated-c-Jun in synoviocytes *(33,34)*.

Bioactive cell-permeable peptide inhibitors of JNK were engineered by linking the minimal 20-amino acid inhibitory domains of the IB proteins (the islet-brain [IB] 1 and 2 proteins, which inhibit JNK signaling) to the 10-amino acid HIV-TAT sequence that rapidly translocates peptides into cells. Addition of the peptides to the insulin-secreting betaTC-3 cell line resulted in a marked inhibition of interleukin-1 (IL-1)-induced c-jun and c-fos expression, indicating inhibition of JNK signaling *(35)*.

1. Wash the treated cells twice with 10 mL of ice-cold PBS. Aspirate PBS completely after the second wash (*see* **Note 4**).
2. Add 0.5 mL of lysis buffer to the cells and incubate on ice for 20 min with occasional swirling.
3. Scrape cell lysate gently off the plate with a cell lifter and transfer the lysate to a sterile ice-cold 1.5-mL microcentrifuge tube. Disrupt cell lysate in a 2-mL Potter Elvehjem tissue grinder submerged in ice by using twenty up and down strokes (*see* **Note 5**).
4. Vortex 30 s and centrifuge in a microcentrifuge at 14,000 rpm at 4°C for 10 min.
5. Transfer the supernatant to a new ice-cold 1.5-mL Eppendorf tube and determine the protein concentration.
6. Adjust 300–600 µg of cell lysate in 500 µL of total lysis buffer.
7. Add 2 µg anti-JNK antibody and incubate at 4°C for 1 h on a rotating navigator.
8. Add 25 µL Protein-A agarose slurry and continue incubation at 4°C for 2 h or overnight on a navigator.
9. Pellet agarose beads for 10 min at 4°C by centrifugation in a microfuge at 1500*g*.
10. Remove supernatant and wash twice with 0.8 mL of ice-cold lysis buffer and twice with 0.5 mL cold kinase buffer.
11. Pellet agarose beads by centrifugation at 4°C for 10 min at 1500*g* in a microcentrifuge.
12. Remove the supernatant and suspend the pellets in 30 µL kinase buffer containing 200 µ*M* ATP and 2 µg of ATF2 fusion protein.
13. Incubate for 30 min at 30°C on a shaking incubator.
14. Repeat **step 10**.
15. Wash the pellet three times with 0.8 mL of ice-cold cell lysis buffer.
16. Add 30 µL cell lysis buffer, 6 µL 6X protein loading buffer, boil at 95°C for 5 min, then cool on ice for 2 min (*see* **Note 3**).

17. Centrifuge at 14,000 rpm for 5 min in a microcentrifuge to collect the beads.
18. Electrophorese the supernatant fraction on to a 7% SDS-PAGE gel.
19. Transfer proteins from the gel onto nitrocellulose membrane. Detect the phospho-ATF2 signal by Western blotting using the phospho-ATF2 antibodies (*see* **Notes 6–10**).

3.4. In Vitro p38 MAPK Assay (15)

The p38 MAPKs are activated by inflammatory cytokines as well as by many other stimuli, including hormones, ligands for G protein-coupled receptors, stresses, and during activation of the immune response. Because the p38 MAPKs are key regulators of inflammatory cytokine expression, they appear to be involved in human diseases such as asthma and autoimmunity *(5)*. Many inhibitors targeting p38 kinase have been developed, including SB203580, SB202190, SB242235, SB239063, SB220025, SB202474, SC68376, and FR167653 *(36–44)*.

SB203580, a pyridinylimidazole compound, is a selective inhibitor of p38 MAP kinase that acts by competitive binding in the ATP-binding pocket. The p38 MAP kinase inhibitors are efficacious in several disease models, including inflammation, arthritis, septic shock, and myocardial injury *(45)*. p38 MAPK is activated significantly in nitric oxide (NO)- or peroxynitrite-induced cell death in a time-dependent manner. Cell death and caspase-3 activation are markedly inhibited by SB203580 *(46)*.

1. Same as in **Subheading 3.3**.
2. Add 0.3 mL of ice-cold cell lysis buffer to the 10 cm cell-culture dish (add 0.6 mL for 15 cm dishes) and incubate for 20 min on ice with occasional swirling.
3. Same as in **Subheading 3.3**.
4. Same as in **Subheading 3.3**.
5. Same as in **Subheading 3.3**.
6. Same as in **Subheading 3.3**.
7. Add anti-p38 MAP kinase antibody (2 µg/reaction) precoupled to a 20 µL mixture of Protein A- and Protein G-agarose beads and incubate at 4°C for 2–3 h with constant rotation. Antibody-coupled beads are washed twice with ice-cold PBS and once with ice-cold lysis buffer before use.
8. Pellet agarose beads for 10 min at 4°C at 1500 g in a microcentrifuge.
9. Remove supernatant and wash beads four times with 1 mL of wash buffer and twice with 1 mL of kinase buffer.
10. Pellet agarose beads for 10 min at 4°C at 1500 g in a microcentrifuge.
11. Remove the supernatant and suspend the pellets in 30 µL of reaction mixture (kinase buffer containing 5 µM ATP, 2 µCi of (δ-^{32}P)-ATP).

12. Incubate the reaction with 2 µg of ATF-2 fusion protein for 30 min at 30°C with constant agitation.
13. Pellet agarose beads for 10 min at 4°C at 1500 g in a microcentrifuge.
14. Transfer 30 µL of the supernatant onto a 2.1-cm diameter Whatman P81 cellulose phosphate filter circles.
15. Wash circles four times for 10 min with 3 mL of 175 mM phosphoric acid and once with 3 mL distilled water for 5 min.
16. Air-dry filters and then measure radioactivity in a liquid scintillation counter (*see* **Note 11–15**).

3.5. Cyclin D1 Reporter Assay (16)

The cyclin D1 gene encodes a labile growth factor and oncogene-inducible regulatory subunit of the holoenzyme that phosphorylates and inactivates the pRb protein. The abundance of cyclin D1 is rate-limiting in the induction of DNA synthesis by diverse mitogenic stimulus *(47)*. The cyclin D1 gene is transcriptionally induced by mitogenic stimuli, including *Ras*, Src, ErbB2, and activated ERK, suggesting that the cyclin D1 promoter is a useful reporter of mitogenic intracellular signaling activity *(16–50)*.

3.5.1. Preparation of Cells for Transient Transfection

1. Plan the transfection experiment. For example, in this experiment, we will examine the effect of Her2/Neu signaling on the cyclin D1 transcription in MCF-7 breast cancer cells. We also want to know if Her2/Neu regulates cyclin D1 through a MAPK pathway and will examine this possibility by using the MEK inhibitor, PD098059 (*see* **Note 16**). In this protocol, we will use a reporter assay to address these questions. Cells will be transiently transfected with the mammalian expression vector of NeuT with cyclin D1 reporter construct, -1745 cyclin D_1-LUC. *c-fos*-LUC will be used as a positive reporter control because it is known that Her2/Neu signaling upregulates c-fos expression *(51)*. Cyclin A-LUC will be used as a negative control. All transfections will be done in triplicate. The transfection plan is shown **Fig. 1** (*see* **Notes 17–19**).
2. Subculture MCF-7 cells for transfection. A day before transfection, seed 0.4×10^5 MCF-7 cells per well in 400 µL DMEM supplemented with 10% FBS into 24-well plates. By the time of transfection, the cells should reach 50–70% confluence (*see* **Note 20**).

3.5.2. Transient Transfection

1. For each well, dilute 1.2 µg reporter plasmid DNA along with either pSV2 control vector (75 ng) or expression vector pSV2-NeuT (100 ng) in 50 µL

MCF-7 cells

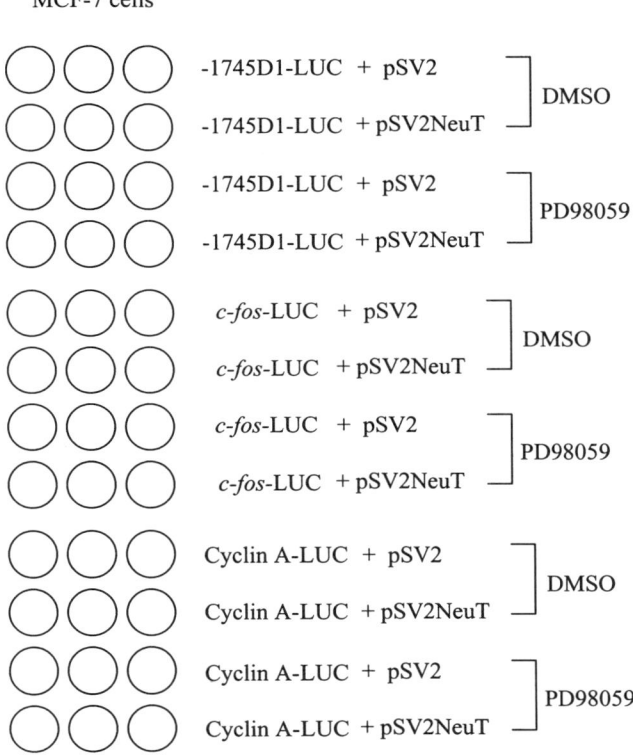

Fig. 1. Example of the transfection plan.

DMEM (serum and antibiotic free medium). Mix well and sit at room temperature for 2 min.

2. Dilute 3–5 µL SuperFect reagent for each well in 50 µL DMEM (serum- and antibiotic-free medium). Mix well and sit at room temperature for 2 min.

3. Combine the diluted DNA with diluted SuperFect, mix by pipeting up and down three to five times, and incubate at room temperature for 10 min to allow formation of the DNA-SuperFect complex.

4. Add 100 µL of the transfection complexes directly into each well containing cell-growth medium. For SuperFect reagent, it is not necessary to change the cell-growth medium to serum- and antibiotic-free medium at this point. However, consult the manufacturer's manual for transfection conditions with different reagents. Mix well by shaking the cell culture plate gently. Incubate the cells in a CO_2 incubator for 24 h.

3.5.3. Treatment of the Cells With PD98059

Twenty-four hours after transfection, replace the medium with 500 µL fresh culture medium containing either DMSO (negative control) or 10 µM PD98059. Incubate the cells for another 24 h (*see* **Note 21–23**).

3.5.4. Luciferase Assay

1. Lyse cells by aspirating the medium from the culture plate and adding 100 µL of cell-extraction buffer. Rotate or shake the cells on a shaking platform at room temperature for 5–10 min.
2. For each sample, add 300 µL of ATP assay buffer into luciferase assay tubes. Prepare six extra tubes as blank controls.
3. Transfer 100 µL of cell lysate into the tube containing the ATP assay buffer and mix.
4. Load the samples onto the luminometer and put the substrate injector into the luciferin container (protected from light with aluminum foil). Make sure that the injector is submerged into the luciferin solution.
5. Measure the integrated light output for 10–60 s. At the end, wash the tubing of the luminometer with distilled water. (If renillar luciferin is used as an internal control, wash the tubing with 70% ethanol six times with six wash tubes and then repeat wash again with distilled water *[52]*).
6. Analyze the data statistically and graph as shown in **Fig. 2**.

4. Notes

1. Safety warnings and precautions: Because the experiments described here involve the use of radioactive (γ-^{32}P)-ATP, be sure to follow your institutional regulations relating to the handling, usage, storage and disposal of such materials. Always use protective barriers.
2. Alternatively, stop the reaction by adding 20 µL of 100 mM EDTA, pH 7.5, centrifuge briefly, and spot 40 µL of each supernatant onto phosphocellular paper. The papers are then washed six times (5–10 min each) in 10% phosphoric acid, soaked briefly in 100% ethanol, and air-dried before analysis in a liquid scintillation counter *(11)*.
3. When heating samples on the heating block, make sure that the microcentrifuge tubes are closed tightly. Place another heating block on top of the tubes will prevent the tops from popping open.
4. When harvesting the cell lysate, be sure to aspirate the PBS buffer completely from the plates. Residual PBS will dilute the concentration of the protein inhibitors in the cell lysis buffer.
5. Keep reconstituted lysis buffer on ice at all times.

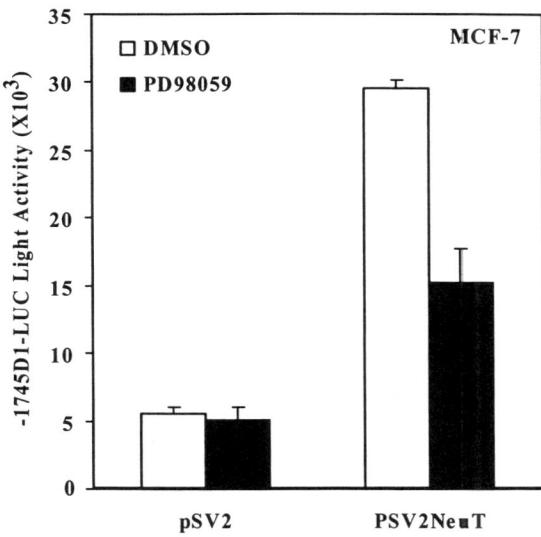

Fig. 2. Regulation of Cyclin D1 by NeuT. 0.4×10^5 MCF-7 cells are seeded into 24-well cell-culture plates and the cells are allowed to grow for 24 h to reach 70% confluence. 1.2 µg of -1745D1-LUC and 100 ng of expression vector of NeuT and 75 ng of control vector are transfected as indicated using the proper transfection reagent. The cells are treated with 5 µ*M* PD98059 for 24 h and the luciferase activity is then measured.

6. Alternatively, incubate 300–600 µg cell lysates with immobilized c-Jun (Cell Signaling Technology, cat. no. 9811) overnight at 4°C.
7. Pellet the agarose beads. Wash the immunoprecipitated products twice with the cell lysis buffer and twice with kinase buffer (**Subheading 2.3.**).
8. Resuspend the pellets in the kinase buffer containing 100 µ*M* ATP.
9. Incubate the reaction for 30 min at 30°C in a shaking incubator.
10. Perform Western blot to detect the phospho-c-Jun signal.
11. Alternatively, stop the reaction by adding 30 µL of 2X Laemmli sample buffer and heat for 5 min at 95°C (*see* **Note 3**).
12. Centrifuge at 12,000 rpm for 5 min.
13. Take 40 µL of the supernatant and resolve on 13% SDS-PAGE gel.
14. Transferred onto polyvinylidene fluoride (PVDF) membranes.
15. Expose the PVDF membrane to a phosphorimager cassette and quantify the amount of radiolabeled phosphate substrate by using a molecular dynamics phosphorimager system.

16. **Table 1** summarizes the various inhibitors for different signaling pathways. Consult the literature for a more extensive list of inhibitors available for a particular pathway *(5,28,33,34,36,37,39–44,46,53–79)*.

17. Different cell types may have a different genetic background. This is particularly true for the cancer cell lines. Careful selection of the cell type for studying a particular signal pathway is very important. For example, to study the effect of a signal pathway on the regulation of p53 expression, consider whether the p53 gene is expressed in the cell type you have chosen and whether or not the p53 gene is mutated. Also, consider whether the status of the particular pathway you are studying is altered (defective or constitutively active) in your chosen cells.

18. Transfection efficiency varies with cell types and different transfection reagents. Several internal control plasmids have been described, but these controls may independently affect activity of the promoter being assessed *(52)*. We suggest using a green flourescent protein (GFP)-expressing vector as a monitor of transfection efficiency.

19. If different cell types are used, a reporter control such as renillar luciferase (or β-galactosidase) should be included in order to adjust the data and make comparisons between different cell types.

20. Because the size of different cell types varies, the number of the cells to be seeded for transfection also varies. In general, the cells should reach 50–80% confluence by the time of transfection. Check the manufacture's manual for special requirements.

21. PD98059 is dissolved in DMSO or methanol. We use DMSO to make a 10 mM stock solution and store in small aliquot at −20°C. PD98059 should be protected from light. Always include the vehicle used to dissolve the inhibitor as a control.

22. For experiments where the ligand of hormone receptors, such as dihydrotestosterone (DHT) or estradiol are used, phenol-free medium and charcoal-stripped serum should be used when the cells are treated with hormones.

23. Treatment usually occurs after 24 h of transfection. The duration of the treatment varies depending on the reagents used and the targeted proteins or signaling pathways. Time-course and dose curves might be necessary.

Acknowledgments

We apologize to the investigators whose work was not cited owing to space limitations. This work was supported in part by awards from the Susan G. Komen Breast Cancer Foundation; Breast Cancer Alliance Inc., and research grants R01CA70896, R01CA75503, R01CA86072, R01CA86071 from NIH (R.G.P.), and 1 R21 DK065220-01 from NIDDK (M.F.).

References

1. Hanahan, D. and Weinberg, R. A. (2000) The hallmarks of cancer. *Cell* **100,** 57–70.
2. Heldin, C. H. (2001) Signal transduction: multiple pathways, multiple options for therapy. *Stem Cells* **19,** 295–303.
3. Lobbezoo, M. W., Giaccone, G., and Van Kalken, C. (2003) Signal transduction modulators for cancer therapy: from promise to practice? *Oncologist* **8,** 210–213.
4. Kyriakis, J. M. (1999) Making the connection: coupling of stress-activated ERK/MAPK (extracellular-signal-regulated kinase/mitogen-activated protein kinase) core signalling modules to extracellular stimuli and biological responses. *Biochem. Soc. Symp.* **64,** 29–48.
5. Johnson, G. L. and Lapadat, R. (2002) Mitogen-activated protein kinase pathways mediated by ERK, JNK, and p38 protein kinases. *Science* **298,** 1911–1912.
6. Pestell, R. G., Albanese, C., Watanabe, G., et al. (1995) Epidermal growth factor and c-Jun act via a common DNA regulatory element to stimulate transcription of the ovine P-450 cholesterol side chain cleavage (CYP11A1) promoter. *J. Bio. Chem.* **270,** 18,301–18,308.
7. Pestell, R. G., Hollenberg, A. N., Albanese, C., and Jameson, J. L. (1994) c-Jun represses transcription of the human chorionic gonadotropin alpha and beta genes through distinct types of CREs. *J. Bio. Chem.* **269,** 31,090–31,096.
8. Lin, W., Albanese, C., Pestell, R. G., and Lawrence, D. S. (2002) Spatially discrete, light-driven protein expression. *Chem. Biol.* **9,** 1347–1353.
9. Fantin, V. R., Berardi, M. J., Scorrano, L., et al. (2002) A novel mitochondriotoxic small molecule that selectively inhibits tumor cell growth. *Cancer Cell* **2,** 29–42.
10. Golub, T. R. (2003) Mining the genome for combination therapies. *Nat. Med.* **9,** 510–511.
11. Wu, W., Graves, L. M., Jaspers, I., et. al. (1999) Activation of the EGF receptor signaling pathway in human airway epithelial cells exposed to metals. *Am. J. Physiol.* **277,** L924–L931.
12. Samarakoon, R. and Higgins, P. J. (2002) MEK/ERK pathway mediates cell-shape-dependent plasminogen activator inhibitor type 1 gene expression upon drug-induced disruption of the microfilament and microtubule networks. *J. Cell Sci.* **115,** 3093–3103.
13. Marques, S. A., Dy, L. C., Southall, M. D., et al. (2002) The platelet-activating factor receptor activates the extracellular signal-regulated kinase mitogen-activated protein kinase and induces proliferation of epidermal cells through an epidermal growth factor-receptor-dependent pathway. *J. Pharmacol. Exp. Ther.* **300,** 1026–1035.
14. Hayakawa, J., Depatie, C., Ohmichi, M., and Mercola, D. (2003) The activation of c-Jun NH2-terminal kinase (JNK) by DNA-damaging agents serves to promote drug resistance via activating transcription factor 2 (ATF2)-dependent enhanced DNA repair. *J. Biol. Chem.* **278,** 20,582–20,592.
15. Sweeney, G., Somwar, R., Ramlal, T., et al. (1999) An inhibitor of p38 mitogen-activated protein kinase prevents insulin-stimulated glucose transport but not glucose transporter translocation in 3T3-L1 adipocytes and L6 myotubes. *J. Biol. Chem.* **274,** 10,071–10,078.

16. Lee, R. J., Albanese, C., Fu, M., et al. (2000) Cyclin D1 is required for transformation by activated Neu and is induced through an E2F-dependent signaling pathway. *Mol. Cell Biol.* **20,** 672–683.

17. Sebolt-Leopold, J. S., Dudley, D. T., Herrera, R., et al. (1999) Blockade of the MAP kinase pathway suppresses growth of colon tumors in vivo. *Nat. Med.* **5,** 810–816.

18. Dai, Y., Yu, C., Singh, V., Tang, L., et al. (2001) Pharmacological inhibitors of the mitogen-activated protein kinase (MAPK) kinase/MAPK cascade interact synergistically with UCN-01 to induce mitochondrial dysfunction and apoptosis in human leukemia cells. *Cancer Res.* **61,** 5106–5115.

19. Dai, Y., Landowski, T. H., Rosen, S. T., et al. (2002) Combined treatment with the checkpoint abrogator UCN-01 and MEK1/2 inhibitors potently induces apoptosis in drug-sensitive and -resistant myeloma cells through an IL-6-independent mechanism. *Blood* **100,** 3333–3343.

20. Alessi, D. R., Cuenda, A., Cohen, P., et al. (1995) PD 098059 is a specific inhibitor of the activation of mitogen-activated protein kinase kinase in vitro and in vivo. *J. Biol. Chem.* **270,** 27,489–27,494.

21. Dudley, D. T., Pang, L., Decker, S. J., et al. (1995) A synthetic inhibitor of the mitogen-activated protein kinase cascade. *Proc. Natl. Acad. Sci. USA* **92,** 7686–7689.

22. Watanabe, G., Howe, A., Lee, R. J., et al. (1996) Induction of cyclin D1 by simian virus 40 small tumor antigen. *Proc. Natl. Acad. Sci. USA* **93,** 12,861–12,866.

23. Watanabe, G., Lee, R. J., Albanese, C., et al. (1996) Angiotensin II activation of cyclin D1-dependent kinase activity. *J. Biol. Chem.* **271,** 22,570–22,577.

24. Watanabe, G., Pena, P., Albanese, C., et al. (1997) Adrenocorticotropin induction of stress-activated protein kinase in the adrenal cortex in vivo. *J. Biol. Chem.* **272,** 20,063–20,069.

25. Reuter, C. W., Morgan, M. A., and Bergmann, L. (2000) Targeting the *Ras* signaling pathway: a rational, mechanism-based treatment for hematologic malignancies? Blood **96,** 1655–1669.

26. Duncia, J. V., Santella, J. B., 3rd, Higley, C. A., et al. (1998) MEK inhibitors: the chemistry and biological activity of U0126, its analogs, and cyclization products. *Bioorg. Med. Chem. Lett.* **8,** 2839–2844.

27. Ge, X., Fu, Y. M., and Meadows, G. G. (2002) U0126, a mitogen-activated protein kinase kinase inhibitor, inhibits the invasion of human A375 melanoma cells. *Cancer Lett.* **179,** 133–140.

28. Torres, C., Li, M., Walter, R., and Sierra, F. (2000) Modulation of the ERK pathway of signal transduction by cysteine proteinase inhibitors. *J. Cell. Biochem.* **80,** 11–23.

29. Force, T., Pombo, C. M., Avruch, J. A., et al. (1996) Stress-activated protein kinases in cardiovascular disease. *Circ. Res.* **78,** 947–953.

30. Kyriakis, J. M. and Avruch, J. (1996) Protein kinase cascades activated by stress and inflammatory cytokines. *Bioessays* **18,** 567–577.

31. Albanese, C., Johnson, J., Watanabe, G., et al. (1995) Transforming p21ras mutants and c-Ets-2 activate the cyclin D1 promoter through distinguishable regions. *J. Biol. Chem.* **270,** 23,589–23,597.

32. Albanese, C., D'Amico, M., Reutens, A. T., et al. (1999) Activation of the cyclin D1 gene by the E1A-associated protein p300 through AP-1 inhibits cellular apoptosis. *J. Biol. Chem.* **274,** 34,186–34,195.

33. Vincenti, M. P. and Brinckerhoff, C. E. (2001) The potential of signal transduction inhibitors for the treatment of arthritis: Is it all just JNK? *J. Clin. Invest.* **108,** 181–183.

34. Han, Z., Boyle, D. L., Chang, L., et al. (2001) c-Jun N-terminal kinase is required for metalloproteinase expression and joint destruction in inflammatory arthritis. *J. Clin. Invest.* **108,** 73–81.

35. Bonny, C., Oberson, A., Negri, S., et al. (2001) Cell-permeable peptide inhibitors of JNK: novel blockers of beta-cell death. *Diabetes* **50,** 77–82.

36. Lahti, A., Kankaanranta, H., and Moilanen, E. (2002) P38 mitogen-activated protein kinase inhibitor SB203580 has a bi-directional effect on iNOS expression and NO production. *Eur. J. Pharmacol.* **454,** 115–123.

37. Nishikori, T., Irie, K., Suganuma, T., et al. (2002) Anti-inflammatory potency of FR167653, a p38 mitogen-activated protein kinase inhibitor, in mouse models of acute inflammation. *Eur. J. Pharmacol.* **451,** 327–333.

38. Ward, K. W., Proksch, J. W., Azzarano, L. M., et al. (2001) SB-239063, a potent and selective inhibitor of p38 map kinase: preclinical pharmacokinetics and species-specific reversible isomerization. *Pharmacol. Res.* **18,** 1336–1344.

39. Barone, F. C., Irving, E. A., Ray, A. M., et al. (2001) SB 239063, a second-generation p38 mitogen-activated protein kinase inhibitor, reduces brain injury and neurological deficits in cerebral focal ischemia. *J. Pharmacol. Exp. Ther.* **296,** 312–321.

40. Jackson, J. R., Bolognese, B., Hillegass, L., et al. (1998) Pharmacological effects of SB 220025, a selective inhibitor of P38 mitogen-activated protein kinase, in angiogenesis and chronic inflammatory disease models. *J. Pharmacol. Exp. Ther.* **284,** 687–692.

41. Badger, A. M., Roshak, A. K., Cook, M. N., et al. (2000) Differential effects of SB 242235, a selective p38 mitogen-activated protein kinase inhibitor, on IL-1 treated bovine and human cartilage/chondrocyte cultures. *Osteoarthritis Cartilage* **8,** 434–443.

42. Badger, A. M., Griswold, D. E., Kapadia, R., et al. (2000) Disease-modifying activity of SB 242235, a selective inhibitor of p38 mitogen-activated protein kinase, in rat adjuvant-induced arthritis. *Arthritis Rheum.* **43,** 175–183.

43. Karahashi, H., Nagata, K., Ishii, K., and Amano, F. (2000) A selective inhibitor of p38 MAP kinase, SB202190, induced apoptotic cell death of a lipopolysaccharide-treated macrophage-like cell line, J774.1. *Biochim. Biophys. Acta* **1502,** 207–223.

44. Wadsworth, S. A., Cavender, D. E., Beers, S. A., et al. (1999) RWJ 67657, a potent, orally active inhibitor of p38 mitogen-activated protein kinase. *J. Pharmacol. Exp. Ther.* **291,** 680–687.

45. Lee, J. C., Kassis, S., Kumar, S., et al. (1999) p38 mitogen-activated protein kinase inhibitors: mechanisms and therapeutic potentials. *Pharmacol. Ther.* **82,** 389–397.

46. Sarker, K. P., Nakata, M., Kitajima, I., et al. (2000) Inhibition of caspase-3 activation by SB 203580, p38 mitogen-activated protein kinase inhibitor in nitric oxide-induced apoptosis of PC-12 cells. *J. Mol. Neurosci.* **15,** 243–50.

47. Brown, J. R., Nigh, E., Lee, R. J., et al. (1998) Fos family members induce cell cycle entry by activating cyclin D1. *Mol. Cell Biol.* **18,** 5609–5619.

48. Hulit, J., Lee, R. J., Russell, R. G., and Pestell, R. G. (2002) ErbB-2-induced mammary tumor growth: the role of cyclin D1 and p27Kip1. *Biochem. Pharmacol.* **64,** 827–836.

49. D'Amico, M., Hulit, J., Amanatullah, D. F., et al. (2000) The integrin-linked kinase regulates the cyclin D1 gene through glycogen synthase kinase 3beta and cAMP-responsive element-binding protein-dependent pathways. *J. Biol. Chem.* **275,** 32,649–32,657.

50. Watanabe, G., Albanese, C., Lee, R. J., et al. (1998) Inhibition of cyclin D1 kinase activity is associated with E2F-mediated inhibition of cyclin D1 promoter activity through E2F and Sp1. *Mol. Cell Biol.* **18,** 3212–3222.

51. Sistonen, L., Holtta, E., Lehvaslaiho, H., et al. (1989) Activation of the neu tyrosine kinase induces the fos/jun transcription factor complex, the glucose transporter and ornithine decarboxylase. *J. Cell Biol.* **109,** 1911–1919.

52. Amanatullah, D. F., Zafonte, B. T., Albanese, C., et al. (2001) *Ras* regulation of cyclin D1 promoter. *Methods Enzymol.* **333,** 116–127.

53. Baselga, J. (2000) New therapeutic agents targeting the epidermal growth factor receptor. *J. Clin. Oncol.* **18,** 54S–59S.

54. Baselga, J. (2000) Current and planned clinical trials with trastuzumab (Herceptin). *Semin. Oncol.* **27,** 27–32.

55. Kim, E. S., Khuri, F. R., and Herbst, R. S. (2001) Epidermal growth factor receptor biology (IMC-C225). *Curr. Opin. Oncol.* **13,** 506–513.

56. Overholser, J. P., Prewett, M. C., Hooper, A. T., et al. (2000) Epidermal growth factor receptor blockade by antibody IMC-C225 inhibits growth of a human pancreatic carcinoma xenograft in nude mice. *Cancer* **89,** 74–82.

57. Normanno, N., Maiello, M. R., and De Luca, A. (2003) Epidermal growth factor receptor tyrosine kinase inhibitors (EGFR-TKIs): simple drugs with a complex mechanism of action? *J. Cell Physiol.* **194,** 13–19.

58. Moasser, M. M., Basso, A., Averbuch, S. D., and Rosen, N. (2001) The tyrosine kinase inhibitor ZD1839 ("Iressa") inhibits HER2-driven signaling and suppresses the growth of HER2-overexpressing tumor cells. *Cancer Res.* **61,** 7184–7188.

59. Monia, B. P., Sasmor, H., Johnston, J. F., et al. (1996) Sequence-specific antitumor activity of a phosphorothioate oligodeoxyribonucleotide targeted to human C-raf kinase supports an antisense mechanism of action in vivo. *Proc. Natl. Acad. Sci. USA* **93,** 15,481–15,484.

60. Hilger, R. A., Scheulen, M. E., and Strumberg, D. (2002) The *Ras*-Raf-MEK-ERK pathway in the treatment of cancer. *Onkologie* **25,** 511–518.

61. Strumberg, D., Voliotis, D., Moeller, J. G., et al. (2002) Results of phase I pharmacokinetic and pharmacodynamic studies of the Raf kinase inhibitor BAY 43-9006 in patients with solid tumors. *Int. J. Clin. Pharmacol. Ther.* **40,** 580–581.

62. El-Kholy, W., Macdonald, P. E., Lin, J. H., et al. (2003) The phosphatidylinositol 3-kinase inhibitor LY294002 potently blocks K(V) currents via a direct mechanism. *Faseb. J.* **17,** 720–722.

63. Vanhaesebroeck, B. and Waterfield, M. D. (1999) Signaling by distinct classes of phosphoinositide 3-kinases. *Exp. Cell Res.* **253,** 239–254.

64. Sakoff, J. A., Ackland, S. P., Baldwin, M. L., et al. (2002) Anticancer activity and protein phosphatase 1 and 2A inhibition of a new generation of cantharidin analogues. *Invest. New Drugs* **20,** 1–11.

65. Kisselev, A. F. and Goldberg, A. L. (2001) Proteasome inhibitors: from research tools to drug candidates. *Chem. Biol.* **8,** 739–758.

66. Rusbridge, N. M. and Beynon, R. J. (1990) 3,4-Dichloroisocoumarin, a serine protease inhibitor, inactivates glycogen phosphorylase b. *FEBS Lett.* **268,** 133–136.

67. Elofsson, M., Splittgerber, U., Myung, J., et al. (1999) Towards subunit-specific proteasome inhibitors: synthesis and evaluation of peptide alpha', beta'-epoxyketones. *Chem. Biol.* **6,** 811–822.

68. Loidl, G., Musiol, H. J., Groll, M., et al. (2000) Synthesis of bivalent inhibitors of eucaryotic proteasomes. *J. Pept. Sci.* **6,** 36–46.

69. Lu, Y. P., Lou, Y. R., Xie, J. G., et al. (2002) Topical applications of caffeine or (-)-epigallocatechin gallate (EGCG) inhibit carcinogenesis and selectively increase apoptosis in UVB-induced skin tumors in mice. *Proc. Natl. Acad. Sci. USA* **99,** 12,455–12,460.

70. Marks, P. A., Richon, V. M., Breslow, R., and Rifkind, R. A. (2001) Histone deacetylase inhibitors as new cancer drugs. *Curr. Opin. Oncol.* **13,** 477–483.

71. Richon, V. M., Zhou, X., Rifkind, R. A., and Marks, P. A. (2001) Histone deacetylase inhibitors: development of suberoylanilide hydroxamic acid (SAHA) for the treatment of cancers. *Blood Cells. Mol. Dis.* **27,** 260–264.

72. Henderson, C., Mizzau, M., Paroni, G., et al. (2003) Role of caspases, Bid, and p53 in the apoptotic response triggered by histone deacetylase inhibitors trichostatin-A (TSA) and suberoylanilide hydroxamic acid (SAHA). *J. Biol. Chem.* **278,** 12,579–12,589.

73. Ailenberg, M. and Silverman, M. (2003) Differential effects of trichostatin A on gelatinase A expression in 3T3 fibroblasts and HT-1080 fibrosarcoma cells: implications for use of TSA in cancer therapy. *Biochem. Biophys. Res. Commun.* **302,** 181–185.

74. Klisovic, M. I., Maghraby, E. A., Parthun, M. R., et al. (2003) Depsipeptide (FR901228) promotes histone acetylation, gene transcription, apoptosis and its activity is enhanced by DNA methyltransferase inhibitors in AML1/ETO-positive leukemic cells. *Leukemia* **17,** 350–358.

75. Jung, M. (2001) Inhibitors of histone deacetylase as new anticancer agents. *Curr. Med. Chem.* **8,** 1505–1511.

76. Lee, B. I., Park, S. H., Kim, J. W., et al. (2001) MS-275, a histone deacetylase inhibitor, selectively induces transforming growth factor beta type II receptor expression in human breast cancer cells. *Cancer Res.* **61,** 931–934.

77. Jaboin, J., Wild, J., Hamidi, H., et al. (2002) MS-27-275, an inhibitor of histone deacetylase, has marked in vitro and in vivo antitumor activity against pediatric solid tumors. *Cancer Res.* **62,** 6108–6115.
78. Bitterman, K. J., Anderson, R. M., Cohen, H. Y., et al. (2002) Inhibition of silencing and accelerated aging by nicotinamide, a putative negative regulator of yeast sir2 and human SIRT1. *J. Biol. Chem.* **277,** 45,099–45,107.
79. Bedalov, A., Gatbonton, T., Irvine, W. P., et al. (2001) Identification of a small molecule inhibitor of Sir2p. *Proc. Natl. Acad. Sci. USA* **98,** 15,113–15,118.
80. Lee, Y. J., Kim, J. H., Chen, J., and Song, J. J. (2002) Enhancement of metabolic oxidative stress-induced cytotoxicity by the thioredoxin inhibitor 1-methylpropyl 2-imidazolyl disulfide is mediated through the ASK1-SEK1-JNK1 pathway. *Mol. Pharmacol.* **62,** 1409–1417.

3

Two-Dimensional Gel Electrophoresis for the Identification of Signaling Targets

Yukihito Kabuyama, Kirsi K. Polvinen, Katheryn A. Resing, and Natalie G. Ahn

Summary

Two-dimensional electrophoresis (2-DE) is a powerful technique to differentially display patterns of protein expression and posttranslational modifications, providing a good strategy to monitor molecular responses induced by the activation or inactivation of specific signaling pathways. In this chapter, optimized protocols for 2-DE using extracts from tissue culture are provided. Protocols for in-gel digestion of gel-resolved proteins, which allow protein identification by mass spectrometry are also discussed.

Key Words: Two-dimensional electrophoresis; signal transduction.

1. Introduction

Two-dimensional electrophoresis (2-DE) is a powerful technique for resolving complex protein mixtures. This method provides the ability to separate and quantify up to 9000 protein forms from unfractionated cell lysates *(1)*. Many parameters and experimental conditions affect the quality of protein resolution on 2-DE, such as temperature, voltage, quality of reagents, and instrumentation, causing variability in quantitative analysis of gel-resolved proteins. However, introduction of immobilized pH gradients for isoelectric focusing (IEF) has led to significant improvements in resolution and reproducibility of

From: *Methods in Molecular Biology, vol. 284:*
Signal Transduction Protocols
Edited by: R. C. Dickson © Humana Press Inc., Totowa, NJ

protein separation, allowing researchers greater accessibility to 2-DE for protein expression profiling. Furthermore, improvements in methods for in-gel digestion, combined with database search algorithms that enable peptide mass fingerprinting and sequencing, allow rapid identification of gel resolved proteins by mass spectrometry (MS).

The ability of 2-DE to differentially display patterns of protein expression and posttranslational modifications provides a good strategy to monitor molecular responses induced by the activation or inhibition of specific signaling pathways. Applications of this method include the analysis of cellular targets downstream of mitogen activated protein kinase (MAPK) *(2)*, transforming growth factor (TGF-β) *(3)*, endothelin 1 *(4)*, *Fas (5)*, and PhoP/Q two component *(6)* signaling pathways. Each of these examples successfully identified novel signaling targets and revealed new functions of each pathways, clearly demonstrating the utility of 2-DE for protein profiling. However, careful methods are required to obtain resolution on 2-DE in order to yield satisfactory protein profiling.

This chapter will describe optimized protocols for 2-DE in ranges of pI from 4 to 7 and 6 to 11, which have been modified from the methods of Gorg et al. *(7)* and Hoving et al. *(8)*. The WM35 human melanoma cell line is used as an example. Protocols for in-gel digestion of gel-resolved proteins are also described.

2. Materials

2.1. Equipment

1. Horizontal electrophoresis apparatus with electrode tray (Multiphor II, Amersham-Pharmacia, Piscataway, NJ).
2. Immobiline drystrip tray (Amersham-Pharmacia).
3. Immobiline drystrip reswelling tray (Amersham-Pharmacia).
4. Thermostatic circulator (e.g., VWR Heated/Refrigerated Circulators).
5. Vertical electrophoresis units (e.g., Protean II xi cell, Bio-Rad).
6. Power supply (EPS3500XL, Amersham-Pharmacia).
7. Slab gel electrophoresis unit (e.g., Bio-Rad Protean II xi).
8. Gradient maker (e.g., Hoeffer SG30 and peristaltic pump (e.g., Rainin Dynamax Model RP-1).

2.2. Reagents

1. Cell lysis buffer A: 7 M urea, 2 M thiourea, 4% (3-[3-cholamidopropyl) dimethylammonio]-1-propane-sulfonate) (CHAPS), 1% IPG buffer (pH 4.0–7.0, Amersham-Pharmacia), 1 mM benzamidine, 25 µg/mL leupeptin, 20 µg/mL pepstatin-A, 20 µg/mL aprotinin, 1 mM sodium vanadate, 1 µM

microcystin-LR, 20 m*M* dithiothreitol (DTT). This buffer can be stored at −80°C up to 3 mo. DTT should be added to a final concentration of 20 m*M* just before use.

2. Cell lysis buffer B: 7 *M* urea, 2 *M* thiourea, 4 % CHAPS, 2 % IPG buffer (pH 6.0–11.0, Amersham-Pharmacia), 1 m*M* benzamidine, 25 μg/mL leupeptin, 20 μg/mL pepstatin-A, 20 μg/mL aprotinin, 1 m*M* sodium vanadate, 1 μ*M* microcystin-LR, 20 m*M* DTT. This buffer can be stored at −80°C up to 3 mo. DTT should be added to 20 m*M* just before use.

3. Rehydration buffer A: 7 *M* urea, 2 *M* thiourea, 4 % CHAPS, 3 % IPG buffer (pH 4.0–7.0, Amersham-Pharmacia), 20 m*M* DTT, 10 μg/mL of bromophenol blue. This buffer can be stored at −80°C up to 3 mo. DTT should be added to 20 m*M* before use.

4. Rehydration buffer B: 7 *M* urea, 2 *M* thiourea, 4% CHAPS, 2% IPG buffer pH 6.0–11.0, Amersham-Pharmacia, 160 m*M* DTT, 10% isopropanol, 5% glycerol, 10 μg/mL bromophenol blue. This buffer can be stored at −80°C up to 3 mo.

5. Rehydration buffer C: Rehydration buffer B with 230 m*M* DTT (instead of 160 m*M* DTT). This buffer can be stored at −80°C up to 3 mo.

6. Equilibration solution A (reduction): 6 *M* urea, 30% glycerol, 2% sodium dodecyl sulfate (SDS), 0.05 *M* Tris-HCl buffer, pH 6.8, 60 m*M* DTT.

7. Equilibration solution B (alkylation): 6 *M* urea, 30% glycerol, 2% SDS, 0.05 *M* Tris-HCl buffer, pH 6.8, 2% iodoacetamide.

8. Immobiline drystrip pH 4.0–7.0, 18 cm (Amersham-Pharmacia). This should be stored at −20°C and thawed immediately before use.

9. Immobiline drystrip pH 6.0–11.0, 18 cm (Amersham-Pharmacia). This should be stored at −80°C and thawed immediately before use.

10. Light mineral oil (Fisher, Piscataway, NJ, cat. no. 0121-1).

11. Light acrylamide solution: 8% acrylamide, 0.21% bisacrylamide, 0.38 *M* Tris-HCl, pH 8.8, 0.03% ammonium persulfate, 0.3% N,N,*N'*,*N'*-tetramethyl-ethylenediamine (TEMED) (added just before use).

12. Heavy acrylamide solution: 18% acrylamide, 0.48% bisacrylamide, 0.38 *M* Tris-HCl, pH 8.8, 0.03% ammonium persulfate, 0.3% TEMED (added just before use).

13. Gel running buffer: 0.1% SDS, 25 m*M* Tris base, 190 m*M* glycine, pH 8.3.

14. Gel fixation solution: 40% ethanol, 12% acetic acid, 0.02% formaldehyde.

15. Sensitivity enhancing solution: 0.02% sodium thiosulfate.

16. Silver-stain solution: 0.2% silver nitrate, 0.03% formaldehyde.

17. Development solution: 6% sodium carbonate, 0.02% formaldehyde, 0.0004% sodium thiosulfate.

18. Phosphate-buffered saline (PBS): 8 g NaCl, 0.2 g KCl, 1.15 g $Na_2HPO_4 \cdot 7H_2O$, 0.2 g KH_2PO_4 per liter, pH approx 7.3.

3. Method

3.1. Sample Preparation

1. Plate 10^6 cells in 10-cm² cell-culture dishes with RPMI-1640 medium containing 10% fetal bovine serum, (FBS) and incubate at 37°C for 48 h.
2. Perturb specific signaling pathways by various methods (e.g., by expressing constitutively active mutants of signaling effector molecules, or by treating cells with reagents that specifically activate or inactivate signaling molecules).
3. Wash cells twice with ice cold PBS, and remove residual PBS. To each dish add 0.7 mL of lysis buffer A (for pI 4–7 separations) or lysis buffer B (for pI 6–11. separations).
4. Harvest each cell lysate by scraping, transfer to a 1.5-mL microcentrifuge tube, and incubate at room temperature for 1 h, with occasional vortex mixing. Centrifuge (200,000 g) at 21°C for 1 h to remove insoluble material.
5. Determine protein concentration by Bradford assay (PIERCE, Rockford, IL). Before assay, the lysate should be diluted with five volumes of deionized water to prevent interfering effects of lysis buffer components.

3.2. Immobilized pH Gradient First Dimension

3.2.1. Isoelectric Focusing for pI 4–7 Range (see **Note 1**)

1. In a 1.5-mL microcentrifuge tube, mix the protein sample (150 µg) with lysis buffer A to a final volume of 175 µL.
2. Add 175 µL of rehydration buffer A and incubate at room temperature for 20 min.
3. Pipet the sample mixture into one channel of the drystrip reswelling tray.
4. Remove an Immobiline drystrip pH 4.0–7.0 gel from the freezer and peel off the plastic backing.
5. Gently lay the drystrip gel (gel side facing down) on the sample mixture in each slot.
6. Using forceps, gently lift and lower the pointed end (anodic or low pI end) of the gel strip and slide it back and forth along the surface of the sample mixture, in order to remove bubbles and evenly spread the sample under the strip.
7. Repeat **step 6**, lifting the flat end (cathodic end) of the gel strip.
8. Repeat **step 6**, lifting the pointed end.
9. Overlay each strip with 2 mL light mineral oil.
10. Repeat **steps 3–9** for each sample.
11. Place the reswelling tray in a chamber at 25°C (e.g., onto the Multiphor flatbed apparatus) and allow the gel strips to hydrate for 16 h.
12. Prepare the Multiphor II apparatus by preequilibratring to 10°C with a circulating water bath. Have ready the electrode tray, the electrodes, the plastic strip aligner, two sheets (15 × 30 cm) of Whatman 3MM paper, two IEF

electrode strips each cut to a length of 100 mm (Amersham-Pharmacia), and light mineral oil.

13. After gel rehydration, remove the gel strips from the reswelling tray, gently rinse with deionized water, and place them onto a 15 × 30 cm sheet of Whatman 3MM paper (plastic side down). Wet another 3MM sheet with deionized water, blot briefly to remove excess water, and place gently onto the surface of the hydrated gel strips to remove excess buffer.
14. Immediately transfer gel strips into the grooves of the plastic strip aligner (Amersham-Pharmacia).
15. Soak the two IEF electrode strips with 500 μL deionized water and remove excessive water by blotting on 3MM paper.
16. Place each IEF electrode strip perpendicular to the aligned gel strips at both cathodic and anodic ends, leaving about 1 mm of gel uncovered at each end.
17. Pipet 20 mL of light mineral oil into the electrode tray.
18. Place the gel strips, electrode strips, and plastic strip aligner (set up in **step 13**) into the electrode tray.
19. Position the electrodes onto the tray and press them down to contact the electrode strips. Precise electrode alignment and contact at this step is important for good IEF resolution.
20. Pipet 20 mL of light mineral oil onto the cooling plate of the Multiphor flatbed apparatus.
21. Place the assembled electrode tray onto the cooling plate.
22. Pipet mineral oil (approx 60 mL) into the electrode tray to completely cover the gel strips.
23. Begin IEF according to the protocol in **Table 1**.

Table 1. First Dimension (IEF) Electrophoresis Conditions for Immobilized pH Gradient Drystrips

	pI 4–7			pI 6–11		
Temperature	20°C			25°C		
Maximum current	0.25 mA / strip			0.5 mA / strip		
Maximum power	1 W / strip			0.5 W / strip		
Electrophoresis	V_{start}	V_{end}	Time	V_{start}	V_{end}	Time
	0V	300V	0.01/h	0V	300V	0.01/h
	300V	300V	1/h	300V	300V	3/h
	300V	3500V	9/h	300V	1400V	6/h
	3500V	3500V	18/h	1400V	1400V	10/h
				1400V	3500V	3/h
				3500V	3500V	2/h

24. At the end of IEF, remove the gel strips from the tray, wrap them between folded sheets of Saran Wrap, and store in a paper envelope at −80°C until equilibration for the second-dimension step.

3.2.2. Isoelectric Focusing for pI 6–11 Range (see **Note 2**)

1. Pipet 350 μL of rehydration buffer B into each channel of the Immobiline drystrip reswelling tray.
2. Remove an Immobiline drystrip pH 6.0–11.0 gel from the freezer and peel off the plastic backing.
3. Gently lay the drystrip gel (gel facing down) onto the rehydration buffer in each slot.
4. Using forceps, gently lift and lower the pointed end of the strip and slide it back and forth along the surface of the sample mixture to remove bubbles.
5. Repeat **step 4**, lifting the flat end of the strip.
6. Repeat **step 4**, lifting the pointed end.
7. Overlay each strip with 2 mL light mineral oil.
8. Repeat **steps 1–7** for each sample.
9. Place the drystrip reswelling tray at 30°C (e.g., onto the Multiphor flatbed apparatus) and allow the gel strips to hydrate for 16 h.
10. Prepare the Multiphor II apparatus by preequilibrating to 10°C with a circulating water bath. Have ready the samples, the electrode tray, the plastic strip aligner, two sheets (15 × 30 cm) of Whatman 3MM paper; IEF electrode strips cut to lengths of 25, 50, and 100 mm (Amersham-Pharmacia), and light mineral oil.
11. After gel rehydration, remove the gel strips from the reswelling tray, gently rinse with deionized water, and place them on a 3MM sheet. Wet another sheet with deionized water, blot briefly to remove excessive water, and place gently onto the surface of the hydrated gel strips to remove excessive buffer.
12. Immediately transfer the gel strips into the grooves of the plastic strip aligner.
13. Cut two IEF electrode strips into 25 mm and 50 mm long pieces, and trim them at one end to form an arrowhead (**Fig. 1**). Soak the 50 mm piece with 250 μL of rehydration buffer C, and place them onto an aligned gel strips at the (square) cathodic ends (basic side). Repeat for each sample.
14. Mix the protein sample (100 μg) with lysis buffer B to a final volume of 100 μL.
15. Soak the 25 mm pieces with each protein sample mixture and place them on the (pointed) anodic ends (acidic side) of each aligned gel strip (**Fig. 1**). Repeat for each sample.

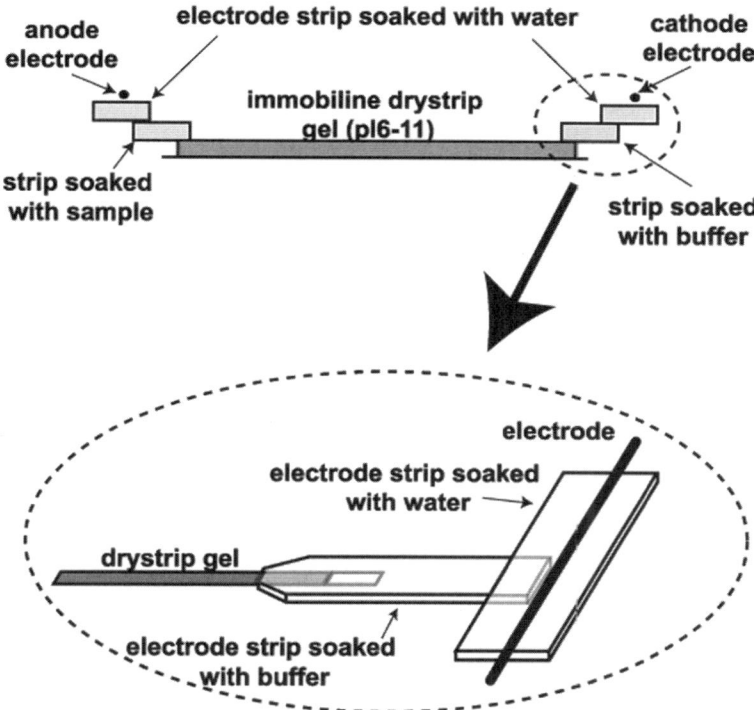

Fig. 1. Configuration of sample application for IEF with pI 6–11 strips Paper IEF electrode strips (25 and 50 mm long) are trimmed at one end to form arrowheads. For Sample application, soak the 25 mm piece with diluted protein sample and place onto the anodic end of the gel strip (pointed low pI end). The 50 mm strips are soaked with rehydration buffer and placed at the cathodic end (square high pI end) of the gel strip.

16. Soak the 100 mm electrode strips with 500 μL deionized water and remove excess water by blotting with 3MM paper. Place the electrode strips perpendicular to and on top of the short electrode strip pieces placed on the ends of aligned gels (**Fig. 1**).

17. Pipet 20 mL of light mineral oil into the electrode tray.

18. Place the set-up (strip aligner, gel strips, strip pieces, and electrode strips assembled at **step 16**) into the electrode tray.

19. Position the electrodes onto the tray and press them down to contact the electrode strips. Precise electrode alignment and contact at this step is important for good IEF resolution.

20. Pipet 20 mL of light mineral oil onto the cooling plate of the Multiphor flatbed apparatus.

21. Place the assembled electrode tray onto the cooling plate.

22. Overlay each strip with approx 60 mL light mineral oil to completely cover the gel strips.
23. Begin IEF according to the protocol given in **Table 1**.
24. At the end of IEF, remove the gel strips from the tray, wrap them between folded sheets of Saran wrap, and store in a paper envelope at −80°C until equilibration for the second-dimension step.

3.3. Equilibration of IEF Gel Strips

1. Add 20 mL of equilibration solution A to glass screw cap tubes (PYREX), cat. no. 9825). Remove strips from −80°C storage, place each into a separate tube, and rock for 20 min.
2. Add 20 mL of equilibration solution B to separate glass screw-cap tubes. Transfer strips from solution A to solution B and rock the tubes for 20 min.
3. After the second equilibration, remove excess solution by placing the strips onto 3MM paper, bending the strips slightly to allow them to stand on their side edges.

3.4. SDS-PAGE Second Dimension

1. Assemble glass plates for SDS-polyacrylamide gel electrophoresis (PAGE) in gel casting stands (e.g., 20 × 20 × 0.1 cm plates and spacers for the Bio-Rad Protean II xi system, Bio-Rad, Hercules, CA).
2. Assemble and appropriate gradient maker (e.g., *Hoeffer SG30*, for a 30 mL gradient) in line with a peristaltic pump (e.g., *Rainin Dynamax Model RP-1*, 1-mm diameter tubing) set to a flow rate of 2.5 mL/min. Attach the flow from the pump to the gel plates using a 18½-gauge needle inserted into the peristaltic tubing.
3. Mix light- and heavy-acrylamide solutions, and add to the gradient maker chambers. The volume of each solution is half the amount needed to fill each gel sandwich (e.g., 17.5 mL of each solution for 20 × 20 × 0.1 cm gels). Add the heavy solution to the mixing chamber with a magnetic stir bar and the light solution to the reservoir chamber. Place the gradient maker onto a magnetic stirrer.
4. Start the mixer, turn on the peristaltic pump, open the gradient maker flow valves, and allow the gel solution to fill the gel sandwich from the top of the gel.
5. After the gel solution has been poured, overlay with water-saturated isobutanol to exclude air bubbles and ensure a level surface at the top of the gel.
6. Allow gel polymerization for 1 h at room temperature, then replace the isobutanol with deionized water. Store the polymerized gels at 4°C until use.
7. Assemble the slab gel electrophoresis unit, precooling the unit and the running buffer to 10°C using a circulating water bath.

8. Replace the deionized water overlay with SDS-PAGE running buffer, and equilibrate for 10 min.
9. Remove the running buffer from the top of the gels, and blot the excess buffer with 3MM paper.
10. Immediately transfer the IPG gel strip equilibrated in equilibration buffer B onto the top of the gel. Use a flat plastic edge (e.g., a thin plastic ruler) to remove air bubbles from underneath the gel strip. Overlay the strip with 2–3 mL of melted 0.5% agarose solution (dissolved in running buffer) taking care to avoid bubbles.
11. Assemble the gels into the electrophoresis chamber and begin electrophoresis, running at constant current of 7mA/gel, and constant temperature of 10°C. The total running time for 20 cm × 20 cm gels is approx 18 h.

3.5. Silver Staining (see Note 3)

All steps are carried out at room temperature with gentle agitation on a rotating platform.

1. After the second-dimension electrophoresis, remove each gel from the glass plates and incubate overnight in 150 mL of gel fixation solution.
2. Rinse gels with 150 mL of 50% ethanol in water for 2 × 10 min.
3. Rinse gels with 150 mL of 30% ethanol in water for 10 min.
4. Soak gels in 100 mL of sensitivity enhancing solution for 1 min.
5. Rinse gels with 200 mL of deionized water for 1 min. Repeat the rinse.
6. Impregnate gels in 150 mL of silver-stain solution for 20 min.
7. Rinse gels with water for 10 s. Rinse again for 10 s.
8. Develop stained proteins with 100 mL of development solution. It takes approx 2–4 min until spot intensities reach maximal levels.
9. Stop development by adding 40% ethanol, 12% acetic acid and incubating for 10 min.
10. Rinse gels with at least four changes of deionized water over 2 h.
11. Scan gel images and analyze spot changes using appropriate computer software. Vendors include GeneBio (Melanie, GeneBio. Geneva, Switzerland), Compugen (Z3, Jamesburg, NJ), Bio-Rad (PDQuest, Hercules, CA), Progenesis (Phoretix, Durham, NC). We have found software programs that allow analysis by image flickering *(8,9)* (e.g., Melanie) to be particularly useful.
12. Stained gels can be stored in water at 4°C until excision and in gel digestion of protein spots.

3.6. Preparation of Samples for Mass Spectrometry

3.6.1. Proteolytic Digestion of Gel-Resolved Proteins

All steps are carried out in a clean hood using gloves to minimize contamination from hands and dust. High-performance liquid chromatography

(HPLC)-grade reagents should be used to make all solutions and bags of plastic tubes and pipet tips should be handled to minimize dust contamination.

1. Precisely excise protein spots from gels and transfer to 1.5-mL microcentrifuge tubes (*see* **Note 4**).
2. Add 500 μL of 50 m*M* sodium thiosulfate, 15 m*M* potassium ferricyanide to gel slices. Incubate for 2 min at room temperature by vortex mixing. Continue until the brown protein staining disappears.
3. Rinse with 1.5 mL water for 5 min. Repeat twice.
4. Add 500 μL of 100 m*M* ammonium bicarbonate and incubate at room temperature for 30 min by vortex mixing.
5. Remove the buffer and add 500 μL of 50% acetonitrile, and 50 m*M* ammonium bicarbonate. Incubate at room temperature for 30 min by vortex mixing.
6. Transfer gel pieces into a clean 0.65-mL microcentrifuge tube and remove as much buffer as possible using a micropipettor.
7. Add 50 μL of 100% acetonitrile and incubate at room temperature for 10 min to shrink gel pieces.
8. Remove the acetonitrile solution from the gel pieces by drying to completion in a clean speedvac concentrator (e.g., for 20 min).
9. Dilute concentrated trypsin solution (e.g., Promega Modified Porcine Trypsin; Princeton Separations Endopeptidase Modified Porcine Trypsin) with 25 m*M* ammonium bicarbonate to a final concentration of 20 μg/mL.
10. Reswell dehydrated gel pieces with 20 μg/mL Trypsin by adding solution directly onto gel pieces in 3–5 μL aliquots, waiting 10 min between each addition. Continue until gel pieces are fully hydrated (typically approx 10 μL).
11. After gel pieces are hydrated, add 25 m*M* ammonium bicarbonate to cover the gel pieces (approx 20 μL). Incubate at room temperature for 20 min.
12. When fully swelled, add 25 m*M* ammonium bicarbonate until gel pieces are covered at the top by approx 1 mm of buffer.
13. Incubate gel pieces at 37°C for 24 h with gentle agitation.
14. After digestion, add 1 μL of 88% formic acid to quench trypsinization and sonicate in a bath sonicator for 20 min.
15. Transfer the supernatant to new 0.65 μL microcentrifuge tube and reduce the volume to 5–10 μL with a speedvac concentrator. Do not Speed-Vac the sample to dryness.

3.6.2. Peptide Concentration and Purification

Concentrate the peptide digest and remove salts and contaminants using a reversed phase resin (e.g., Millipore Zip Tip μ-C18 columns). This step is critical for successful matrix-assisted laser desorption/ionization-time-of-flight mass spectrometry (MALDI-TOF MS).

1. Using a micropipetor, wet the ZipTip resin with 4 × 10 μL passes of 100% methanol. (One pass is equivalent to drawing the solvent up into the Zip Tip, then expelling it completely.)
2. Rinse the resin with 4 × 10 μL passes of 1% (v/v) trifluoroacetic acid (TFA).
3. Adsorb peptides to the resin with 15–20 passes of the peptide digest solution.
4. Wash the resin with 4 × 10 μL passes of 1% formic acid.
5. Aliquot 2 μL of 60% acetonitrile, 1% formic acid to a second 0.65-mL microcentrifuge tube. Elute the peptides from the resin with 20 passes of the acetonitrile solution into the second tube. This eluate is ready for MALDI-TOF MS analysis. For liquid chromatography based MS, the purified sample must be dried by Speed-Vac centrifugation and resuspended in 0.1% formic acid.

4. Notes

1. The protocol described here represents a modification of the *in-gel rehydration* method described by Gorg et al. *(6)*. Modifications include the temperature regulation during gel rehydration and the electrophoresis conditions for protein focusing. In our experience, resolution and reproducibility of protein separation were significantly improved by rehydrating at a constant temperature of 25°C, and by extending the focusing time to approx 80,000v/h.
2. The protocol described here represents a modification of the *paper-bridge application* method described by Hoving et al. *(7)*. The main modification is the regulation of temperature at gel-rehydration step, where in our experience, the penetration of proteins into the gel strip is improved by rehydrating the strips at a constant temperature of 30°C prior to sample application.
3. This protocol was originally reported by Blum et al. *(11)*, where we have incorporated several modifications. This method provides high sensitivity of detection, allowing approx 5000 protein spots to be visualized upon combining pI 4–7 with pI 6–11 2-DE gels (**Fig. 2**). However, the method leads to high fixation of proteins which precludes efficient peptide recovery. Therefore, we often visualize protein changes using the high-fixation method, but recover proteins from replicate gels stained with lower fixation silver-staining protocol described by Shevchenko et al. *(12)*.
4. Protein spots can be cut manually using P1000 pipet tips cut with a razor blade to the appropriate diameter, then beveled using an X-acto knife to create a sharp edge at the tip. Insert the sharpened pipet tip onto a plastic transfer pipet

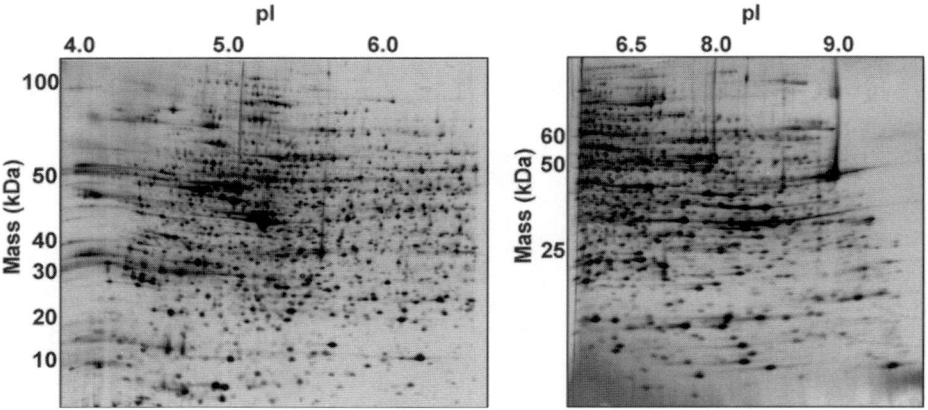

Fig. 2. Silver-stained 2-DE gels of cultured human melanoma cells (WM35) using immobilized pH gradient gel strips. Second dimension SDS-PAGE was carried out using 8–18 % acrylamide gradient gels. **(A)** 150 µg of WM35 cell lysate was resolved on a pI 4–7 gel strip (18 cm) by in-gel rehydration (**Subheading 3.2.1.2.**). **(B)** 100 µg of WM35 cell lysate was resolved on a pI 6–11 gel strip (18 cm) by the paper bridge application method (**Subheading 3.2.2.2.**). Approximately 5000 protein spots can be visualized on these gels after silver staining by the method in **Subheading 3.5.**

(Sarstedt no. 86-1171), and use the resulting instrument to excise the proteins from gels immersed in water. The gel piece can be easily drawn into the pipet tip and transferred to a storage tube. In some cases, multiple gels should be harvested to ensure peptide detection. The number of protein spots required for the identification of unknowns vary from 2–4 for high-abundance proteins (>10 ng) to 20–30 for low-abundance proteins (1–2 ng per spot). Typically, 500 fmol is adequate for protein identification.

References

1. Klose, J., Nock, C., Herrmann, M., et al. (2002) Genetic analysis of the mouse brain proteome. *Nat. Genet.* **2,** 385–393.
2. Lewis, T. S., Hunt, J. B., Aveline, L. D., et al. (2000) Identification of novel MAP kinase pathway signaling targets by functional proteomics and mass spectrometry. *Mol. Cell* **6,** 1343–1354.
3. Kanamoto, T., Hellman, U., Heldin, C. H., and Souchelnytskyi, S. (2002) Functional proteomics of transforming growth factor-beta1-stimulated Mv1Lu epithelial cells: Rad51 as a target of TGFbeta1-dependent regulation of DNA repair. *EMBO J.* **21,** 1219–1230.
4. Predic, J., Soskic, V., Bradley, D., and Godovac-Zimmermann, J. (2002) Monitoring of gene expression by functional proteomics: response of human lung fibroblast cells to stimulation by endothelin-1. *Biochemistry* **41,** 1070–1078.

5. Gerner, C., Frohwein, U., Gotzmann, J., et al. (2000) The Fas-induced apoptosis analyzed by high throughput proteome analysis. *J. Biol. Chem.* **275,** 39,018–39,026.

6. Adams, P., Fowler, R., Howell, G., et al. (1999) Defining protease specificity with proteomics: a protease with a dibasic amino acid recognition motif is regulated by a two-component signal transduction system in Salmonella. *Electrophoresis* **11,** 2241–2247.

7. Gorg, A., Obermaier, C., Boguth, G., et al. (2000) The current state of two-dimensional electrophoresis with immobilized pH gradient. *Electrophoresis* **21,** 1037–1053.

8. Hoving, S., Gerrits, B., Voshol, H., et al. (2002) Preparative two-dimensional gel electrophoresis at alkaline pH using narrow range immobilized pH gradients. *Proteomics* **2,** 127–134.

9. Lemkin, P. F. (1999) Comparing 2-D electrophoresis gels across internet databases. *Methods Mol. Biol.* **112,** 393–410.

10. Appeal, R. D. and Hochstrasser, D. F. (1999) Computer analysis of 2-D images. *Methods Mol. Biol.* **112,** 363–381.

11. Blum, H., Beier, H., and Gross, H. J. (1987) Improved silver staining of plant proteins, RNA and DNA in polyacrylamide gels. *Electrophoresis* **8,** 93–99.

12. Shevchenko, A., Wilm, M., Vorm, O., and Mann, M. (1996) Mass spectrometric sequencing of proteins from silver -stained polyacrylamide gels. *Anal. Chem.* **68,** 850–858.

4

A High-Throughput Mammalian Cell-Based Transient Transfection Assay

Daniel J. Noonan, Kenneth Henry, and Michelle L. Twaroski

Summary

In eukaryotic organisms gene expression is regulated through a variety of upstream transacting factors (transcription factors) whose primary function appears to be the targeting of coregulatory protein complexes, which interact with basal transcription machinery to define the relative rate of transcription for a specific gene. Understanding the regulatory forces mediating transcription factor activity has been the focus of both academic and industrial research efforts over the past 15 yr, and in this time frame a variety of methodologies have been developed for reconstituting and assaying transcription factor activities in mammalian cell environments. Presented here is a high-throughput version of one of these methodologies that can be readily adapted to the screening of a variety of transcription factors. This technology utilizes co-transfection of mammalian expression and luciferase reporter plasmids to reconstitute transcription events in a mammalian host cell. Included is a detailed protocol for the use of a 96-well plate format, along with a variety of cost-effective measures that can be implemented to facilitate the use of the technology in the average low budget academic laboratory.

Key Words: Nuclear receptor; luciferase reporter; $CaPO_4$ precipitation; 96-well plate; plate-reading luminometer.

1. Introduction

In eukaryotic cells, a class of transcription regulatory proteins known as upstream-transcription factors function to regulate gene expression through

From: *Methods in Molecular Biology, vol. 284:*
Signal Transduction Protocols
Edited by: R. C. Dickson © Humana Press Inc., Totowa, NJ

binding-specific DNA sequences in the regulatory regions (promoters) of the genes whose activity they are to modulate *(1)*. The primary purpose of this event appears to be targeting specific coregulatory complexes to the promoter region, which in turn facilitates or inhibits the transcriptional activities mediated by the RNA polymerase complex *(1)*. The activity of many of these upstream-transcription factors is directly or indirectly linked to a signaling pathway initiated by some ligand-receptor interaction, and many of the ligand receptor interactions have been linked to vital metabolic functions and disease processes. The simplest example of this would be the steroid/nuclear receptor family of proteins, wherein the intracellular receptors for such lipophilic compounds as estrogens, androgens, corticosteroids and prostaglandins are receptor proteins converted into upstream transcription factors upon binding their ligand *(2–5)*. These compounds and their receptors regulate metabolic activities and eukaryotic development in response to a variety of internal and external signals. Furthermore, misregulation of their function has been associated with a variety of diseases and cancer *(6)*. For this reason many of these receptors have been the targets for large drug-discovery efforts in a variety of pharmaceutical companies. In this regard, methods for reconstituting transcription factor activity in mammalian cell culture systems have been developed and perfected over the past two decades *(7,8)*. This technology has proven to be a valuable tool for defining structure and function relationships between transcription factors, the signaling pathways that activate them, and their mechanism for modulating gene expression. Commercially, these assays have also proven to be valuable tools for screening for small-molecule agonists and antagonists of transcription-factor activity *(7,8)*. Over the past decade a variety of high throughput screens have been developed that function through the reconstitution of the transcription factor activity in some foreign host cell. Two basic screening approaches have been developed. One attempts to identify modulators of ligand receptor-mediated events *(7)*, and the other attempts to identify modulators of transcription factor–coregulatory complex interactions *(9)*. In this methods article, we will present a modified approach to high-throughput screening first developed back in the early 1990s for the screening of small molecule agonists and antagonists of steroid receptor-mediated transcription events. This technology utilizes mammalian expression and luciferase reporter plasmids to reconstitute steroid and nuclear receptor-mediated transcription events in a host cell. Furthermore, it converts a cumbersome 10-cm plate assay into an assay that is performed in 96-well plates. Although this technology was originally developed for high throughput screening in a pharmaceutical company environment, we have found it to be both cost effective and highly utile in our academic pursuits.

2. Materials

In this section we have listed the reagents and reagent preparations necessary to implement the co-transfection assay we utilize in our laboratory, and the one detailed in the Methods section. Generic laboratory chemicals (e.g., NaCl, $MgCl_2$, KH_2PO_4, $K_2H_2PO_4$, $NaHPO_4$, NaH_2PO_4, Tris-HCl, EGTA, glycerol, BSA, etc.) are standardly purchased at their highest reagent grade from a reputable distributor such as Sigma or Fisher. Many variations and sources can also apply to several of the specialty reagents used in this protocol, but we will list our sources of these specialty reagents, and where applicable, we will also attempt to amplify on possible alternatives. Transient transfection assays are not cheap to run, but there are places that one can economize the assay substantially. Using the "home brew" approach outlined here, it can be estimated that, for disposable materials alone, the assay ranges from $5 to $7 per 96-well plate. Depending on the kit used, this could prove to be a substantial savings over the available "kit science" approaches, which can range anywhere from $30 to $80/plate. Beyond the reagents and luminometer listed here, two other fairly nonstandard laboratory purchases that are basic necessities in the high throughput technology are a good multichannel pipettor and disposable reagent reservoirs. These items can be purchased from any of the major distributors of scientific equipment and supplies.

Unless otherwise indicated, all reagents should be prepared with the highest quality double-distilled water (ddH_2O) available. We have been more detailed in describing the preparation of some of these reagents, owing both to their importance and in hopes that this will facilitate the process for the reader.

1. 0.2 *M* Phenylmethanesulfonyl fluoride (PMSF) solution: Dissolve 0.348 g PMSF in 10 mL isopropyl alcohol. *Caution: PMSF is toxic.* Use gloves and a mask when weighing.
2. Ligand base: Into 80 mL of ddH_2O add 0.851 g Trizma phosphate, 1.0 g bovine serum albumin (BSA), 0.152 g EGTA, 2.0 g CHAPS (Amresco, cat. no. 0465; *see* **Note 1**), 1.0 g L-Lecithin (Sigma, St. Louis, MO, cat. no. P-9671), 15 mL glycerol. Adjust pH 7.8 $+/-.02$, QS to 100 mL filter-sterilize and store at 4°C.
3. 2X HBS Buffer: Into 950 mL ddH_2O add 10 g HEPES and 16 g NaCl. Adjust the pH to 7.10 $+/- 05$ (*see* **Note 2**), QS to 1 L, filter-sterilize, and store at room temperature.
4. 100X Phosphate: Into 95 mL of ddH_2O dissolve 0.994 g Na_2HPO_4 and 0.966 g $NaH_2PO_4 \cdot H_2O$. QS to 100 mL, filter-sterilize, and store at room temperature.
5. 2 *M* $CaCl_2$: Into 95 mL of ddH_2O dissolve 36.61 g $CaCl_2 \cdot 4H_2O$ (EM Sciences, Fort Washington, PA, cat. no. 2384-2). QS to 100 mL, filter-sterilize, and store at room temperature.

6. Phosphate-buffered ATP Solution (KHPO$_4$-ATP): Into 80 mL of ddH$_2$O dissolve 0.116 g KH$_2$PO$_4$ and 1.59 g K$_2$HPO$_4$. Adjust pH to 7.80 +/− .05. Add 0.221 g ATP (disodium salt, Sigma cat. no. A5394). Adjust pH to 7.80 +/− .05, QS to 100 mL, filter-sterilize, divide into 25-mL aliquots and store at −20°C. Good for 6 mo.

7. 0.1 *M* Potassium phosphate buffer: Into 900 mL of ddH$_2$O dissolve 1.16 g KH$_2$PO$_4$ and 15.9 g K$_2$HPO$_4$. Adjust pH to 7.8 +/−.05, QS to 1 L, filter–sterilize, and store at room temperature.

8. 10X Luciferin stock solution: Dissolve 100 mg of D-Luciferin (Promega, Madison, WI, cat. no. PBI 1300B; *see* **Note 3**) in 33.1 mL 0.1 *M* potassium phosphate buffer, pH 7.8 (*see* **step 7**). Divide into 5-mL aliquots, cover individual aliquots with aluminum foil and store at −20°C. Dilute to 1X in 0.1 *M* potassium phosphate buffer, pH 7.8, just prior to use. *Keep shielded from light at all times.*

9. 1*M* MgCl$_2$-50 m*M* Tris-HCl: Into 80 mL of ddH$_2$O, dissolve 20.33 g MgCl$_2$, 053 g Tris-HCl, and 0.20 g Tris base. Adjust pH to 7.8 +/−.1, filter sterilize, and store at room temperature.

10. *O*-Nitrophenyl D-Galactopyranoside (ONPG) solution (*see* **Note 4**): Into 80 mL of ddH$_2$O, dissolve 0.2 g ONPG (Sigma cat. no. N-1127), 0.117 g NaH$_2$PO$_4$·H$_2$O and 1.30 g Na$_2$HPO$_4$. Adjust pH to 7.8 +/− .05, QS to 100 mL, filter-sterilize, divide into 50 mL aliquots and store at −20°C. Good for 6 mo.

11. Dithiothreitol (DTT) 0.10 *M*: Into 10 mL of ddH$_2$O, dissolve 0.155 g of DTT. Divide into 0.5-mL aliquots and freeze at −20°C. Good for 1 yr.

3. Methods

3.1. Choosing an Expression Vector

One of the first and often overlooked considerations when instituting the transient transfection assay is choice of vector system to be used. Almost every manufacturer of molecular biology reagents has its own mammalian expression system and most have multiple permutations of their system. There are far too many to mention, which is not the object of this review, so we will simply overview some of the considerations one needs to reflect on when choosing a vector.

Perhaps the most important consideration is the promoter system driving your gene or transcription factor of interest. As with latex gloves, these generally come in small, medium, and large. This of course refers to the level of expression generated by the specific promoter elements. Most often researchers want high levels of transcription-factor expression. This generally is to increase sensitivity and to overcome shortfalls in transfection efficiency. One needs to be cautious about choosing an extremely high level of expression vector, especially when attempting to investigate some question of physiological relevance with respect to the transcription factor being studied. On the other hand, if one is screening chemical compounds for transcription-factor activation, sensitivity may be of great importance. Perhaps the most popular and most active of these expression vectors are the cytomegalovirus (CMV)-driven expression systems. Less active but sometimes more uniform in their expression profiles are the Rous sarcoma virus (RSV) and simian virus (SV-40)-driven promoters.

A second concern that we, and others, have encountered serendipitously with the various mammalian expression vectors used in these studies is the presence of cryptic enhancers and transcription factor-binding sites *(10–12)*. This appears to be especially true for many of the CMV-driven vectors and some reporter vectors. It is therefore judicious to run a vector-only control with every assay, and, if examining the effects of some other expression construct on the activity of your transcription factor, make sure it is not the vector producing the effects.

3.2. Choosing a Reporter Vector

Here again there are a variety of systems to choose from. Firefly luciferase has been our reporter of choice simply because of its sensitivity and the availability of its substrate. A variety of other luminescent enzymes have been cloned more recently along with their available substrates *(13–15)*. In addition there are a variety of fluorescing proteins that have been, or can be, adapted to function as reporters *(15–17)*. Use of these in high-throughput assays will require a plate-reading fluorimeter, but might be worth the investment if one wishes to develop scenarios that incorporate some form of fluorescence energy transfer (FRET) into their readout *(9,17)*. Finally, there are reporters that use either chromophores (e.g., β-galactosidase) or radioactive isotopes (e.g., chloramphenicol acetyltransferase [CAT]). These reporters are generally thought to be either too cumbersome or not sensitive enough for high-throughput screening protocols.

3.3. Choosing a Host

The primary concerns when selecting a host cell line for transient transfections are its transfectability, its endogenous level of transcription factors, and its biological relevance to your transcription factor. As discussed next, there are several approaches available for transfecting DNA into mammalian cells. Several cell lines appear to be highly receptive to this process, whereas most primary cells are very refractive to it. Cell lines we have had reasonable success with using the least expensive technology available ($CaPO_4$ precipitation) include the human embryonic kidney (HEK) 293 cell line, the human hepatocellular carcinoma cell line

Hep G2, and the African Green Monkey kidney cell line COS-7. Our research utilizes this assay to reconstitute and evaluate steroid receptor-mediated transcription events *(18,19)*. One complication we have encountered with cell lines is background noise from endogenous receptors. For this reason, reporter-only controls are necessary for all transfections. Identifying a host cell system that fits your needs is a combination of literature research and trial and error.

3.4. Choosing a Transfection System

Three basic approaches—$CaPO_4$ precipitation, liposome technologies and electroporation—have usually been employed to transfect DNA into mammalian cells. Academic scientists with low budgets find it important to identify cost-effective measures to implement research of this nature. The transfection system we will describe in this article is the most economical we could find: the $CaPO_4$ precipitation methodology *(18,20)*. We might also point out that, like all of the technologies available, it has its strengths and weaknesses. The $CaPO_4$ precipitation technology is the cheapest and is fairly reproducible once the system is defined, but of the three it is the lengthiest and the most limited in host cell choices. The liposome technology is easy, quick, and useful with a large number of host cells, but from a reagent perspective it the most expensive and struggles at times with reproducibility issues. Electroporation is moderately cheap once one has made the purchase of the machinery, requires very little time, and can be used with a variety of cell types, but can suffer from reproducibility and cell viability issues, which can be detracting in high-throughput studies.

3.5. Choosing a Plate-Reading Luminometer

Choosing the right plate-reading luminometer can also be a serious consideration. To begin with, they are not inexpensive. Plate-reading luminometers range in price from $10,000 to $90,000, but we would caution anyone serious about doing these types of studies to try to get the best machine they can find for the type of studies they hope to implement. This doesn't mean buying the $90,000 machine, but rather purchasing the highest-quality machine for the type of assay they wish to perform. Contact vendors, set up demos, and talk to other investigators performing these types of assays. The complexities of these luminometers vary considerably and the software packages that come with them can either decrease or increase your workload substantially. One consideration that affects both the type and the price of the luminometer chosen is the choice of reporter and reporter system one hopes to use. If reading luciferin-flash kinetics or doing dual luciferase assays, we have found that luminometers that have dual-injector systems seem to give more reproducible results. If simply reading glow kinetics as generated by Renilla luciferase, then plate readers without injectors work fine. In our laboratory, we are currently running a Molecular Devices Lmax plate reading luminometer with dual injectors. This luminometer allows us to do either single luciferase or dual luciferase assays and the software package and service support are very good. This is not to say there are not comparable or better systems out there, so shop around before you buy.

3.6. Robotics or Manual

If you are serious about implementing a high-throughput assay to screen a large number of chemical compounds and have the money available, you may wish to consider including robotic workstations into your assay to reduce human error and streamline your operation. At first this may seem impractical for something like the $CaPO_4$ precipitation protocol, but when we were designing this 96-well plate assay back in the early 1990s, we had programmable Bio-Mek workstations set up in laminar flow hoods that created all of the ligand dilutions, distributed the transfected cells into 96-well plates, prepared the luminometer plates, prepared the β-galactosidase plates, distributed the lysates, and read the β-galactosidase. This relegated the scientists to doing the transfection and operating the computers.

3.7. Kit Science or Home Brew

We believe our preference here (Home Brew) should be clear to the reader, but I feel it important to point out that there are many commercial luciferase assay systems available now (*see* http://www.biocompare.com/molbio.asp?catid= 1620 for a comparison of the various systems available), some of which have published track records to support their credibility. The upside of these systems is that you simply follow the manufacturer's protocol; the downsides are the price and some of the innate limitations on troubleshooting the systems.

3.8. Preparing Plasmids

The use of clean plasmid preps can be critical to obtaining reproducible results with transfection technologies. Two pieces of advice relating to the preparation of plasmids for transfection are: check the integrity of the plasmid preparation prior to using it in the assay and, if storing a plasmid for over 1 mo, freeze it at $-20°C$. We normally use double cesium-banded plasmid preparations. Others have used commercial kits (e.g., Qiagen, Valencia, CA) for plasmid preparation with some success, but it is important that these plasmid preparations be examined on an agarose gel for contamination with genomic DNA and RNA, which will occur occasionally if the capacity of the technology is pushed to its upper limits. One final note of interest that we have observed while using the $CaPO_4$ technology is that the salts used in the final precipitation of the plasmid preparation can impact the $CaPO_4$ precipitate formation. When potassium salts (e.g., reprecipitating with 1/10th volume of solution 3 of the alkaline lysis procedure) (*21*) are used in the final precipitation of plasmid DNA, the $CaPO_4$ precipitate formed in the transfection protocol generally is more robust and flocculent, whereas if NH_4 acetate or sodium salts are used, the precipitate is finer. This can be good or bad depending on the cell type one is transfecting into. For example, transfection efficiency into Hep G2 cells appears to be improved substantially by the flocculent precipitate, whereas the finer precipitate appears to work better in the COS 7 cells. Again, this is a trial and error type of technology so our advice is to test the water and do pilot studies before jumping into the deep end of the pond.

3.9. Preparing Cells

Transfection of mammalian cells is still somewhat of a mystery science with regard to defining the correct conditions for transfection. Depending on the technology used, a variety of factors appear to impact the efficiency of transfection. These include pH, salts, cell cycle, and integrity of one's DNA preparation. We have also found that using a logarithmic growing population of cells generally gives us our highest transfection efficiency. For this purpose, we routinely passage cells on consecutive days, plating at approx 70% confluency each day. In addition, newly thawed cells need to be passaged for a week or so before they begin to transfect with any efficiency. Finally, efficiency also seems to decline as cells reach higher passage numbers (e.g., 40–50 passages for the cells used here).

3.10. Our Home Brew Protocol

This is the protocol used on a daily basis in our laboratory to analyze ligand activation of steroid receptor systems reconstituted in mammalian cells. Our mammalian expression plasmid of choice is a pRSV vector (*22*) and our DNA binding elements specific for our receptor plasmid are cloned into a pBL luciferase vector (*22*) that utilizes a thymidine kinase minimal promoter for facilitation of transcription initiation. Using this protocol without any robotic assistance and analyzing a single set of plasmids, a trained technician can easily analyze 10–96 well plates in 1 wk.

3.10.1. Day 1

1. Choose one plate of cells (e.g., COS-7 cells) that is >90% confluent.
2. Wash cells with phosphate-buffered saline (PBS).
3. Trypsinize cells with 1 mL Trypsin.
4. Incubate 37°C, 5% CO_2, 5–15 min (*see* **Note 5**).
5. Resuspend cells with DMEM 10% fetal bovine serum (FBS) in a total volume of 5 mL per plate.
6. Split the above plate 1:5 to create a sufficient log growth–phase stock of cells (*see* **Note 6**).

3.10.2. Day 2

1. Wash cells with PBS.
2. Trypsinize cells with 1 mL Trypsin.
3. Incubate 37°C, 5% CO_2, 5–15 min.
4. Resuspend cells with Dulbecco's Modified Eagle's Medium (DMEM) 10% FBS in a total volume of 5 mL per plate.
5. Combine the cells from all the 10-cm plates into a 50-mL centrifuge tube.
6. Vortex the tube for 10–20 s. Remove 50 µL for counting.
7. Add the 50 µL of cells to 450 µL 0.4% Trypan Blue and count cells using a hemocytometer (*see* **Note 7**).
8. Plate cells at 8.5×10^5 cells/plate in a total volume of 8 mL/plate.

3.10.3. Day 3

1. Determine the number of plates to be transfected and the amount of each plasmid needed.
2. We commonly use the following four plasmids as a plasmid set:
   a. Luciferase reporter.
   b. β-Galactosidase expression plasmid for normalizing data.
   c. Transcription factor plasmid in a mammalian expression vector.
   d. pUC19 carrier plasmid.
3. We use 5 µg of reporter, 5 µg normalizing plasmid, 1 µg of receptor, and 9 µg of carrier for each plate, holding the concentration at 20 µg per plate (*see* **Note 8**). Determine the amount of the following reagents needed:
   a. 2X HBS: 500 µL/plate.
   b. 100X Phosphate: 10 µL/plate.
   c. ddH_2O: 420 µL/plate.
   d. 2 *M* $CaCl_2$: 60 µL/plate.
4. Perform the following in a laminar flow hood. The order of addition is important.
   a. Begin by labeling two tubes per transfection:
      i. Tube A1-An (sterile Eppendorf tube).
      ii. Tube B1-Bn (sterile 15 mL Falcon polypropylene tube).
   b. Add the 2X HBS (500 µL) and 100X phosphate (10 µL) per plate to each Tube B.
   c. Add the sterile ddH_2O (420 µL) to each Tube A.
   d. Add each plasmid to their respective Tube A.
   e. Add the 60 µL of 2 *M* $CaCl_2$ to tube A1 and mix gently by pipetting.
   f. While gently vortexing (e.g., setting 4), add dropwise the contents in Tube A1 to Tube B1.

g. Start the timer and continue with Tube A2 and so forth.
5. Incubate for 30 min at room temperature.
6. After 30 min, add the solution from B1 dropwise to cells. Observe precipitate (*see* **Note 9**). Continue with B2-Bn.
7. Incubate for 6–7 h, 37°C, 5% CO_2.
8. One h prior to the end of the incubation period, begin preparing ligand dilutions (20 µL/well) using DMEM + 10% FBS (*see* **Note 10**). Add ligands to appropriate wells of a 96-well plate.
9. Remove plates from incubator and wash each plate twice with PBS.
10. Trypsinize to release cells.
11. Resuspend each plate in a total volume of 10 mL with DMEM + 10% FBS (*see* **Note 10**).
12. Transfer cell suspensions into centrifuge tubes. If more than one plate of the same plasmid set was made, combine cell suspensions.
13. Vortex centrifuge tube. Remove 50 µL and count cells using Trypan Blue (*see* **Note 7**).
14. Determine dilutions needed based on 180 µL per well at a concentration of 15,000 cells/well.
15. Dilute cells and add 180 µl into appropriate wells of your 96-well plate containing ligand(s).
16. Return each plate to the incubator when finished.

3.10.4. Day 4

Continue incubation of transfected cells (*see* **Note 11**).

3.10.5. Day 5

1. Determine the volume of reagents needed (*see* **Note 12**).
   a. Lysis buffer: Per 96-well plate mix 5.645 mL ligand base, 0.046 mL 1 M $MgCl_2$-50 mM Tris-HCl, 0.058 mL 100 mM dithiothreitol (DTT), and 0.012 mL 0.2 M PMSF. Keep at room temperature.
   b. Potassium-Phosphate $KHPO_4$-ATP + $MgCl_2$ solution: Per 96-well plate mix 11.52 mL $KHPO_4$-ATP solution with 0.247 mL 1M $MgCl_2$-50 mM Tris-HCl. Store covered and on ice until use.
   c. β-Galactosidase buffer: Per 96-well plate mix 23.04 mL ONPG solution with 0.075 mL 14.2M β-mercaptoethanol. Store at room temperature. Good for 1 d.
   d. Luciferin solution: Per 96-well plate, mix 9.504 mL of 0.1 M potassium phosphate buffer with 1.056 mL 10X luciferin. Store covered and at room temperature until use. Good for 1 d.

2. Remove plates from incubator and examine for viability (*see* **Note 13**). Aspirate off media.
3. Add 50 µL of lysis buffer to each well. Incubate the plate 30 min at room temperature.
4. Aliquot 100 µL/well KHPO$_4$-ATP + MgCl$_2$ into luminometer plates.
5. At the end of 30 min, add 20 µL of the cell lysate to appropriate luminometer plate wells containing 100 µL KHPO$_4$-ATP + MgCl$_2$.
6. Submit luminometer plate to the luminometer for luciferase activity readings (*see* **Note 14**).
7. To remaining 30 µL of lysate, add 200 µL of ONPG+βME to each well. Begin timer when completing plate.
8. Place plates at 37°C and check for yellow color formation every 5–10 min for a maximum total time of 150 min (*see* **Note 15**).
9. Remove large bubbles from the surface of the plate by rapidly passing the flame of a Bunsen burner over the surface of the 96-well plate. Record the time elapsed. Determine absorbance at 415 nm on an Eliza plate reader. An optimum sample has an absorbance value of 0.30–1.00 at 415 nm.
10. Determined normalized results: Luciferase value/(β-gal at A415 nm/time).

3.11. Tabulating the Data

Most luminometers and plate readers available today come with software packages that allow either direct manipulation of the data generated in a luciferase assay, or the ability to download that data into another file or program. For our assays, we have found it convenient to download both the luciferase and β-galactosidase data into a Microsoft Excel program. With a little preparatory work, one is able to set up Macros in Excel wherein the β-galactosidase data can be merged with the luciferase data in a single spreadsheet. From there it is fairly straightforward to create Macros that: (1) incorporate the time it took the β-galactosidase to develop and calculate a β-galactosidase rate, (2) divide the luciferase data by the β-galactosidase rate to establish a normalized luciferase value, and (3) average the repeats for each treatment to establish an average for the normalized data (*see* **Note 16**). In addition, Excel allows you to graphically display these data, so simple Macros can also be developed to facilitate this process.

4. Notes

1. These are detergents that we have tested for efficient and gentle cell lysis of mammalian cells and work optimal in our system. They are slightly more expensive then using Triton X-100 or NP-40, but we feel they are worth it. We have only tested this detergent combination for its impact on luciferase and β-galactosidase activity, so if the use of another reporter (e.g., GFP or RFP) should be desired it would be advisable to re-evaluate.

2. The pH of the HBS solution can have significant effects on the efficiency of transfections, and although we list pH 7.1 here, it is possible that the optimum pH for your cells may be different. If problems occur with transfection efficiency that are not traceable to plasmid preparations, you may want to try playing with the pH of your HBS solution (e.g., performing test transfections while varying the pH by 0.5 pH units between pH 6.8 and 7.2).

3. If using "kit science" the purchase of luciferin and most of the remainder of these reagents is not required. For economical reasons, we buy our Luciferin in 1-g quantities (10×100 mg vials) that we store at $-70°C$ in powder form. One-hundred milligrams is enough luciferin to theoretically evaluate more than thirty-four 96-well plates, but in practice, this number is closer to 30 plates. This totals approx \$4/plate for Luciferin.

4. We generally use flash kinetics for our luciferase assays and normalize for transfection efficiency and nonspecific variations in reporter expression using a β-galactosidase expression plasmid (*18*). The use of dual luciferase approaches and normalizing with firefly luciferase, Renilla luciferase, or green fluorescent protein (GFP) plasmids are also popular systems (*23*). Using alternative normalizing plasmids to the one described here will require some modification to the reagent list given, but may be worth tracking down a bulk supplier of those reagents and developing them yourself.

5. Always monitor detachment of cells using a microscope. Different cell lines have different sensitivities to the trypsinization process and commercial trypsin can vary from lot to lot as well as with age. Occasionally tapping the edge of the plate against the palm of your free hand can facilitate the detachment process.

6. Each plate split 1:5 on Monday should be sufficient to make at least three to five plates for transfection. If more than 25 plates are needed for transfection, combine two to three parent plates and vortex them thoroughly before aliquoting to insure the population of cells is homogeneous. Furthermore, in the final platting for transfection, if your cell line is loosely adherent (e.g., HEK cells) you may want to coat your 10-cm plates with poly-L-lysine. Incubate a 10 cm plate with 1–2 mL (just enough to cover the bottom) of a 50 μg/mL solution of poly-L-lysine (made up in sterile ddH$_2$O), at 37°C for 30 min just prior to plating. Aspirate off the poly-L-lysine solution and add media and cells. The same applies to the 96-well plates used with these cells.

7. We generally count the living cells (those not filled with blue color) in all four outer corners of the hemocytometer grid (each corner contains 16 smaller squares). Multiplying this number by 25,000 gives the number of cells per mL in your sample. If your cell viability (live cells/dead cells) is not greater than 95%, then you may either be trypsinizing the cells too long or handling the cells too roughly.

8. The amount and ratio of your transcription factor plasmid to your reporter plasmid can significantly impact the level of expression of your reporter. If you are having trouble getting good expression, it might be worth the investment of time to test different amounts of transcription factor plasmid (e.g., 0.1 µg–5.0 µg) in the assay. Remember to standardize your total DNA to 20 µg by altering the amount of pUC plasmid used.

9. A precipitate should be formed almost immediately. If your luciferase and β-galactosidase values are low and your precipitate fails to rapidly form upon addition to your cells, then this may be the source of problems. The two most likely culprits would be either one of the solutions or your plasmid DNA preparation. Check the pH on your HBS solution to make certain it has not changed and try reprecipitating plasmid preps.

10. If screening with ligands that might be a constituent in FBS (e.g., growth factors, steroids, and other lipophilic ligands), it might be necessary to change to serum-free media, charcoal-stripped media (HyClone), or delipidated (Sigma) media, all of which are commercially available. Also, note that each ligand will be diluted 1:10 in the well when cells are added and the volume is brought up to 200 µL, therefore, if you want a 10^{-4} M final concentration, add 20 µL of a 10^{-3} M ligand to each well.

11. Incubation of cells can be for anywhere from 18–72 h. If using a strong promoter (*see* **Subheading 3.1.**) on your mammalian expression plasmids, it might be better to process your transfections 24–36 h after transfecting DNA into cells, especially if you feel there might be some toxicity issue with one of your plasmids.

12. The calculation of these volumes as presented includes an extra 20% to accommodate for pipeting errors and luminometer priming should it be necessary.

13. If cells look sick, are detaching from the surface, and/or there are a lot of floating cells in the wells, something is either wrong with one of your solutions (the most probable is the ligand, many of which are toxic at high concentrations), your plasmids (e.g., transfection of apoptosis or growth-suppressing proteins) or your media (check pH if using phenol red free-media or serum-free media).

14. To reduce background phosphorescence, allow the luminometer plate to incubate a minute or so in the darkness of the luminometer prior to starting the reading.

15. This is a factor of the transfection efficiency, the length of time transfected cells were incubated, and the promoter driving the β-galactosidase gene. If incubating moderately poor transfecting cells for relatively short periods of time (18–36 h), use a pCMV-β-gal plasmid. If incubating good transfecting cells for longer periods of time (36–72 h), use a pRSV-β-gal plasmid.

Generally, moderately good transfecting cell lines with the pRSV-β-gal plasmid take approx 30 min to develop.

16. One final note for consideration is potential complications with normalizing plasmids. We have observed that when analyzing specific modulators of transcription factor activity (e.g., coregulatory molecules) that these proteins can also modulate the levels of our β-galactosidase expression, making it difficult to normalize the luciferase data. In those cases, it is first necessary to define this modulation as a proliferation, protein degradation, or transcription effect. If it turns out to be a transcription effect, it may be necessary to normalize to some other parameter (e.g., cell number, DNA, or total protein). Alternatively, there are other assays available wherein the readout is based on protein: protein interactions (**9**).

References

1. Gill, G. (2001) Regulation of the initiation of eukaryotic transcription. *Essays Biochem.* **37**, 33–43.
2. Beato, M. and Klug, J. (2000) Steroid hormone receptors: an update. *Hum. Reprod. Update* **6**, 225–236.
3. Hatina, J. and Reischig, J. (2000) Hormonal regulation of gene transcription: nuclear hormone receptors as ligand-activated transcription factors. *Cesk Fysiol* **49**, 61–72.
4. Kato, S. (2000) Nuclear receptor-mediated signaling pathway. *Nippon Yakurigaku Zasshi* **116**, 133–140.
5. Aranda, A. and Pascual, A. (2001) Nuclear hormone receptors and gene expression. *Physiol. Rev.* **81**, 1269–1304.
6. Wiseman, H. and Duffy, R. (2001) New advances in the understanding of the role of steroids and steroid receptors in disease. *Biochem. Soc. Trans.* **29**, 205–209.
7. Jones, T. K., Pathirana, C., Goldman, M. E., (1996) Discovery of novel intracellular receptor modulating drugs. *J. Steroid Biochem. Mol. Biol.* **56**, 61–66.
8. Silverman, L., Campbell, R., and Broach, J. R. (1998) New assay technologies for high-throughput screening. *Curr. Opin. Chem. Biol.* **2**, 397–403.
9. Zhou, G., Cummings, R., Li, Y., et al. (1998) Nuclear receptors have distinct affinities for coactivators: characterization by fluorescence resonance energy transfer. *Mol. Endocrinol.* **12**, 1594–1604.
10. Boshart, M., Kluppel, M., Schmidt, A., et al. (1992) Reporter constructs with low background activity utilizing the cat gene. *Gene* **110**, 129–130.
11. Thirunavukkarasu, K., Miles, R. R., Halladay, D. L., and Onyia, J. E. (2000) Cryptic enhancer elements in luciferase reporter vectors respond to the osteoblast-specific transcription factor Osf2/Cbfa1. *Biotechniques* **28**, 506–510.
12. Ibrahim, N. M., Marinovic, A. C., Price, S. R., et al. (2000) Pitfall of an internal control plasmid: response of Renilla luciferase (pRL-TK) plasmid to dihydrotestosterone and dexamethasone. *Biotechniques* **29**, 782–784.
13. Stables, J., Scott, S., Brown, S., et al. (1999) Development of a dual glow-signal

firefly and Renilla luciferase assay reagent for the analysis of G-protein coupled receptor signalling. *J. Recept. Signal Transduct. Res.* **19,** 395–410.

14. Dyer, B. W., Ferrer, F. A., Klinedinst, D. K., and Rodriguez, R. (2000) A noncommercial dual luciferase enzyme assay system for reporter gene analysis. *Anal. Biochem.* **282,** 158–161.

15. Wang, Y., Yu, Y. A., Shabahang, S., et al. (2002) Renilla luciferase-Aequorea GFP (Ruc-GFP) fusion protein, a novel dual reporter for real-time imaging of gene expression in cell cultures and in live animals. *Mol. Genet. Genomics* **268,** 160–168.

16. Zhang, J., Campbell, R. E., Ting, A. Y., and Tsien, R. Y. (2002) Creating new fluorescent probes for cell biology. *Nat. Rev. Mol. Cell Biol.* **3,** 906–918.

17. Lippincott-Schwartz, J. and Patterson, G. H. (2003) Development and use of fluorescent protein markers in living cells. *Science* **300,** 87–91.

18. Henry, K., O'Brien, M. L., Clevenger, W., et al. (1995) Peroxisome proliferator-activated receptor response specificities as defined in yeast and mammalian cell transcription assays. *Toxicol. Appl. Pharmacol.* **132,** 317–324.

19. Mukherjee, R., Jow, L., Noonan, D., and McDonnell, D. P. (1994) Human and rat peroxisome proliferator activated receptors (PPARs) demonstrate similar tissue distribution but different responsiveness to PPAR activators. *J. Steroid Biochem. Mol. Biol.* **51,** 157–166.

20. Wigler, M., Pellicer, A., Silverstein, S., et al. (1979) DNA-mediated transfer of the adenine phosphoribosyltransferase locus into mammalian cells. *Proc. Natl. Acad. Sci. USA* **76,** 1373–1376.

21. Birnboim, H. C. and Doly, J. (1979) A rapid alkaline extraction procedure for screening recombinant plasmid DNA. *Nucleic Acids Res.* **7,** 1513–1523.

22. Giguere, V., Hollenberg, S. M., Rosenfeld, M. G., and Evans, R. M. (1986) Functional domains of the human glucocorticoid receptor. *Cell* **46,** 645–652.

23. Day, R. N., Kawecki, M., and Berry, D. (1998) Dual-function reporter protein for analysis of gene expression in living cells. *Biotechniques* **25,** 848–856.

5

Determining Protein Half-Lives

Pengbo Zhou

Summary

Controlling the stability of cellular proteins is a fundamental way by which cells regulate growth, differentiation, survival, and development. Measuring the turnover rate of a protein is often the first step in assessing whether or not the function of a protein is regulated by proteolysis under specific physiological conditions. Over the years, procedures to determine the half-life of proteins in cultured eukaryotic cells have been well-established. This chapter describes in detail the two most frequently used methods, pulse-chase analysis and cycloheximide blocking, to determine a protein's half-life in yeast and cultured mammalian cells.

Key Words: Protein degradation; proteasome; pulse-chase; turnover.

1. Introduction

Proteolysis has recently emerged as an essential regulatory mechanism underlying virtually any cellular processes, including the cell cycle, signal transduction, apoptosis, and embryonic development. Owing to the irreversible nature and the profound efficiency of various cellular protein-degradation pathways, regulated proteolysis has evolved as the most efficient means cells exploit to rapidly reprogram cellular processes in response to alterations of growth conditions or environmental cues. In most cases, regulated proteolysis is carried out via the ubiquitin-proteasome pathway, which constitutes the bulk of cytoplasmic proteolysis (reviewed in **ref. 1**). Other cellular protein destruction apparatus, such as calpain and lysosome, also play significant roles in posttranslational processing and degradation of specific cellular proteins (reviewed in

From: *Methods in Molecular Biology, vol. 284:*
Signal Transduction Protocols
Edited by: R. C. Dickson © Humana Press Inc., Totowa, NJ

refs. *2*,*3*). Determining the stability of a given cellular protein is one of the first steps towards understanding whether its cellular abundance and activity are subjected to proteolytic control. The availability of specific inhibitors for different proteolytic enzymes further allows for rapid identification of specific protein degradation pathway(s) that are involved.

Among the approaches developed to measure protein half-lives, pulse-chase analysis is the most frequently used method because it imposes minimal disruption or interference with normal cell growth and metabolism. Typically, the protein of interest is first metabolically labeled in rapidly growing cells for a short period of time with a radioactive precursor (e.g; ^{35}S-labeled methionine and/or cysteine). During the subsequent chase period, an excess of nonradioactive precursor molecules are added to the culture to prevent further incorporation of the radiolabel into proteins. At different times during the chase period, samples of the cells are lysed and immunoprecipitated with antibody against the target protein. Radiolabeled protein is subjected to sodium dodecyl sulfate-polyacrylamide gel electrophoresis (SDS-PAGE) and quantified by phosphorimage analysis or similar procedure. The half-life of a protein is defined as the time it takes for the concentration of the radiolabeled target protein to be reduced by 50% relative to the level at the beginning of the chase.

Another method to determine the turnover rate of a given protein is referred to as cycloheximide blocking. Protein synthesis is inhibited by cycloheximide and the decay of a target protein over time is determined by SDS-PAGE and immunoblot analysis. One caveat for the cycloheximide block approach is that the protein half-life is measured when overall protein synthesis is abrogated and, thus, may not reflect the actual turnover rate under normal growth conditions. The stability and abundance of the proteolytic enzymes themselves might also be affected, which further complicates the accurate measurement of protein turnover rate. However, cycloheximide blockage is a useful alternative approach in the event that the target protein is refractory to metabolic labeling. In yeast cells, target genes expressed under the control of inducible promoters, such as the galactose-inducible *GAL1* promoter, can be turned off rapidly (within minutes) after switching cells to glucose-containing medium (reviewed in **ref. *4***). Therefore, the turnover rate of a protein can be determined by following its decay after the silencing of the inducible *GAL1* promoter. This chapter will describe commonly used methods to determine the half-lives of proteins in mammalian and yeast cells.

2. Materials

2.1. Pulse-Chase Analysis in Mammalian Cells

1. Dulbecco's Modified Eagle's Medium (DMEM) containing 1 g/L glucose and 4 m*M* glutamine, supplemented with 10% fetal calf serum (FCS).

2. DMEM free of L-methionine and L-cysteine (Invitrogen, Carlsbad, CA, cat. no. 21013024, or Mediatech, Herndon, VA, cat. no. 17-204-CI).
3. Phosphate-buffered saline (PBS): 1 L is made from 10 g NaCl, 0.25 g KCl, 1.5 g Na_2HPO_4, 0.25 g KH_2PO_4, pH 7.2. The solution is autoclaved and stored at room temperature or 4°C.
4. Trypsin-EDTA solution (Invitrogen, cat. no. 25200056).
5. Express-^{35}S protein labeling mix containing both ^{35}S-methionine and ^{35}S-cysteine (Applied Biosystems, Foster City, CA, cat. no. NEF-772).
6. 200 mM L-methionine in sterile water (Sigma, St. Louis, MO., cat. no. M2893).
7. 200 mM L-cysteine in sterile water (Sigma, cat. no. C7880).
8. RIPA lysis buffers: 50 mM Tris-HCl, pH 7.5, 150 mM NaCl, 1% Nonidet P-40 (NP-40), 0.5% deoxycholate (DOC), and 0.1% SDS.
9. Protease inhibitor cocktail: 2.5 µg/mL chymostatin, 2.5 µg/mL pepstatin A, 2.5 µg/mL leupeptin, and 2.5 µg/mL antipain. Phenylmethylsulfonyl fluoride (PMSF) is made fresh in 95% ethanol and added to lysis buffer at a final concentration of 1 mM just prior to use.
10. Bradford protein assay kit (Bio-Rad, Richmond, CA, cat. no. 77432A).
11. Protein-A or Protein-G sepharose slurry: wash the Protein-A or Protein-G sepharose suspensions (Amersham Biosciences, Piscataway, NJ, cat. no. 17-0974-01 or 17-0618-01) three times in RIPA buffer. Resuspend in RIPA buffer to the original suspension volume and store at 4°C.
12. Sterile microfuge tubes (1.5 and 2 mL).

2.2. Pulse-Chase and Promoter Turn-Off Analysis in Yeast Cells

1. YPD medium: 1% yeast extract, 2% peptone, and 2% dextrose (glucose).
2. Synthetic complete (SC) medium (1 L): 1.3 g complete supplement mixture (CSM) dropout powder (Qbiogene, Carlsbad, CA), 1.7 g yeast nitrogen base without amino acids and ammonium sulfate, 5 g $(NH_4)_2SO_4$, and 20 g dextrose or raffinose.
3. 20% Dextrose (glucose).
4. 20% Galactose.
5. CSM medium without L-methionine (Qbiogene, cat. no. 4510-712).
6. Acid-washed glass beads (425-600 µm) (Sigma, cat. no. G8772).
7. ECL plus™ Western blotting reagents pack (Amersham Bioscience, cat. no. RPN2124), or a regular ECL kit (Western Lightning™ Chemiluminescence Reagent Plus, Applied Biosystems, cat. no. NEL105).
8. 2X Laemmli buffer (10 mL): 0.15 g Tris-HCl base (final concentration is 0.125 M), pH adjusted to 6.8 with HCl, 4 mL 10% SDS (final concentration is 4%), 1 mL glycerol (final concentration is 10%), 20 mg bromophenol blue (final concentration is 0.02%), 0.4 mL β-mercaptoethanol (final concentration is 4%).

9. Sterile 50 mL Falcon tubes.
10. Sterile microfuge tubes.

2.3. Equipment

1. Vibra Cell™ sonicator (VC 130PB) with microtip (Sonics and Materials, Newtown, CT).
2. End-over-end mixer (VWR, West Chester, PA).
3. Microcentrifuge and refrigerated table top centrifuge.
4. Spectrophotometer.
5. Vertical gel electrophoresis system.
6. Western-blotting apparatus.
7. Gel dryer.
8. Phosphorimager and storage phosphor screens.
9. Orbital shaker.

3. Methods

3.1. Measuring Half-Lives of Proteins in Mammalian Cells by Pulse-Chase Analysis

This section describes a standard procedure for measuring half-lives of proteins in adherent tissue culture cells. The same principle applies to nonadherent cells, although the initial metabolic labeling procedure will be slightly different (*see* **Note 1**).

3.1.1. Cell-Culture Preparation

1. For measuring the half-lives of proteins exogenously expressed via transient transfection, plate 2×10^6 HeLa cells on one to two 10 cm dishes, and incubate in CO_2 incubator overnight.
2. Transfect with a plasmid expressing the gene of interest by calcium phosphate or other transfection procedures and incubate cells for 12–16 h.
3. Wash with cells PBS and incubate them in fresh DMEM medium with 10% FCS for 6–8 h.
4. Remove medium, trypsinize, and seed equal aliquots of cell suspensions in five 100-mm dishes. Incubate overnight in a CO_2 incubator at 37°C.

   For measuring half-lives of endogenous proteins, plate equal number of cells in five 100 mm dishes and culture for 1 d. Enough cells should be plated on each dish so they reach 60–80% confluency on the second day, which is optimal for pulse labeling (*see* **Note 2**).

3.1.2. Pulse-Chase

1. Aspirate medium and wash the cells three times with 5 mL prewarmed PBS.
2. Remove PBS completely. Starve cells by adding 1 mL methionine- and cysteine-free DMEM medium containing 10% dialyzed FCS. Incubate at

37°C for 1 h. Shake occasionally so that the cells are fully covered by the medium.

3. Add 100 µCi of Express-^{35}S protein-labeling mix to each 100-mm plate (*see* **Note 3**). Incubate the cells for 30 min at 37°C with shaking every 5–10 min.
4. Remove radioactive medium. Wash cells once with 5 mL of prewarmed PBS.
5. For the "Time 0" plate, proceed to cell lysis and extract preparation in **step 8.**
6. For all other time points, add 3 mL chase medium containing prewarmed DMEM, 10% FCS, 3 mM L-methionine, and 1 mM L-cysteine to each plate. Immediately return plates to the CO_2 incubator. Note that cell permeable protease inhibitors can be included in chase medium in a parallel experiment to assess whether or not a specific proteolytic mechanism is involved in degrading the protein of interest (*see* **Note 4**).
7. Remove medium from plates at desired time points (0.5, 1.0, 2.5, and 5.0 h are good starting points). Wash once with 3 mL PBS and then remove PBS completely.
8. Add 300 µL of ice-cold RIPA lysis buffer containing protease inhibitors to plate. Scrape cells using a disposable cell scraper and transfer them to a fresh microfuge tube. Cells can also be frozen in liquid nitrogen at this step and processed when all time points are collected (*see* **Note 5**). Denaturing lysis buffer can be used if high background is observed following immunoprecipitation and SDS-PAGE analysis (*see* **Note 6**).
9. Sonicate cells with three to five bursts using the microtip of a Vibra-Cell™ VC500 sonicator.
10. Rotate the microfuge tube containing cell extracts on an end-over-end mixer for 30 min at 4°C.
11. Centrifuge at 14,000 rpm for 10 min at 4°C. Transfer the supernatant fluid to a fresh microfuge tube.
12. Determine the protein concentration with the Bradford assay kit. Samples are diluted 200–500 times in Bradford assay buffer to minimize interference to the colorimetric reactions by detergents in RIPA buffer. Alternatively, use the Bio-Rad DC protein assay kit, which is compatible with both ionic and nonionic detergents.

3.1.3. Immunoprecipitation, SDS-PAGE, and Autoradiography

1. Transfer 500 µg to 1 mg each of cell lysate to fresh microfuge tubes. Make up the total volume (300–500 µL) with RIPA buffer.
2. Preclear the lysates by adding 20 µL of pre-washed Protein-A or Protein-G sepharose beads. Rotate at 4°C for 1 h on an end-over-end mixer.
3. Centrifuge samples at 14,000 rpm in a microcentrifuge for 10 s at 4°C. Transfer supernatant fluids to fresh microfuge tubes.

4. Add the appropriate amount of antibody to each sample (*see* **Note 7** and the chapter in this volume on immunoprecipitation). Rotate on an end-over-end mixer at 4°C for 1 h.
5. Centrifuge tubes briefly. Add 50 µL of 50% slurry of prewashed Protein-A or Protein-G sepharose beads. Rotate on an end-over-end mixer for 1 h to overnight at 4°C.
6. Centrifuge samples at top speed for 10 s at 4°C.
7. Remove supernatant fluid. Add 1 mL of RIPA buffer to the pellet and shake up and down to wash. Centrifuge samples at 4°C for 10 s to pellet the sepharose beads containing the bound protein of interest. Aspirate supernatant fluid.
8. Repeat washing two more times.
9. Aspirate all liquid from the sepharose beads. Resuspend in 50 µL of 2X Laemmli buffer (**5**). Heat at 95°C for 5 min.
10. Centrifuge briefly in microcentrifuge. Resolve the eluted proteins by SDS-PAGE.
11. Fix gel in 50% methanol and 10% acetic acid for 30 min to 1 h after electrophoresis, dry under vacuum, and expose to a phosphor screen.
12. Quantitate the intensity of individual bands of the protein of interest using the phosphoimager and plot the percentage of proteins on a logarithmic scale over time. The half-life is the time point at which 50% of the protein remains relative to that at the beginning (Time 0) of the chase.

3.2. Measuring the Half-Life of Yeast Proteins

This section describes two commonly used methods to determine the half-lives of proteins in yeast: pulse-chase analysis and promoter turn-off. Pulse-chase analysis is generally used for determining the turnover rate of endogenous proteins or proteins expressed from constitutive promoters on plasmids. Promoter turnoff is a simple procedure used to measure the stability of proteins exogenously expressed from inducible promoters on plasmids.

3.2.1. Pulse-Chase Analysis

3.2.1.1. Preparation of Yeast Culture

1. Grow yeast cells overnight to stationary phase in either YPD for endogenous proteins or synthetic complete medium lacking the amino acid corresponding to the selectable marker on the transformed plasmid.
2. Dilute yeast cells to an OD_{600nm} of 0.1 in 30 mL medium. Grow to log phase (OD_{600nm} of 0.4–0.6) (*see* **Note 8**).

3.2.1.2. PULSE-CHASE

1. Collect yeast cells by centrifugation in a tabletop centrifuge at 2000 rpm for 5 min at room temperature. Remove supernatant fluid completely.
2. Resuspend cells in 30 mL of synthetic complete medium without L-methionine. Transfer to a sterile flask and shake at 30°C for 1 h in an orbital shaker.
3. Transfer culture to a 50 mL Falcon tube. Centrifuge at 2000 rpm for 5 min to pellet cells and resuspend in 6 mL synthetic complete medium without L-methionine.
4. Add 100–500 µCi of Express-^{35}S protein labeling mix per mL of cell culture and shake at 30°C for 10 min. See **Note 10** for measuring half-life in temperature sensitive mutant yeast cells.
5. Pellet cells and remove radioactive medium carefully.
6. Resuspend cells in 6 mL prewarmed synthetic complete medium containing 1 mg/mL L-methionine and 1 mg/mL L-cysteine (add fresh).
7. Immediately aliquot 1.4 mL yeast culture to a 2 mL screw-cap microfuge tube as the "Time 0" cells. Centrifuge at 14,000 rpm for 15–20 s in the microfuge tube. Remove supernatant fluid and freeze the cell pellet in liquid nitrogen. Keep at −80°C until all the time points are collected.
8. At the desired time points, aliquot 1.4 mL yeast culture, collect cell pellet by centrifugation and freeze in liquid nitrogen as in **step 7**. Typical time points are 0, 10, 20, 40, 60, and 120 min, but shorter or longer time points may be needed in specific cases.

3.2.1.3. PREPARATION OF YEAST EXTRACT

1. Resuspend cell pellets in 300 µL ice-cold RIPA buffer containing protease inhibitors.
2. Add 300 µL acid-washed and ice-chilled glass beads (500 µm) to each tube.
3. Vortex vigorously for 30 s in the cold room. Chill on ice for at least 30 s. Repeat vortexing and chilling cycle six more times (*see* **Note 9**).
4. Centrifuge at 14,000 rpm for 10 min at 4°C. Transfer supernatant fluid to a fresh microfuge tube. Centrifuge for another 10 min. Transfer supernatant fluid to a fresh microfuge tube.
5. Determine the protein concentrations by the Bradford assay or the DC protein assay (Bio-Rad) (*see* **step 11** of **Subheading 3.1.2.**).
6. Proceed to **step 7** or freeze samples in liquid nitrogen for later analysis.
7. Perform immunoprecipitation, SDS-PAGE, and autoradiography to determine the half-life of target protein as described in **Subheading 3.1.3.**

3.2.2. Promoter Turn-Off

To determine the turnover rate of a target protein exogenously expressed from inducible promoters, such as *GAL1*, a promoter turn-off procedure is often the method of choice.

3.2.2.1. TRANSIENT INDUCTION AND PROMOTER TURN-OFF

1. Yeast cells transformed with plasmid expressing target gene are grown to early log phase (OD_{600} of 0.4–0.6) in 30 mL synthetic complete dropout medium containing 2% raffinose. The carbon source must be one that does not repress expression of the *GAL1* promoter, which is strongly repressed by glucose.
2. Add sterile galactose to cell culture at a final concentration of 2% to induce target gene expression for 30 min.
3. Transfer cells to a 50 mL sterile Falcon tube. Centrifuge at 2000 rpm for 5 min. Remove supernatant fluid.
4. Wash cells by resuspending in 30 mL of sterile PBS. Centrifuge as above to pellet cells. Remove supernatant fluid.
5. Resuspend cells in 6 mL synthetic complete dropout medium containing 2% raffinose and 2% glucose (*see* **Note 11**).
6. Immediately transfer 1 mL of cells to a fresh microfuge tube as the "Time 0" sample. Centrifuge in a microcentrifuge at 14,000 rpm for 15 s. Remove supernatant fluid and freeze the cell pellet in liquid nitrogen.
7. At appropriate time points (*see* **step 8** of **Subheading 3.2.1.2.**), remove 1 mL of cells, centrifuge and freeze cell pellet in liquid nitrogen.

3.2.2.2. YEAST EXTRACT PREPARATION

Prepare yeast cell extracts as described in **Subheading 3.2.1.3.**

3.2.2.3. SDS-PAGE AND IMMUNOBLOTTING

1. 100–200 µg each of total protein extract is subjected to SDS-PAGE and immunoblotting using standard procedures (*5*).
2. Visualize the protein bands of interest using the ECL plus reagent. Expose the blot to a phosphorimager screen to determine the intensity of bands over time using a phosphorimager. If regular ECL reagents are used, multiple exposures to X-ray films are necessary to determine the range in which the response of the film is linear. Densitometers can be used to scan the exposure within the linear range to measure the band intensity.

4. Notes

1. For metabolic labeling of suspension cultures, collect cells by centrifugation at 400 g for 5 min. Wash once with prewarmed PBS and centrifuge as

before. Resuspend cells at approx10^7 cells/mL in prewarmed medium containing 10% dialyzed FCS, lacking L-methionine and L-cysteine. Incubate for 1 h at 37°C. Add Express-^{35}S protein labeling mix at 10 mCi/mL and incubate for 30 min to 1 h at 37°C. Mix occasionally. Collect cells by centrifugation and suspend in cold chase medium containing excess L-methionine (3 mM) and L-cysteine (1 mM). Collect cells at appropriate time points as in **Subheading 3.1.2.**

2. Confluency of cells required for efficient ^{35}S labeling will depend on the growth rate of individual cells. HeLa cells are fast growing and efficient labeling can be achieved at approx 60% confluency. For slower growing cells, a higher density of 75–85% confluency is required.

3. Proteins with only a few amino acid residues of the radioactive precursor, or those with a slow turnover rate might not be labeled efficiently. Using Express-^{35}S protein labeling mix, which contains both radioactive methionine and cysteine, often results in increased labeling of the target protein compared to that of ^{35}S-methionine alone. If both methionine and cysteine are rare in the target protein, other radioactive precursors, such as ^{35}S-leucine, can be used as an alternative. Proteins that have slow metabolic turnover rate should be labeled for a longer period of time prior to chase. If the aforementioned methods fail, cycloheximide blocking procedure can be considered as an alternate approach for half-life measurement (*6*).

4. Proteasome, calpain, or lysosomal inhibitors can be included in the chase medium in parallel experiments to assess whether degradation of the protein of interest is mediated by a specific proteolytic mechanism. Frequently used proteasome inhibitors include lactacystine (12.5 µM), MG132 (50 µM), and LLnL (50 µM). Lactacystin specifically inhibits the 26S proteasome activity, whereas the more economical MG132 and LLnL abrogate the functions of both calpain and the proteasome. Other inhibitors used in the literature include calpeptin (30–60 µM) or calpain inhibitor II (5 µM) for calpain, and E64 (50 µM) for cysteine proteases found in organelles including lysosome (*6*). Note that many protease inhibitors often block the function of several proteases. The result from half-life studies using protease inhibitors should only serve as a starting point to estimate cellular degradation pathways that may or may not be involved in the turnover of the protein of interest.

5. Cells can be scraped into 0.5 mL ice-cold PBS. Transfer to microfuge tube and centrifuge at 14,000 rpm for 1 min at 4°C. Remove the supernatant fluid and freeze the cell pellet in liquid nitrogen. Store at −80°C until all time points are collected and then proceed to cell lysis and extract preparation.

6. If high background is encountered, cells can be lysed under denaturing conditions for immunoprecipitation: Scrape cells from a 100-mm dish are

scraped in 350 μL of preboiled 1% SDS lysis buffer (50 mM Tris-HCl, pH 7.5, 0.5 mM EDTA, 1% SDS, 1 mM dithiothreitol[DTT]). If the lysate is too viscous owing to high-molecular-weight DNA, sonicate to break the DNA. Boil for 10 min, then dilute 1:10 in 0.5% NP-40 lysis buffer (50 mM Tris, pH 7.5, 150 mM NaCl, 0.5 % NP-40, 50 mM NaF, 1 mM DTT, 1 mM NaVO$_3$, and protease inhibitors) and perform the immunoprecipitation as in **Subheading 3.1.3.**

7. The amount of antibody and cell extracts used should be determined on a case-by-case basis *(5)*. For 500 μg of total protein extracts, 2–3 μL of crude antiserum or 2.5–5 μg of affinity-purified antibody are good starting points, but it is best to titrate the sample with increasing amounts of antibody to find the amount needed to immunoprecipitate the protein of interest.

8. Yeast cells can be synchronized with α-factor (3 μg/mL) at G1 phase, hydroxyurea (10 mg/mL) in S phase, or 15 μg/mL nocodazole in M phase of the cell cycle. The half-life of the target protein at different cell cycle stages can then be measured *(3,9)*.

9. To determine whether degradation of a given yeast protein is dependent on a specific protease, yeast strains carrying temperature-sensitive mutations of the corresponding enzyme can be used to measure the half-lives at nonpermissive temperature. These measurements can then be compared to the half-lives determined at permissive temperature *(9)*. Grow temperature-sensitive mutant yeast to early log phase (OD$_{600}$ of 0.4–0.6) at permissive temperature and shifted to nonpermissive temperature for 1–2 h. Pulse-chase or promoter turnoff experiments are then performed at the nonpermissive temperature using procedures described in **Subheadings 3.2.1.2.** and **3.2.2.2.**

10. To determine if yeast cells are completely lysed by mechanical force, a few microliters of lysis material can be spotted on a glass slide and observed under the light microscope. Additional vortexing is necessary if a large number of yeast cells are still intact.

11. For proteins translated from short-lived mRNAs (e.g., Cln3, t$_{1/2}$ approx 3.5 min), glucose addition is sufficient to block the biosynthesis of the target protein *(9)*. For those synthesized from stable mRNAs, 1 mg/mL of cycloheximide can be included in chase medium containing glucose to block both transcription and translation *(7–9)*.

Acknowledgments

The author would like to thank Josie Siegel and Maurizio DiLiberto for critical reading of the manuscript, Xiaoai Chen and Yue Zhang for optimizing these methods, and James Miller and Danial DiBartolo for editing. Studies in our laboratory in this area are supported by the Sidney Kimmel Foundation for Cancer

Research, the AMDeC Foundation, the Mary Kay Ash Charitable Foundation, the New York Academy of Medicine, the Susan G. Komen Breast Cancer Foundation, the Dorothy Rodbell Cohen Foundation for Sarcoma Research, and the National Institute of Health.

References

1. Hershko, A. and Ciechanover, A. (1998) The ubiquitin system. *Annu. Rev. Biochem.* **67,** 425–479.
2. Carafoli, E. and Molinari, M. (1998) Calpain: A protease in search of a function? *Biochem. Biophys. Res. Commun.* **247,** 193–203.
3. Pillay, C. S., Elliott, E., and Dennison, C. (2002) Endolysosomal proteolysis and its regulation. *Biochem. J.* **363,** 417–429.
4. Johnston, M. (1987) A model fungal gene regulatory mechanism: the *GAL* genes of *Saccharomyces cerevisiae. Microbiol. Rev.* **51,** 458–476.
5. Harlow, E. and Lane, D. (1988) *Antibodies: A Laboratory Manual.* Cold Spring Harbor Laboratory Press, Cold Spring Harbor, NY.
6. Patrick, G. N., Zhou, P., Kwon, Y. T., et al. (1998) p35, The neuronal-specific activator of cyclin-dependent kinase 5 (Cdk5) is degraded by the ubiquitin-proteasome pathway. *J. Biol. Chem.* **273,** 24,057–24,064.
7. Amon, A., Irniger, S., and Nasmyth K. (1994) Closing the cell cycle circle in yeast: G2 cyclin proteolysis initiated at mitosis persists until the activation of G1 cyclins in the next cycle. *Cell* **77,** 1037–1050.
8. Yaglom, J., Linskens, M. H., Sadis, S., et al. (1995) p34 Cdc28-mediated control of Cln3 cyclin degradation. *Mol. Cell Biol.* **15,** 731–741.
9. Zhou, P. and Howley, P. M. (1998) Ubiquitination and degradation of the substrate recognition subunits of SCF ubiquitin-protein ligases. *Mol. Cell* **2,** 571–580.

6

Assaying Protein Kinase Activity

Jan Brábek and Steven K. Hanks

Summary

Protein kinases, encoded by approx 2% of eukaryotic genes, represent one of the major classes of cell-regulatory molecules. Assessment of the catalytic activity of a specific protein kinase can be an important step in elucidating signal-transduction pathways that affect cell behavior. As an example of approaches taken to measure protein kinase activity, this chapter presents methods useful for determination of the activity of the oncogenic protein-tyrosine kinase v-Src. Included are protocols for heterologous expression of the kinase in yeast *Saccharomyces cerevisiae*, immunoaffinity purification from yeast cell lysates, kinase reactions using incorporation of ^{32}P into peptide substrates, and quantifying protein kinase activity. The Notes section discusses alternative approaches for assaying the activity of Src recovered from vertebrate cells and it gives recommendations for assaying the activity of the other protein kinases with respect to the substrate specifity and the composition of kinase reaction buffer.

Key Words: Src; protein-tyrosine kinase; kinase assay; peptide substrate; heterologous protein expression; *Saccharomyces cerevisiae*.

1. Introduction

Phosphorylation by protein kinases is a major signal-transduction mechanism used by eukaryotic cells to regulate proliferation, gene expression, metabolism, motility, membrane transport, and virtually every other activity that defines their phenotypic behavior. Given their diverse cellular roles, it is not surprising that protein kinases are encoded by a substantial portion of eukaryotic genes. The genome of the budding yeast *S. cerevisiae* contains 130 distinct

From: *Methods in Molecular Biology, vol. 284:*
Signal Transduction Protocols
Edited by: R. C. Dickson © Humana Press Inc., Totowa, NJ

genes encoding protein kinases, representing approx 2% of all genes *(1)*. The human genome carries 518 protein kinase genes (approx 1.7% of all genes) *(2)*. A large majority (approx 90%) of these protein kinases belong to the eukaryotic protein kinase (EPK) superfamily defined on the basis of a homologous kinase catalytic domain *(3,4)*. The EPK domain interacts with both ATP and protein (peptide) substrates and functions to transfer the γ-phosphate of ATP onto the hydroxyl group of a serine, threonine, or tyrosine amino-acid residue within the peptide substrate.

Based on catalytic domain relatedness, the EPK superfamily has been broadly subdivided into seven major groups: (1) AGC kinases, (2) CAMK-related kinases, (3) CMGC kinases, (4) STE kinases, (5) type I casein kinases, (6) tyrosine kinases, and (7) "tyrosine kinase-like" kinases *(4,5)*. The latter two groups are absent from yeast, reflective of their functions in signaling pathways associated with metazoan complexity. Each major group is composed of many distinct families whose individual members have highly-similar EPK domains and exhibit additional homology outside the EPK domain core. Other EPK families fall outside the seven major groups. **Table 1** shows the number of human EPKs within each family (and group) that are known or likely to have protein kinase activity. Notably, this classification scheme groups together protein kinases that have common specificities in peptide recognition and phosphorylation. The obvious example is the tyrosine kinase group composed of EPKs that specifically phosphorylate tyrosine residues. However, protein kinases that fall within the other major groups also tend to have similar serine/threonine specificity determinants *(6–8)*. Thus, kinases of the AGC group (which includes the PK<u>A</u>, PK<u>G</u>, and PK<u>C</u> families) and the closely-related CAMK group (which includes families of calmodulin-regulated kinases) tend to be basophilic; that is, they frequently phosphorylate serine/threonine residues residing near basic residues. Similarly, members of the casein kinase I group are acidophilic. The tyrosine kinases also tend to be acidophilic. Many CMGC group members (including the namesake <u>C</u>DK, <u>M</u>APK, <u>G</u>SK, and <u>C</u>LK families) are highly selective for serine/threonine residues lying immediately N-terminal to a proline residue.

To successfully assay protein kinase phosphotransfer activity, the investigator must first obtain a pure preparation of the protein kinase of interest. Immunoprecipitation of the endogenous protein kinase from nondenaturing cell lysates (described in another chapter of this book), is commonly used to achieve this goal. However, this approach should be taken with caution (and the use of proper controls), because it is possible the immunoprecipitate will not only contain the kinase of interest, but other protein kinases that either coprecipitate, cross-react with the antibody, or nonspecifically bind to the affinity matrix. In many cases, it is useful to carry out the assays using the most highly purified

Table 1. Human EPK Groups and Families [a]

Group	Families (# members)
Tyrosine kinases (84 total)	Ack (2), Abl (2), Csk (2), FAK (2), Fer (2), JAK (4), Src (11), Syk (2), Tec (5), Alk (2), Axl (3), DDR (2), EGFR (3), Eph (12), FGFR (4), InsR (3), Lmr (3), Met (2), Musk (1), PDGFR/VEGFR (8), Ret (1), Ror (2), Sev (1), Tie (2), Trk (3),
AGC kinases (61 total)	PK<u>A</u> (5), PK<u>G</u> (2), PK<u>C</u> (9), AKT (3), DMPK (7), GRK (7), MAST (5), NDR (4), PKB (1), PKN (3), RSK (9), SGK (3), YANK (3).
CAMK-related (65 total)	CAMK1 (5), CAMK2 (4), CAMKL (20), DAPK (5), DCAMKL (3), MAPKAPK (5), MLCK (3), PHK (2), PIM (3), PKD (3), PSK (1), RAD53 (1), Trio (4), TSSK (5), CAMK-Unique (1).
CMGC kinases (61 total)	<u>C</u>DK (20), <u>M</u>APK (14), <u>G</u>SK (2), <u>C</u>LK (4), CDKL (5), DYRK (10), RCK (3), SRPK (3)
STE kinases (45 total)	STE7 (7), STE20 (28), STE11 (8), STE-Unique (2).
Tyrosine kinase-like (37 total)	RAK (2), LISK (4), LRRK (2), MLK (9), RAF (3), RIPK (5), STKR (12).
Type I casein kinases (11 total)	CK1 (7), TTBK (2), VRK (2).
"Other" kinases (63 total)	Aur (3), BUB (1), Bud32 (1), CAMKK (2), CDC7 (1), CK2 (2), IKK (4), IRE (2), MOS (1), NAK (4), NEK (11), NKF1 (3), NKF2 (1), NKF4(2), PEK (4), PLK (4), TLK (2), TOPK (1), TTK (1), ULK (4), VPS15 (1), WEE (3), Wnk (4), Other-Unique (2).

[a] Only EPK family members with known or likely kinase activity are included. About 50 additional EPKs that lack one or more key residues important for phosphotransfer activity are not represented. Family names are those used in the KinBase searchable database found in **ref. 27**.

protein kinase preparation that can be obtained through heterologous expression in bacteria, yeast, or insect cells. It is also necessary to obtain an appropriate protein or peptide substrate that can be efficiently phosphorylated by the protein kinase of interest. The Protein Kinase Factsbook *(6)* provides much information regarding physiological substrates and specificity determinants for many protein kinases. The PhosphoBase database *(8)* provides phosphorylation site data for >50 well-characterized protein kinases and is another useful resource for identifying suitable substrates for utilization in kinase reactions. For poorly characterized protein kinases of unknown specificity, identifying an appropriate substrate presents a challenge. However, it is anticipated that proteomic approaches to define protein kinase peptide specificity determinants *(9,10)* will soon provide information on suitable protein or peptide substrates for all protein kinases.

This chapter illustrates commonly used approaches to assay protein kinase activity using the oncogenic Src tyrosine kinase as an example. First, we present a protocol for obtaining purified v-Src expressed in the yeast *Saccharomyces cerevisiae*, which has proven to be a useful heterologous expression system because nonspecific copurification of endogenous tyrosine kinases is minimized *(10–13)*. The system enables rapid production of a large amount of the kinase with specific activity comparable to that obtained with kinases expressed in vertebrate cells *(14,15)*. To assay the activity of a purified serine/threonine kinase, the only major change that would need to be made to these basic protocols is the substitution of a suitable peptide substrate for the kinase reaction.

2. Materials

1. Basic molecular biology reagents and equipment including equipment for agarose gel electrophoresis, thermocycler and thermostable polymerase for polymerase chain reaction (PCR), restriction endonucleases, T4 DNA ligase, *Escherichia. coli* strains and growth media, ampicillin, and so on.
2. Equipment for SDS-PAGE (sodium dodecyl sulfate-polyacrylamide gel electrophoresis) and Western blotting.
3. v-*src* Gene (Prague C variant) in plasmid pATV-8 (American Type Culture Collection).
4. Oligonucleotide primers for PCR amplification of the v-*src* gene:
   upstream: 5′-gtcggatccatgggtagtagcaagagcaagc-3′
   downstream: 5′-gccgaattcttactcagcgacctccaacac-3′.
5. Plasmid pYES2 (Invitrogen, Carlsbad, CA).
6. *S. cerevisiae* strain EGY48 (Invitrogen).
7. YPAD media for standard yeast culture: 20 g/L peptone (Difco, Detroit, MI), 10 g/L yeast extract (Difco), 100 µg/L adenine hemisulfate (Sigma, St. Louis, MO), 2% glucose.
8. SD-U media for yeast selection: 6.7 g/L yeast nitrogen base without amino acids (Difco), 0.6 g/L -His/-Leu/-Trp/-Ura DO (dropout) supplement (Clontech; Palo Alto, CA) 20 mg/L L-histidine HCl monohydrate, 100 mg/L L-leucine, 20 mg/L L-tryptophan. For solid media, include 18 g/L agar.
9. Additional reagents for yeast selection, induction, transformation and, lysis (all can be obtained from Sigma): raffinose, galactose (glucose-free), lithium acetate, polyethylene glycol 3350 average MW, salmon sperm DNA, 0.5 mm glass beads for cell lysis.

10. Lysis buffer LB1: 50 mM HEPES, pH 7.4, 0.5% Nonidet P40, 5% glycerol, 100 mM sodium chloride, 0.5 mg/mL Pefabloc (Roche, Indianapolis, IN), 5 μg/mL leupeptin (Sigma), 5 μg/mL aprotinin (Sigma), 0.5 mM sodium orthovanadate, 1 mM sodium fluoride.
11. Anti-v-Src monoclonal antibody MAb 327 (Calbiochem, San Diego, CA).
12. Antimouse IgG agarose (Sigma).
13. Stock solutions for protein kinase assay buffer (*see* **Notes 1** and **2**): 500 mM HEPES, pH 7.4, 800 mM MgCl$_2$, 200 mM MnCl$_2$, 2 mM ATP, 10 mM sodium orthovanadate, 500 μg/mL leupeptin, and 500 μg/mL aprotinin.
14. [γ-^{32}P]ATP (3000 Ci/mmol; Amersham Biosciences, Piscataway, NJ).
15. Tyrosine kinase substrate peptide RRLIEDAEYAARG (Sigma).
16. 21-mm Diameter circles of P81 cellulose phosphate paper (Whatman, Clifton, NJ).
17. Trichloroacetic acid (TCA).
18. Vacuum manifold for washing filters.
19. Random amino acid copolymer poly (Glu, Tyr) (4:1) (Sigma).
20. Liquid-scintillation counter.
21. Device for scanning densitometry.

3. Methods

The methods described below outline: (1) expression of v-Src kinase in yeast *S. cerevisiae*, (2) immunoaffinity purification of the expressed kinase, (3) assay of Src tyrosine kinase activity, and (4) quantification of kinase activity.

3.1. Expression of Src in S. cerevisiae

This section presents a brief overview of the steps involved in subcloning the v-Src cDNA into a yeast expression plasmid, introducing the expression plasmid into *S. cerevisiae*, and inducing expression of v-Src in the yeast cells. The investigator should be familiar with basic molecular biology methods, which are not described in detail.

The v-*src* gene is amplified from the pATV-8 plasmid by standard PCR using the oligonucleotide primers, digested with *Bam*HI and *Eco*RI restriction enzymes, cloned into the *Bam*HI and *Eco*RI sites of the galactose-inducible vector pYES2, and verified by sequencing. The resulting pYES2-Src plasmid is then transformed into *S. cerevisiae* strain EGY48 by standard Li-acetate method *(16)* and selected on SD-U agar plates. The transformed yeast cells are grown overnight (*see* **Note 3**) (to saturation density) in 30 mL SD-U medium supplemented with 2% raffinose, pelleted by centrifugation (2500 *g*, 3 min, 20°C), and then grown for an additional 4 h in 50 mL fresh SD-U medium containing 2% raffinose and 2% galactose to induce v-Src expression.

3.2. Purification of Expressed v-Src Protein

This section gives detailed protocols for preparing yeast cell extracts and recovering the induced v-Src protein by immunoprecipitation.

3.2.1. Preparing Yeast Cell Extracts

1. After cooling the 50 mL induced yeast culture on ice, pellet the cells by centrifugation (2500g, 3 min, 4°C).
2. Resuspend the cell pellet in 2 mL ice-cold lysis buffer LB1 (*see* **Note 4**). Transfer the suspension equally into three precooled 10-mL glass centrifuge tubes.
3. Pellet the cells again by centrifugation (2000 g, 3 min, 0°C), carefully aspirate the supernatant, then add 1.5 mL of washed, dried, and precooled 0.5 mm glass beads to each tube. Wrap the tubes with parafilm and lyse the cells by vortexing vigorously for 5 min at 4°C (in cold room).
4. Add 1 mL of lysis buffer LB1 to each of the three tubes and mix with the lysed cell/bead slurry by pipetting with a 1-mL volume tip. Collect the crude extract solutions (without the beads) into two 1.5-mL microfuge tubes.
5. Centrifuge (15,000g, 15 min, 0°C) to remove any insoluble cellular material, nonlysed cells, or contaminating beads. Pool the supernatants. Aliqouts can be stored at −80°C or used for the next step.

3.2.2. Immunoaffinity Purification of Src From Yeast Extracts

1. Aliqout 400 µL of the yeast cell extract to a 1.5-mL microtube then add 1 µg/mL (400 ng) of anti-v-Src monoclonal antibody 327. Vortex to mix, then incubate for 3 h at 4°C to allow the antigen/antibody complex to form (*see* **Note 5**).
2. During the aforementioned incubation step, prepare a 50% slurry of anti-mouse IgG1 agarose beads by washing three times with at least five volumes of 50 mM HEPES, pH 7.4. After the last wash, add 1 bead-volume of the same buffer.
3. After the 3 h incubation of *step 1*, add 40 µL of the 50% washed slurry to the tube. Gently rock the tube on a rotator for 1 h at 4°C, allowing the antigen-antibody complexes to become bound to the beads.
4. Wash the beads (immunoprecipitates) three times with at least 50 vol of LB1 buffer, then once with 50 mM HEPES, pH 7.4. The immunoprecipitates should be kept on ice during these steps. For each wash, the beads are resuspended by gentle vortexing and spun down by centrifugation (3000g, 3 min, 0°C). At the final wash step, divide the beads equally into two 0.5 mL tubes. The final bead volume will be approx 10 µL per tube. One tube is used to assess the recovery of v-Src in the immunoprecipitates

(**step 5**). The other tube is used for the kinase assays described in **Subheading 3.3.**, and should be kept on ice in the final 50 mM HEPES wash buffer until ready for use.

5. Add 20 μL 2X SDS-PAGE sample buffer to one tube, mix with the bead by vortexing, heat for 5 min at 100°C. Then separate the sample by SDS-PAGE and analyze by Western blot to ascertain that the Src protein was properly expressed and affinity-purified. The anti-v-Src 327 MAb can also be used for the Western analysis.

3.3. Assaying Src Protein-Tyrosine Kinase Activity

Two different protocols are presented below to assay the activity of purified Src kinase using incorporation of ^{32}P into different peptide substrates (*see* **Note 6**).

3.3.1. Assaying Tyrosine Kinase Activity Using Substrate-Based Oligopeptide

The first protocol makes use of a small synthetic peptide substrate. Such peptides can be designed as optimized substrates for any protein kinase of interest, thus enabling precise determination of kinetic parameters *(17)*. The oligopeptide includes basic residues allowing its separation from nucleotides and free phosphate present in the kinase reaction mixture through its ability to bind tightly to phosphocellulose paper *(18)*.

1. For each reaction, prepare 10 μL of 3X concentrated protein kinase assay buffer (3XPKB): 50 mM HEPES, pH 7.4, 24 mM MgCl$_2$, 6 mM MnCl$_2$, 300 μM Na$_3$VO$_4$, 60 μM ATP, 15 μg/mL leupeptin, 15 μg/mL aprotinin, and 5 μCi [γ-^{32}P]ATP (*see* **Notes 1** and **2**). Use the kinase assay buffer stock solutions and mix thoroughly by pipeting after adding each compound. Be sure to account for radioactive decay when adding the [γ-^{32}P]ATP. The [γ-^{32}P]ATP should be added last; from this step you must exercise radioactive safety precautions.
2. Add 10 μL 3XPKB to the immunoprecipitates (*see* **Note 7**).
3. Start the kinase reaction by adding 10 μL of the RRLIEDAEYAARG peptide substrate from a 3.6 mM solution prepared in 50 mM HEPES (1.2 mM final concentration) (*see* **Note 8**).
4. Incubate the kinase reaction tube for 20 min at 30°C with gentle shaking (i.e., rotate in a hybridization oven). For kinetic analysis, the reaction time and/or enzyme concentration should be varied to achieve kinetics in the linear range. If time course experiments are carried out, a large reaction can

be prepared. At the desired time points, aliquots are withdrawn and processed as described later.

5. Terminate the reaction by adding 45 µL of ice cold 5% TCA, then incubate 5 min on ice.

6. Centrifuge the sample (12,000g, 1 min, 4°C). Spot 20 µL of the supernatant (in triplicate) on 21 mm diameter P81 cellulose phosphate filter circles.

7. Wash the P81 filter circles three times with 100 mL 0.5% phosphoric acid using a filter-washing device such as a vacuum manifold (approx 1 min per wash). Unincorporated ATP will be washed off the paper. Progress of the washing steps can be followed by including a P81 filter circle for a blank reaction (lacking the peptide substrate) and monitoring with a Geiger counter.

8. Transfer papers into 5-mL polypropylene scintillation vials with 2.5 mL of deionized H_2O. Measure the incorporation of [γ-^{32}P]ATP into substrate bound to filters in a liquid-scintillation counter using the Cerenkov method. For quantification of the results, see **Subheading 3.4**.

3.3.2. Protein-Tyrosine Kinase Assay Using Synthetic Aminoacid Polymers

This alternative protocol utilizes a random polymer-containing glutamate and tyrosine in a four to one ratio. This inexpensive aminoacid polymer is a suitable substrate for most, if not all, protein-tyrosine kinases *(19,20)* (*see* **Note 9**). The phosphorylated polymers can be separated from the reaction mixture by either TCA precipitation or gel electrophoresis.

1. Prepare the Src immunoprecipitates and 3X concentrated protein kinase assay buffer (3XPKB) as seen earlier, except substitute 10 µL of synthetic poly (Glu, Tyr) aminoacid polymer (from a 1 mg/mL stock solution prepared in 50 m*M* HEPES) for the RRLIEDAEYAARG peptide.

2 Incubate the kinase reaction tube for 20 min at 30°C with gentle shaking (i.e., rotate in a hybridization oven).

3. Terminate the reaction by adding 30 µL of 2X SDS-PAGE sample buffer. Mix by vortexing and heat for 5 min at 100°C.

4. Mix 10 µL of the sample together with 2 µL of SDS-PAGE marker (prestained broad range) and load onto a 12% polyacrylamide SDS-PAGE minigel for electrophoretic separation. Stop the electrophoresis just as the dye runs off the gel (the unincorporated ATP will also have run out of the gel by this time).

5. Cut the gel lane containing the loaded sample using a razor blade (*see* **Note 10**). The prestained marker proteins enable precise cutting of individual lanes. Transfer each gel slice (using tweezers) into a separate scintillation vial containing 2 mL of deionized H_2O.

6. Measure the incorporation of [γ-^{32}P]ATP into the synthetic poly (Glu, Tyr) substrate in a liquid-scintillation counter using the Cerenkov method.

3.4. Quantification of Results

In the protocols presented earlier, Src kinase activity is measured as incorporation of ^{32}P into either a substrate-based oligopeptide (*see* **Subheading 3.3.1.**) or synthetic random amino-acid polymer (*see* **Subheading 3.3.2.**). The levels of phosphorylated substrate are quantified by determining the ^{32}P-radioactivity in washed phosphocellulose filters (*see* **Subheading 3.3.1.**) or excised gel slices (*see* **Subheading 3.3.2.**). This section discusses approaches that can accurately quantify the results and determine kinetic parameters.

First, it is necessary to obtain a quantitative measurement of the level of Src protein from the Western blot analysis performed in **Subheading 3.2.2., step 5**. This can be achieved by using densitometry scanning of the Western blots, whether they are developed by either alkaline phosphatase reactions on the membrane or by a chemiluminescence exposure of X-ray film. For accurate quantification, it is necessary to quantitate several dilutions of the kinase sample (e.g., 1:2, 1:4, 1:8), which also ensures that the Western blots are not overdeveloped. A normalized kinase activity can then be determined with respect to the amount of Src protein in the reaction mixture relative to the measured incorporation of radioactivity into the substrate. Normalized kinase activities obtained in this way are useful for comparing different kinase isoforms or mutational variants.

If the protein kinase under study is available in a defined concentration, it is possible to determine K_m (for peptide) and V_{max} for the kinase reactions described here. In this case, one would carry out multiple reactions using various amounts of peptide substrate (ranging from 0.1 to 10 mM; the range could vary with different enzymes and substrates), and determine kinetic constants by weighted nonlinear least-squares fit to the hyperbolic velocity versus peptide concentration using iterative programs such as DYNAFIT *(21)*, or KINSIM *(22),* and FITSIM *(23)*.

4. Notes

1. Optimal buffer composition, pH, ionic strength, and divalent cation concentration must be determined for each kinase under study. The most frequently used buffers are HEPES, PIPES, and TRIS-HCl at the concentrations in the range 20–50 mM, with pH in the range 6.8–8.0. Most kinases have Km values for ATP in the range of 5–200 µM; it is necessary to use saturation concentration of ATP when determining K_m and V_{max} for substrate. The optimal concentration of divalent cations is usually between

10–20 mM. In general, serine/threonine kinases prefer $MgCl_2$, whereas tyrosine kinases prefer $MnCl_2$. For dual specifity kinases, specifity may be determined by the nature of the divalent cation bound to ATP; Mg^{2+} results in serine phosphorylation and Mn^{2+} promotes tyrosine phosphorylation (24). The addition of phosphatase inhibitors, e.g., sodium vanadate for tyrosine phosphatases or okadaic acid, and EGTA for serine-threonine phosphatases is necessary if there is a possibility of phosphatase contamination of the kinase preparation (as in immunoprecipitates of endogenous proteins).

2. $MnCl_2$ is light-sensitive and the stock solution should be prepared just before performing the assay.

3. v-Src interferes with cell-cycle progression in yeast cells (25,26) and it is therefore necessary to use the inducible expression system and induce Src expression for a short time (3–4 h) at high cell density.

4. When protein kinases are expressed in mammalian cells for use in kinase assays, the cells are usually lysed in RIPA or NP-40 buffer (lacking SDS), and cell debris is removed by centrifugation before immunoprecipitation. It is essential that the investigator assures that no "contaminating" protein kinases are present in the immunoprecipitate. It is important to carry out control reactions in which the primary antibody incubation step is omitted. If protein kinase activity is associated with the affinity beads obtained from such control lysates, steps must be taken to eliminate the nonspecific activity. The lysates for example can be preincubated with the affinity beads to reduce or eliminate the nonspecific activity.

5. Some kinase antibodies can inhibit or modify kinase activity. Thus, it may be necessary to test several antibodies to identify one suitable for direct assays of immunoprecipitated kinases.

6. Alternatively, proteins can be used as a substrate for protein kinase assays. Commonly used protein substrates include acid-treated enolase, alpha and beta casein, myelin basic protein (MBP), calmodulin, histones H1 and H2B, and angiotensin. After the kinase reaction, proteins are resolved using SDS-PAGE, transferred to membrane, and incorporation of ^{32}P is assessed by autoradiography followed by densitometry scanning.

7. Highly purified active c-Src can be purchased from commercial sources (e.g., Upstate Cell Signaling Solutions [Charlottesville, VA], Calbiochem [San Diego, CA]) and used as an alternative source of enzyme for the kinase assay. In this case, the reaction should be initiated by the addition of kinase to the complete reaction mixture.

8. The peptide can be incorporated directly into the kinase buffer if multiple assays are being performed using the same peptide substrate concentration. The RRLIEDAEYAARG peptide utilized in this protocol is a good substrate for many tyrosine kinases in addition to Src. Regarding assays of serine/

threonine kinases, the investigator will need to identify an appropriate substrate peptide, as discussed in the Introduction.

9. Specificity for individual groups of tyrosine kinases can be increased by the introduction of other amino-acid residues such as alanine or lysine *(19)*.
10. Commercially available synthetic amino-acid polymers are very heterogenous with respect to the range of molecular weight (even each batch of product may differ) and thus are not suitable for reproducible pattern analysis using SDS-PAGE (autoradiography).

Acknowledgments

We wish to thank D. Mojzita for his useful comments on the protocols. Work from the authors' laboratory is supported by Public Health Service grants GM49882 and DK56018.

References

1. Hunter, T. and Plowman, G. D. (1997) The protein kinases of budding yeast: six score and more. *Trends Biochem. Sci.* **22,** 18–22.
2. Manning, G., Whyte, D. B., Martinez, R., et al. (2002) The protein kinase complement of the human genome. *Science* **298,** 1912–1934.
3. Hanks, S. K., Quinn, A. M., and Hunter, T. (1988) The protein kinase family: conserved features and deduced phylogeny of the catalytic domains. *Science* **241,** 42–52.
4. Hanks, S. K. and Hunter, T. (1995). The eukaryotic protein kinase superfamily: kinase (catalytic) domain structure and classification. *FASEB J.* **9,** 576–596.
5. Manning, G., Plowman, G. D., Hunter, T., and Sudarsanam, S. (2002) Evolution of protein kinase signaling from yeast to man. *Trends Biochem. Sci.* **27,** 514–520.
6. Hardie, G. and Hanks, S. K. (1995) *The Protein Kinase Facts Book. Vol. II: Protein-Serine Kinases.* Academic Press, London, UK.
7. Kreegipuu, A., Blom, N., Brunak, S., and Järv, J. (1998) Statistical analysis of protein kinase specificity determinants. *FEBS Lett.* **430,** 45–50.
8. Kreegipuu, A., Blom, N., and Brunak, S. (1999) PhosphoBase, a database of phosphorylation sites: release 2.0. *Nucleic Acids Res.* **27,** 237–239.
9. Songyang, Z., Lu, K. P., Kwon, Y. T., et al (1996) A structural basis for substrate specificities of protein Ser/Thr kinases: Primary sequence preference of casein kinases I and II, NIMA, phosphorylase kinase, calmodulin-dependent kinase II, CDK5, and Erk1. *Mol. Cell Biol.* **16,** 6486–6493.
10. Zhu, H., Klemic, J. F., Chang, S., et al (2000) Analysis of yeast protein kinases using protein chips. *Nature Genet.* **26,** 283–289.
11. Murphy, S. M., Bergman, M., and Morgan, D. O. (1993) Suppression of c-Src activity by C-terminal Src kinase involves the c-Src SH2-domain and SH3-domain-analysis with *Saccharomyces cerevisiae. Mol. Cell Biol.* **13,** 5290–5300.
12. Gonfloni, S., Williams, J. C., Hattula, K., et al (1997) The role of the linker between the SH2 domain and catalytic domain in the regulation and function of Src. *EMBO J.* **16,** 7261–7271.

13. Brábek, J., Mojzita, D., Novotny, M., et al (2002) The n- and RT-loops of the SH3 domain of Src can downregulate its kinase activity in the absence of the SH2 domain-Y527 interaction. *Biochem. Biophys. Res. Commun.* **296,** 664–670.
14. Kornbluth, S., Jove, R., and Hanafusa, H. (1987) Characterization of avian and viral p60src proteins expressed in yeast. *Proc. Natl. Acad. Sci. USA* **84,** 4455–4459.
15. Brábek, J., Mojzita, D., Hamplová, L., and Folk, P. (2002) The regulatory region of Prague C v-Src inhibits the activity of the Schmidt-Ruppin A v-Src kinase domain. *Folia Biol. (Prague)* **48,** 28–33.
16. Schiestl, R. H. and Gietz, R. D. (1989) High efficiency transformation of intact yeast cells using single stranded nucleic acids as a carrier. *Curr. Genet.* **16,** 339–346.
17. Casnellie, J. E. (1998) Assay of protein kinases using peptides with basic residues for phosphocellulose binding, in *Protein Phosphorylation* (Sefton B. M. and Hunter, T., eds.), Academic Press, San Diego, CA, pp. 111–116.
18. Witt J. J. and Roskoski, R. Jr. (1975) Rapid protein kinase assay using phosphocellulose-paper absorption. *Anal. Biochem.* **66,** 253–258.
19. Braun, S., Raymond, W. E., and Racker, E. (1984) Synthetic tyrosine polymers as substrates and inhibitors of tyrosine-specific protein kinases. *J. Biol. Chem.* **259,** 2051–2054.
20. Racker, E. (1998) Use of synthetic amino acid polymers for assay of protein-tyrosine and protein-serine kinases, in *Protein Phosphorylation* (Sefton B. M. and Hunter, T., eds.), Academic Press, San Diego, CA, pp. 111–116.
21. Kuzmic, P. (1996) Program DYNAFIT for the analysis of enzyme kinetic data: application to HIV proteinase. *Anal. Biochem.* **237,** 260–273.
22. Barshop, B. A., Wrenn, R. F., and Frieden, C. (1983) Analysis of numerical methods for computer simulation of kinetic processes: development of KINSIM; a flexible, portable system. *Anal. Biochem.* **130,** 134–45.
23. Zimmerle, C. T. and Frieden, C. (1989) Analysis of progress curves by simulations generated by numerical integration. *Biochem. J.* **258,** 381–387.
24. Yuan, C. J., Huang C. Y., and Graves D. J. (1993) Phosphorylase kinase, a metal ion-dependent dual specificity kinase. *J. Biol. Chem.* **268,** 17,683–17,686.
25. Boschelli, F., Uptain, S. M., and Lightbody, J. J. (1993) The lethality of p60(v-src) in *Saccharomyces cerevisiae* and the activation of p34(cdc28) kinase are dependent on the integrity of the SH2 domain. *J. Cell Sci.* **105,** 519–528.
26. Florio, M., Wilson, L. K., Trager, J. B., et al (1994) Aberrant protein phosphorylation at tyrosine is responsible for the growth-inhibitory action of pp60(v-src) expressed in the yeast *Saccharomyces cerevisiae. Mol. Biol. Cell* **5,** 283–296.
27. Kinase.com [http://kinase.com/]

7

Comparative Phosphorylation Site Mapping From Gel-Derived Proteins Using a Multidimensional ES/MS-Based Approach

Francesca Zappacosta, Michael J. Huddleston, and Roland S. Annan

Summary

Understanding how phosphorylation regulates the behavior of individual proteins is critical to understanding signaling pathways. These studies usually involve knowledge of which amino acid residues are phosphorylated on a given protein and the extent of such a modification. This is often a rather difficult task in that most phosphoproteins contain multiple substoichiometric sites of phosphorylation.

Here we describe the multidimensional electrospray (ES) mass spectrometry (MS)-based phosphopeptide-mapping strategy developed in our laboratory. In the first dimension of the process, phosphopeptides present in a protein digest are selectively detected and collected into fractions during on-line liquid chromatography (LC)/ES/MS, which monitors for phosphopeptide-specific marker ions. This analysis generates a phosphorylation profile that can be used to assess changes in the phosphorylation state of a protein pointing to those phosphopeptides that require further investigation. The phosphopeptide-containing fractions are then analyzed in the second dimension by nano-ES with precursor-ion scan for the marker ion m/z 79. As the final step, direct sequencing of the phosphopeptides is performed by LC/ES/MS/MS. Merits and limitations of the strategy, as well as experimental details and suggestions, are described here.

Key Words: Phosphorylation; protein kinase; mass spectrometry; electrospray; liquid chromatography.

1. Introduction

Protein phosphorylation is surely the most important posttranslational modification by which signals are transmitted within the cell (*1*). Signal transduction

From: *Methods in Molecular Biology, vol. 284:*
Signal Transduction Protocols
Edited by: R. C. Dickson © Humana Press Inc., Totowa, NJ

through reversible phosphorylation of intracellular proteins plays an essential role in controlling many aspects of cell growth, metabolism, and differentiation. Dissecting the biological pathways that lead to these effects often involves understanding how phosphorylation regulates the behavior of individual proteins in the pathway. Key to this process is identifying sites of phosphorylation on selected components in the pathway.

Knowledge of the specific amino acids phosphorylated on a given protein is often critical to understanding how that protein's function is regulated. Recently it has become clear that multisite phosphorylation is quite common (2), but that not every site contributes to regulating a particular function. The extent to which any given site is modified in response to stimulation of the cell or a change in its environment speaks to the physiological relevance of that site.

The challenges for any technique designed to study protein phosphorylation are at least fourfold. First, the modification can occur at one or several (usually) of many potential sites within the protein sequence. Therefore, the analytical technique must sample as much sequence coverage as possible to reasonably ensure that no sites are missed. Second, many phosphoproteins of interest are present in the cell at very low level, thus the technique needs to be quite sensitive. Third, the extent of phosphorylation at the various sites can vary greatly (from 1 to 100%), therefore, the analysis technique needs to have enough selectivity to allow detection of phosphorylated peptides from among the usually very abundant nonphosphorylated peptides. Last, once the phosphorylated peptide has been detected and tentatively identified, there are usually multiple potential amino acids that could be modified, therefore, it is almost always necessary to determine the exact site of phosphorylation by direct sequencing. The two techniques that deal most effectively with these challenges are the Hunter 2D phosphopeptide mapping technique (3) coupled to Edman sequencing and LC coupled to ES/MS/MS (4,5). The major drawback with the first method is the need to incorporate a radioactive label onto the phosphate group. Although this can be conveniently done for in vitro assays, metabolic labelling of growing cells requires large amounts of radioactivity and the prospect of purifying a low copy number protein from a liter or more of radioactive cell-culture material is unappealing to most researchers. For this reason, MS is emerging as the key technology both for phosphosite mapping and for the determination of phosphorylation stoichiometry (5–12).

In our laboratory, we have developed and over the years refined, a multidimensional ES/MS-based strategy that allows for selective and sensitive detection and identification of multisite phosphorylation on proteins. Because the first dimension provides a profile of the phosphorylation state of the entire protein, it can also give a semiquantitative measure of the extent to which phosphorylation at any given site changes in response to different cellular conditions

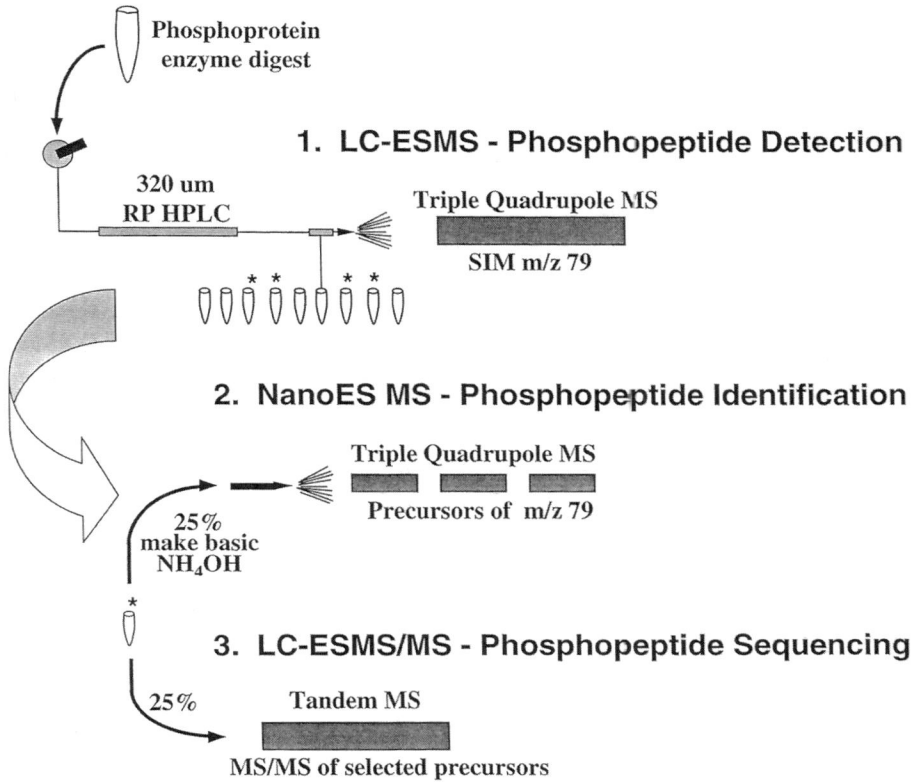

Fig. 1. Schematic diagram of the multidimensional ES/MS method for mapping phosphorylation sites in proteins.

(5). These changes in phosphorylation stiochiometry speak to the functional significance of that site under a certain set of conditions.

The strategy described here utilizes several different MS scanning techniques to detect and identify phosphopeptides and to determine the exact site of phosphorylation. A schematic workflow for the strategy is shown in **Fig. 1**. In the first dimension of the process, phosphopeptides present in the proteolytic digest of a protein are selectively detected and collected into fractions by monitoring for phosphopeptide specific marker ions PO_3^- (m/z 79) and PO_2^- (m/z 63) produced in the ion source of the MS during on-line LC-negative ion-ES/MS *(6)*. In this type of experiment, the marker ions are produced in a region of relatively high pressure located between the ion source and the MS analyzer *(see* **Fig. 2**) via collision-induced dissociation. Energetic collisions between phosphopeptide ions and gas molecules result in cleavage of the phosphate group from the

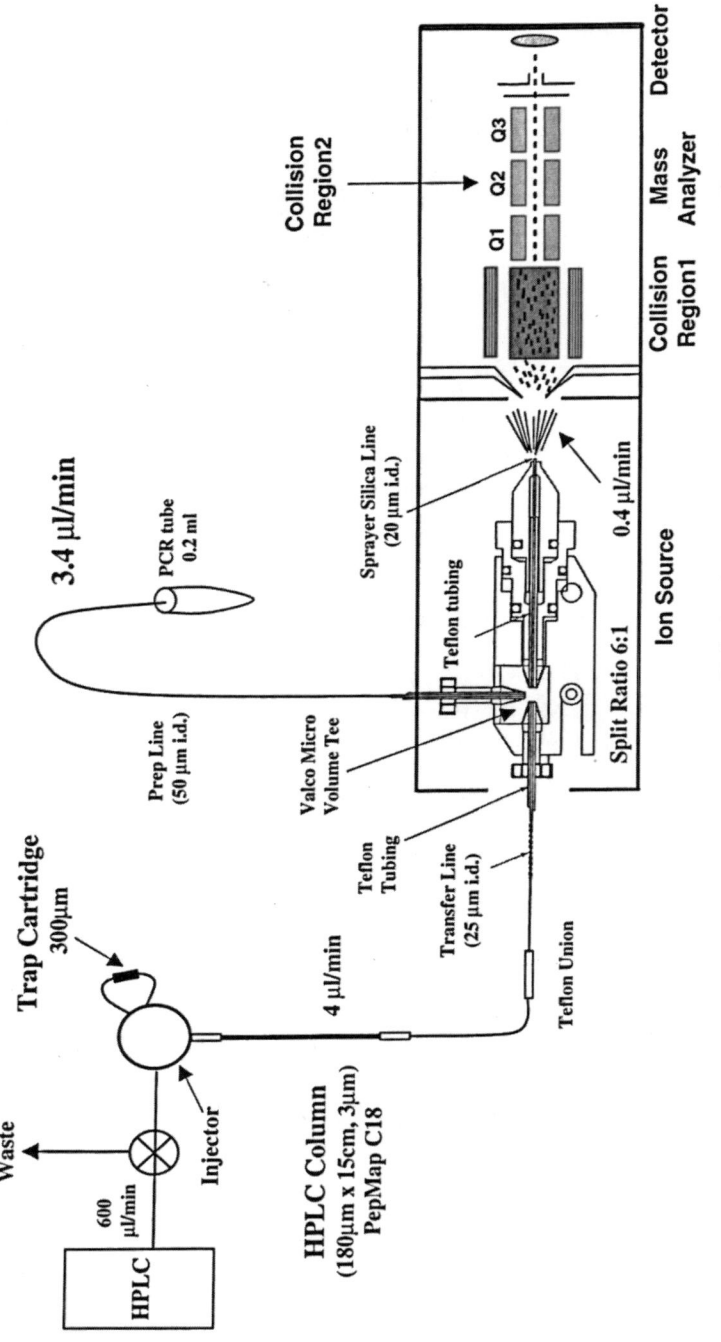

Fig. 2. Configuration of the first dimension LC/ES/MS experiment for the detection and collection of phosphorylated peptides. The mass spectrometer operates in the negative ion mode for this experiment. Elements are not drawn to scale.

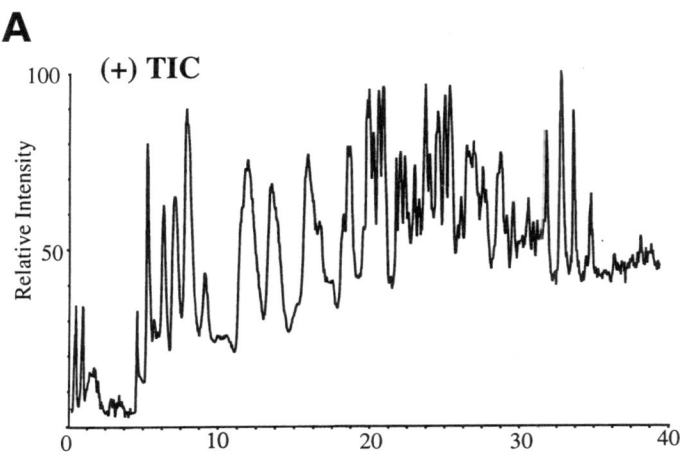

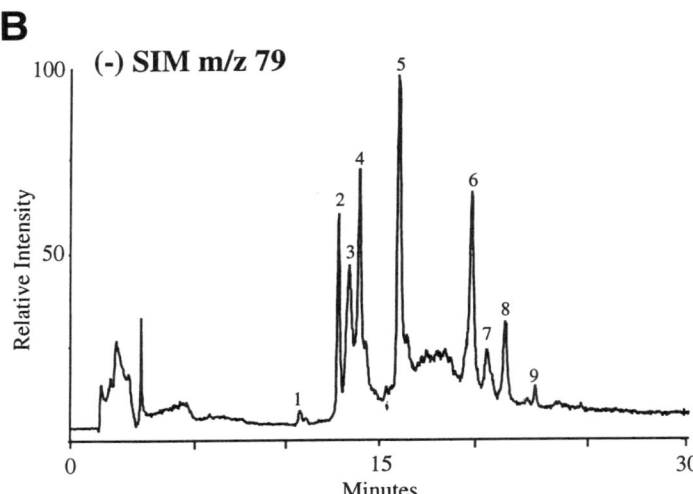

Fig. 3. First dimension LC/ES/MS analysis of phosphoprotein. (**A**) Positive ion, total ion current (TIC) showing the MS signal for all peptides present in a tryptic digest of the protein. (**B**) Negative ion, single ion monitoring for m/z 79, showing the MS signal for the phosphopeptide-specific marker ion PO_3^-.

side chain of serine, threonine, and tyrosine to form PO_3^- and PO_2^-. The instruments ideally suited for this experiment are quadrupole MS because of their ability to detect a single ion (or several) 100% of the time. The readout from this experiment (*see* **Fig. 3B**) resembles the radioactivity trace for the high-performance liquid chromatography (HPLC) separation of a ^{32}P-labeled sample, except there isn't any radioactivity. The inclusion of a chromatographic step is an important component of the overall strategy. Chromatographic separation of the digest prior to phosphopeptide analysis greatly minimizes problems associated with the MS analysis of unfractionated samples such as ion-suppression effects, charge-state overlap, and dynamic-range limitation and helps maximize the sequence coverage. Furthermore, the chromatographic step partitions the total phosphopeptide pool, reducing the complexity of the nonphosphorylated background (compare **Fig. 3A, B**) and provides a substantial measure of sample clean-up, all of which facilitate the identification and sequencing of the specific phosphopeptides. A chromatographic separation prior to phosphopeptide analysis is essential when dealing with a complex phosphorylation profile or very large proteins *(13–16)*.

The column effluent is split just before the ion source (for details *see* **Fig. 2**), with ca. 20% going to the MS for phosphopeptide detection and 80% collected for further analysis. Columns with an internal diameter of 500 μm or larger (optimal flow rate 20–50 μL/min) can be used when coupled with a conventional ES source that has an optimal flow rate at the source after splitting of 2–5 μL/min. Smaller-diameter HPLC columns, however, will provide higher sensitivity owing to the sample concentration effect. Columns with an internal diameter of 180 or 320 μm (optimal flow rate of 3–4 μL/min) can be used if coupled to micro-electrospray sources that operate at an optimal flow rate of 0.2–0.5 μL/min *(5)*. The selected ion monitor (SIM) trace for m/z 79 and 63 indicates peaks that contain phosphorylated peptides. Comparison of the SIM trace in **Fig. 3B** with the corresponding full scan TIC trace from a separate LCMS run in **Fig. 3A** shows the reduced complexity of the peptide profile when monitoring only for phosphate-specific marker ions. This trace also serves as fingerprint for the phosphorylation profile of a protein. Change in the phosphorylation state of a protein would be reflected by a change in the phosphorylation profile pointing to those phosphorylation sites that require further study.

Figure 4A shows the phosphopeptide specific marker-ion profiles for a tryptic digest of the motor protein myosin-V treated with two different cell extracts: one extract prepared from cells in interphase and the other extract prepared from cells in mitosis *(17)*. These data clearly show that phosphorylation of peptides in fractions 7, 8, and 9 is highly cell cycle-dependent, whereas phosphorylation of the remaining phosphopeptides in the profile is not. Because myosin-V-dependent organelle transport had been shown to be cell cycle-dependent and suggested to be regulated by phosphorylation *(18)*, further analysis was carried out on the relevant phosphopeptides in fractions 7–9. In the second dimension of the analysis, the molecular weight of phosphopeptides present in the selected fractions is determined by analyzing each fraction using a precursor ion scan for m/z 79 in the negative ion mode (**Fig 4B**) *(8,9)*. The precursor-ion scan for m/z 79 detects only those ions that can fragment to yield HPO_3^-. This step is needed because the fractions collected in the first-dimension LC analysis almost invariably contain more than one peptide and the phosphopeptides might represent only a small percentage of the total peptide mixture present in the fraction. Using conventional MS techniques such as matrix-assisted laser desorption/ionization (MALDI)-MS or full-scan ES/MS, phosphopeptides are often masked by the more abundant nonphosphorylated peptides present in a sample. Using precursor-ion scanning, phosphopeptides can easily be detected even though they account for less than 1% of the intensity of the largest ion in the spectrum. In the case of the myosin sample, we found that all three fractions, 7–9, contained the same two phosphopeptides with masses 1921.8 and 1982.6.

From the precursor ion-scan data, a tentative assignment of the amino-acid sequence for most of the peptides can usually be made; however, the identity of some peptides will be ambiguous at this point. For instance, the 1921 Da peptides shown in **Fig. 4B** was assigned to the myosin sequence 1648–1664 with one mole of phosphate; however, the 1982 Da peptide could not be assigned to any reasonable sequence. For this peptide and in fact for all identified phosphopeptides, we use direct sequencing to establish a confident identity and assign the exact site of modification. Thus, as the final step in the analytical process, direct sequencing of the phosphopeptides is done by on-line LC, positive-ion tandem MS of selected precursors (for example *see* **Fig. 4C**). This last step confirms the peptide sequence and usually allows determination of the exact site of modification. It also identifies peptides that result from nonspecific enzyme cleavage and that contain other, unanticipated modifications. **Figure 4C** shows the MS/MS spectrum for the doubly charged ion (m/z 961.2) from the 1921 Da phosphopeptide found in Fraction 8. In fact, all three 1921 Da peptides yielded a similar MS/MS spectrum. They showed that in each case phosphorylation was confined to the first three residues. Based on the presence of three chromatograph peaks each containing a phosphopeptide with the same molecular weight, we hypothesized that the protein was phosphorylated on all three residues to some extent. Using several weak fragment ion peaks found in the spectrum shown in **Fig. 4C**, we also suggested that the most abundant phosphorylation site was the third residue Ser1650. Site-directed mutagenesis of all three sites showed these hypotheses to be correct *(17)*. The 1982 Da peptides was found by MS/MS to have an identical sequence to the 1921 Da peptide with an unknown modification to the N-terminal serine residue.

Using the strategy outlined in this chapter, we have shown that from among the many phosphorylation sites present on the myosin-V protein, a single site was significant in regulating its organelle transport function. Using the strategy

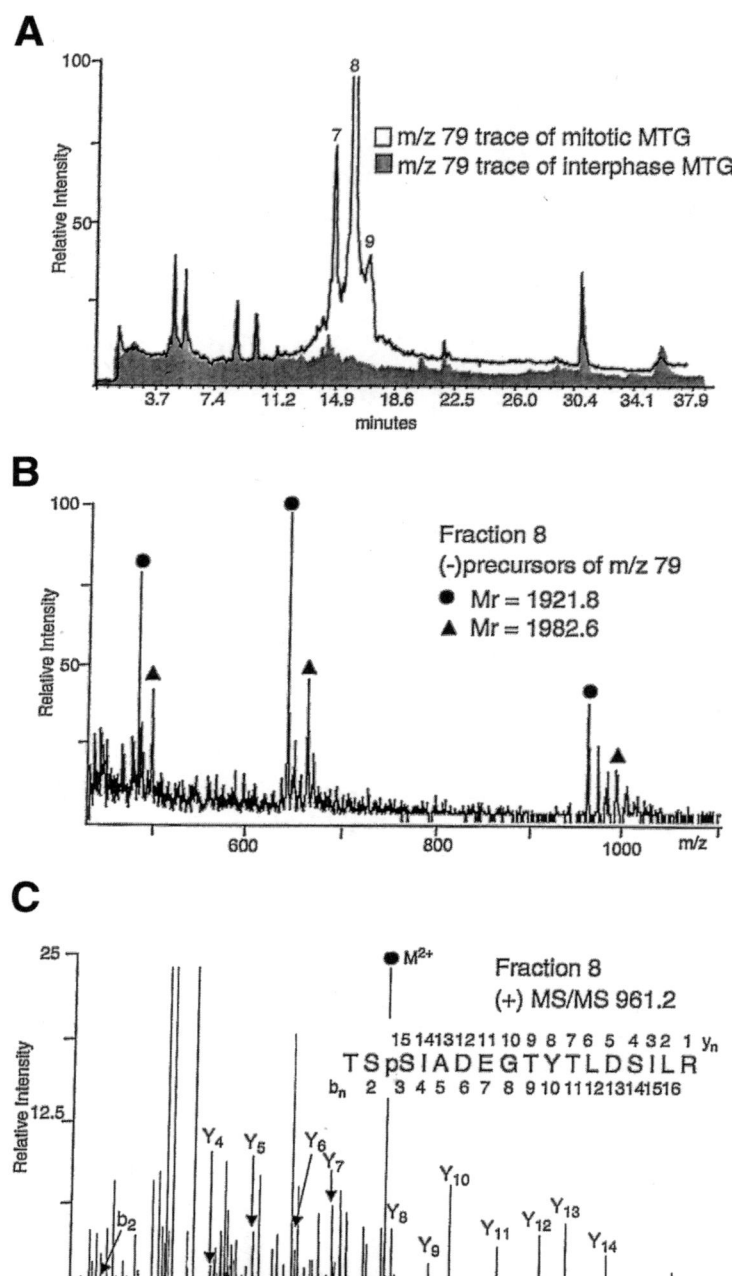

Fig. 4. Comparative phosphorylation site mapping. **(A)** Phosphopeptide-selective profile of myosin-V treated with mitotic (white) or interphase (gray) extract *(17)*. After

outlined here, we were able to identify that single site without resorting to the analysis of the entire phosphorylation complement.

2. Materials

2.1. Protein In-Gel Digestion

1. 100 mM ammonium bicarbonate, pH 8.0.
2. 45 mM dithiothreitol (DTT) in 100 mM ammonium bicarbonate, pH 8.0.
3. 100 mM iodoacetamide in 100 mM ammonium bicarbonate, pH 8.0.
4. Acetonitrile (HPLC-grade).
5. Modified Trypsin (Promega, Madison, WI).
6. Heating block at 37°C.

2.2. Reverse-Phase HPLC Solvents (see Note 1)

1. Solvent A: 2% acetonitrile in water containing 0.1% formic acid + 0.02% trifluoroacetic acid (TFA).
2. Solvent B: 90% acetonitrile in water containing 0.1% formic acid + 0.02% TFA.
3. Sample loading solution: 0.1% formic acid + 0.02% TFA.
4. Autosampler needle wash solution: 40% acetonitrile + 40% 2-propanol.

2.3. HPLC Columns and Trap Cartridges

1. PepMap C18 trap cartridge (300 μm × 5 mm) (Dionex-LC Packings).
2. PepMap C18 capillary column, (300 μm × 15 cm, 3 μm particles) (Dionex-LC Packings).
3. PepMap C18 capillary column (75 μm × 15 cm, 5 μm particles) (Dionex-LC Packings).

---

purification by SDS-PAGE, myosin-V bands were excised and digested *in situ* with trypsin. Tryptic digests of each sample were fractionated by RP-HPLC as described in the text. Fractions 7–9 were analyzed by nano-ES using precursor ion scanning for m/z 79. **(B)** Precursor-ion scan spectrum for fraction 8 showing multiple charge states (2−, 3− and 4−) for two phosphopeptides. Molecular masses are isotope-weighted average values derived from the individual charge states for each peptide. **(C)** MS/MS product ion spectrum of the doubly charged ion (m/z 961.2) from the 1921.8 Da phosphopeptide in Fraction 8. The y_n ion series shows that residues 1651–1664 are not phosphorylated. A weak b_n series is present, showing a b_2 ion corresponding to unmodified Thr[1648] and Ser[1649]. All subsequent b_n ions (b_3–b_{10}) are shifted in mass by 80 Da. For the sake of clarity not all ions are labeled on the spectrum. Actual sequence coverage is indicated on the peptide sequence. Nomenclature is after Biemann *(19)*.

2.4. Nano-ES Solvents for m/z 79 Precursor-Ion Scanning

1. Basic solution: 50% methanol in water containing 10% ammonium hydroxide (made from a 30% stock solution).

2.5. Instrumentation

1. Phosphopeptide selective LC/ES/SIM experiments have been carried out using an API III triple quadrupole MS (Sciex/Applied Biosystem) equipped with a nanoflow source from Micromass (Waters). This experiment has been shown by other laboratories around the world to work equally well on a variety of triple quadrupole instruments from this and other manufacturers. Instrument lens and voltages used to optimize sensitivity for production of the marker ions will vary among the different kinds of triple quads. Single quadrupole instruments can also be used.
2. Precursor-ion spectra are acquired on an API 3000 triple quadrupole MS (Sciex/Applied Biosystem) equipped with a nano ES source (Sciex/Applied Biosystem). These experiments can also be performed on any other triple quadrupole MS system. We use medium size metal-coated capillary nano-ES spray tips available commercially (Proxeon, Odense M Denmark).
3. Targeted LC/MS/MS is performed on a Micromass QTOF instrument equipped with a Micromass nanoflow source. Any type of LC-tandem MS instruments can be used to produce MS/MS data. Sensitivity and resolution will vary with make and model.
4. A conventional Shimadzu HPLC system is used for the capillary LC/ES/MS/SIM experiments. A precolumn flow rate of 4 µL/min was achieved using an LC-Packing Acurate microflow splitter. Postcolumn split is made using a stainless steel or titanium 0.15-mm i.d. micro-volume Valco tee near the source to make the high-voltage connection.
5. An LC Packings Ultimate system is used for the nanobore LC/MS/MS experiments.

3. Methods

3.1. Protein In-Gel Digestion (see Note 2)

1. Excise the band of interest from a Coomassie Blue stained gel (*see* **Note 3**) using a clean razor blade or scalpel. The gel band is placed in a 2.0-mL screw-cap tube.
2. Wash with 500 µL of 100 mM ammonium bicarbonate, pH 8.0, for at least 1 h or until Coomassie blue stain is gone (*see* **Note 4**).
3. Discard the liquid. Add 150 µL of 100 mM ammonium bicarbonate, pH 8.0, and 10 µL of 45 mM DTT in 100 mM ammonium bicarbonate. Incubate for 30 min at 37°C.

4. Add 10 µL of 100 m*M* Iodoacetamide in 100 m*M* ammonium bicarbonate. Incubate for 30 min at room temperature in the dark (*see* **Notes 5** and **6**)
5. Discard the liquid. Wash the gel piece with 1 mL of 50% acetonitrile (v/v) in 100 m*M* ammonium bicarbonate, pH 8.0, for 2 h.
6. Discard the liquid. The gel band is cut in small pieces (1×1 mm) and placed in a 0.5-mL microcentrifuge tube.
7. Add 50 µL of acetonitrile. Let it stand for 15 min at room temperature. Remove the liquid and dry down the gel piece in a Speed-Vac.
8. Gel band is re-swollen in 10 µL of a modified trypsin solution (10 ng/µL in 100 m*M* ammonium bicarbonate, pH 8.0). Allow to re-swell for 15 min and then add the appropriate volume of 100 m*M* ammonium bicarbonate, pH 8.0, to cover the band (usually 15–30 µL) (*see* **Notes 7** and **8**).
9. Incubation at 37°C overnight.
10. Transfer the liquid to a clean 0.5-mL microcentrifuge tube.
11. Add 25 µL of acetonitrile (or enough liquid to cover the band). Let stand for 10 min. Transfer the liquid to the tube containing the digestion solution.
12. Repeat **step 10** (*see* **Note 9**).
13. Concentrate the sample using a Speed-Vac. This step is required both to remove the organic solvent before LC separation as well as to reduce sample volume. Do not let the sample go dry.

3.2. Phosphopeptide Mapping Using a Multidimensional ES/MS-Based Strategy

3.2.1. First-Dimension Analysis: Micropreparative Fractionation of Phosphopeptides Using Phosphate-Specific LC/ES/MS (see Note 10)

1. Prior to analysis of samples the production and detection of the phosphate-specific marker ions by LC/MS should be optimized (*see* **Note 11**). Generally speaking, the optimization of the ion-source conditions for the production of the marker ions needs only to be done once, but sensitivity should be checked prior to any round of experiments. For general tuning and calibration (*see* **Note 12**).
2. After checking the MS sensitivity, the LC is connected to the mass spectrometer and the whole set-up tested by injecting 200 fmol of phosphopeptide standard TRDIYETDYpYRK (Anaspec, San Jose, CA). For this phosphopeptide, under these conditions, a signal-to-noise ratio of ca. 10:1 should be observed.
3. Protein digests to be analyzed are acidified (pH ≤3.0) using an appropriate volume of 0.1% TFA or loading solution.
4. Protein digests are injected onto a PepMap C18 trap cartridge (300 µm × 5 mm) used in place of the sample loop on the HPLC injector (*see* **Note 13**). After washing with loading solution, the peptides are back flushed off of

the cartridge with a 4 µL/min acetonitrile/water gradient (2–50% B in 20–30 min) onto a PepMap C18 capillary column (300 µm × 15cm, 3 µm particles) fitted directly into the injector. HPLC mobile phases contain both 0.1% formic acid and 0.02% TFA (*see* **Note 14**). The column outlet is connected to a Micromass nanoflow ion source via a 25 µm i.d. fused silica line threaded directly into the source (*see* **Fig. 2**). Column flow is split prior to the MS by means of a 0.15 mm i.d. microvolume Valco tee insert, which directs 0.4–0.6 µL/min to a 20 µm i.d. tapered fused silica ES tip from New Objectives (Cambridge, MA) (*see* **Note 15**). The remainder of the flow is sent to the prep line for manual collection into polymerase chain reaction (PCR) tubes. The prep line is PEEK (63 µm i.d., 22.5 cm length) from the source tee to a Valco union (SS, 0.15 mm i.d.) and fused silica (50 µm i.d., 45 cm length) on the other side of the union (*see* **Note 16**).

5. Fractions that have been shown to contain phosphopeptides are manually collected into PCR tubes (*see* **Note 17**), and immediately placed on ice. Fractions are stored at –70°C (*see* **Note 18**).

3.2.2. Second-Dimension Analysis: Determination of the Molecular Masses of the Phosphopeptides by Nano-ES With Precursor Ion Scanning

1. Phosphopeptides are detected in the fractions collected in the first dimension by precursor-ion scanning for m/z 79 (PO_3^-) (*see* **Note 19**). Precursor ion mass spectra are obtained using a nano-ES interface.
2. One-half (or less for abundant phosphopeptides) of each fraction that contains phosphopeptides (or only fractions containing phosphopeptides of interest in cases of differential phosphorylation studies) is made basic by adding one volume of 50/50 methanol/water containing 10% ammonium hydroxide (*see* **Note 20**). 1.5 µL of sample in the basic solution is loaded into the back of a medium, metal-coated ES spray needle using an electrophoresis gel-loading pipet tip. The nanospray needle is positioned in the ion source with the aid of the 2 CCD cameras (*see* **Note 21**).
3. Acquire full-scan negative ion spectra to check instrument status and determine optimal spray flow and positioning (*see* **Notes 22** and **23**).
4. Acquire negative ion m/z 79 precursor-ion spectra to determine the molecular weight of the phosphopeptide(s) present in the fractions (*see* **Note 24**). In some cases, phosphopeptides can be identified at this stage by their molecular weight.

3.2.3. Third-Dimension Analysis: Localization of the Site of Modification by Targeted On-Line LC/MS/MS

Once identified, phosphorylated peptides are sequenced by targeted LC/ES/MS/MS on a Micromass QTOF equipped with a nanoflow ion source. Individual fractions are loaded on a PepMap C18 trap cartridge (300 μm × 5 mm) and, after washing of the cartridge with loading solution, are back-flushed onto a PepMap C18 capillary column (75 μm × 15cm, 5μm particles) at 0.3 μL/min using a 5–50% B gradient in 30 min. HPLC mobile phases contained 0.1% formic acid and 0.02% TFA. MS/MS data are collected on a single precursor or on a set of predefined precursors (*see* **Note 25**) using a 2-s scan with a precursor-ion window of +3 Da. When multiple precursors need to be sequenced (*see* **Note 26**), different precursors can be analyzed according to their retention time in time windows during the LC separation. For closely eluting peptides, the instrument can be set to alternate between precursors after each scan. By targeting a single precursor m/z (or several in a time dependent manner), we are able to integrate the MS/MS signal over the entire elution profile of the peptide, thus obtaining some of the nano-ES signal averaging advantage while taking advantage of the tremendous concentration advantage provided by 75 μm (i.d.) capillary LC columns. The use of the chromatographic step also significantly reduces interference from chemical noise and other peptides of similar m/z (*see* **Note 27**).

Alternatively, if there is no fractionated sample left, targeted LC/ES/MS/MS can be performed on the whole peptide mixture.

4. Notes

1. All solvents must be HPLC-grade. All chemicals must be of the highest purity commercially available.
2. This procedure assumes one SDS-PAGE band from a 1.0-mm thick mini gel. When working with gel-purified sample, gloves must be worn at all times (including staining of the gel, excision of the bands, and protease digestion) and clean laboratory procedures must be applied (i.e., always keep tubes in closed containers, change solutions often, do not touch with bare hands tubes and tips, wash surfaces that will be in contact with the gel with water and methanol). This will help to minimize keratin and dust contamination. Although these precautions do not seem to be as critical in the first two dimensions of the strategy in which only phosphopeptides are detected, it might affect the third-dimension experiment.
3. The procedure reported for the detection and identification of phosphopeptides has not been tested with silver-stained gel bands. Whenever possible, colloidal Coomassie should be used.
4. For heavily Comassie-stained bands, an overnight wash is advisable. Coomassie Blue dye contains a sulfonate group that will generate a SO_3^- ion (m/z 80) both in the first-dimension and in the second dimension experiment. During the LC separation, the dye will elute very late in the gradient, so in most cases it should not interfere with the analysis, however, it is probably advisable to remove it anyway. Extensive washes

are mandatory in cases when no LC separation is used prior to m/z 79 precursor-ion scanning.

5. DTT and iodoacetamide solutions can be prepared and kept at 4°C for a few days. The iodoacetamide solution should be kept in the dark (for instance, by wrapping the tube in foil).

6. Reduction and alkylation is essential for samples separated by first-dimension SDS-PAGE even when the reducing agent was already present in the loading buffer. This step, however, might be skipped if samples are derived from second-dimension-gel electrophoresis in that a reduction and an alkylation step is already performed between the first- and second-dimension analysis.

7. The most commonly used enzyme for generating peptide digests for mass-spectrometric analysis is trypsin. This enzyme, which specifically cleaves after the Lys and Arg residues (except cases when the following residue is a Pro), usually generates peptides ideally suited for MS analysis in size (500–3000 Da) and that present fragmentation patterns well-understood. However, other proteolytic enzyme can be used if the sequence of the protein under investigation requires it. Asp-N, Glu-C, Lys-C, and Arg-C are all useful. Enzymes with less-stringent specificity such as chymotrypsin might still be used, though nearly all peptide assignments will need to be validated by MS/MS sequencing.

8. Modified porcine trypsin (Promega) is usually employed in our laboratory. In this form of the enzyme, the Lys residues have been modified so as to partially inhibit autolysis. A 100 ng/μL trypsin solution is prepared in water, aliquoted (10 μL/aliquots), and immediately stored at –20°C. For each use, 90 μL of 100 mM ammonium bicarbonate, pH 8.0, are added to an aliquot, vortexed, and 10 μL added to each dried gel band. The unused diluted enzyme is discarded.

9. Although we found that organic extraction does not greatly improve the extraction of the peptides from the gel matrix, we still use it to displace all the peptide-containing aqueous solution trapped in the gel pieces.

10. Introducing an HPLC separation step will provide sample clean-up and peptide fractionation. However, very hydrophilic or very hydrophobic (phospho) peptides might not be recovered after the LC separation owing to their chromatographic properties on typical C18 stationary phases. In these cases, an alternative (or additional) proteolytic digestion might be employed in order to change the character of the peptide.

11. Conditions described here are specific to acquisition of ES mass spectra on Sciex/Applied Biosystems quadrupole mass spectrometer API-III (Concord, Ontario, Canada). Instrument conditions (declustering potential, spray position, sample flowrates, gas flows, voltage settings, etc.) for the

production of m/z 63 and 79 are optimized by infusing a 5 pmol/µL solution of a phosphopeptide standard in 30% HPLC mobile phase B containing 0.1% formic acid + 0.02 % TFA. We use KRPpSQRHGSK (University of Michigan Protein and Carbohydrate Structure Facility). The signal for m/z 63 and 79 are simultaneously monitored in real time by SIM (100 ms dwell per ion). The signals are then maximized individually by adjusting the declustering potential for each. On the Sciex API-III, the optimal declustering voltages are ca. −350V for m/z 63, and −300V for m/z 79 using TFA-containing mobile phases. The low mass region of the spectrum containing the phosphate marker ions is shown in **Fig. 5A**. For comparison the solvent background is shown in **Fig. 5B**. The signal-to-background for m/z 63 and 79 is typically ca. 40:1 and 70:1, respectively, by constant infusion of the standards under these conditions.

12. Tuning and calibration is performed using a solution containing a mixture of polypropylene glycol (PPG) 425, 1000, and 2000 (3 × 10-5M, 1 × 10-4M, and 2 × 10-4M, respectively) in 50/50/0.1 water/methanol/formic acid (v/v/v), 1 mM ammonium acetate. The m/z range of 10–2400 is calibrated in the negative-ion mode by multiple-ion monitoring of the isotope clusters of six PPG ion signals and two TFA-related ions, m/z 69 (CF3) and m/z 113 (CF3CO$_2$). The TFA-derived ions are present as background in any mass spectrometer that is regularly exposed to TFA-containing HPLC mobile phases. Mass spectra are recorded at instrument conditions sufficient to resolve the first two isotopes of anion m/z 991.7 (PPG + HCO$_2$)− so that the valley between them is 55% of the height of the second isotope for the singly charged ion.

13. The use of a trap cartridge in place of an injection loop has two advantages: it allows loading of a larger sample volume and provides sample clean-up. Injection volume can be up to several hundred µL. Wash of the trap cartridge after sample loading should be performed in 0.1% TFA or loading solution regardless of the solvent used during the LC run. Under these conditions, retention of the hydrophilic peptides on the trap cartridge is maximized. For best retention of peptides, caution should be used not to load the sample too quickly. On this size trap, we typically load 10 µL in 20–30 s.

14. Although conventional 0.1% TFA-containing mobile phases may be employed, sensitivity for phosphopeptide detection is increased approx 2.5-fold (as measured by signal-to-noise ratio) using a combination of 0.02% TFA and 0.1% formic acid. The degree to which the quality of the chromatographic separation is compromised can range from minimal to significant depending on the specific type of C18 employed. Therefore, columns should be tested with a mixture of standard phosphorylated and nonphosphorylated peptides prior to committing real samples for analysis.

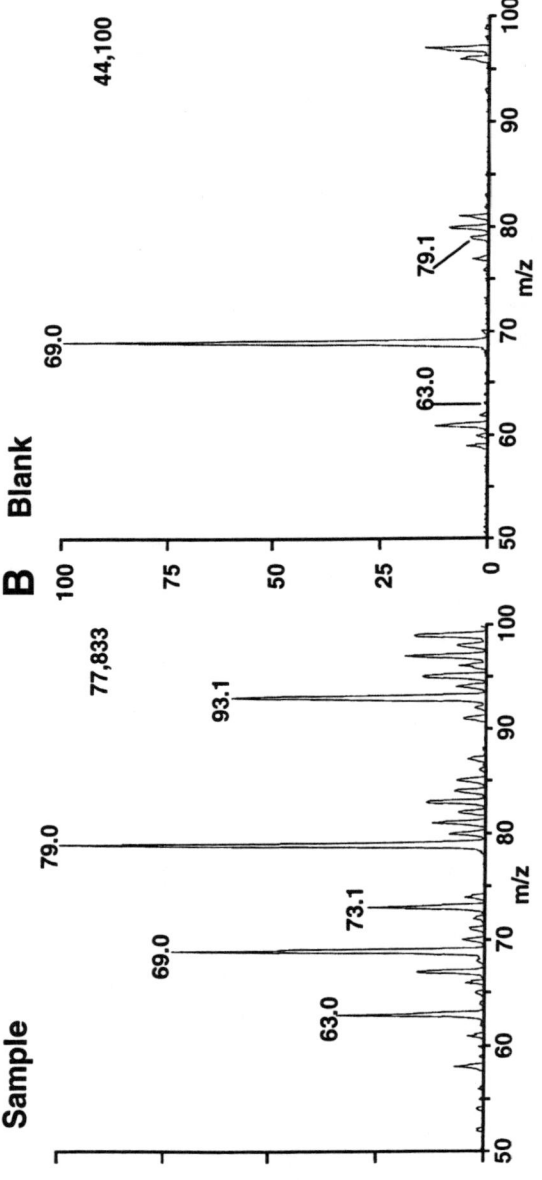

Fig. 5. Typical appearance of the marker ion region in the negative ion mass spectrum of (**A**) 5 pmol/μL solution of a phosphopeptide standard in a TFA-containing mobile phase; (**B**) TFA-containing mobile phases alone. Both samples are infused at the same rate using the same declustering potential (−350V on the orifice).

For maximum sensitivity of peptide analysis by LC/ES/MS positive ion, TFA may be eliminated altogether and replaced with 0.2% formic acid. However, no apparent benefit is seen for phosphate marker ion sensitivity when doing LC/ES/MS SIM with 0.2 % formic acid alone. Some chromatographic peak integrity is sacrificed under these conditions, although certain column packings (such as the PepMap C18 phase, LC-Packings used in our laboratory) can perform very well.

15. To reduce the risk of plugging, immediately following LC/ES/MS, a syringe is connected in place of the column flow and methanol is infused to flush the source and split lines free of acetonitrile and TFA. Following this, the syringe is filled with air and the system flushed until no liquid is seen coming from the ends of the spray and the prep fused silica.

16. The prep line union is grounded to prevent electrospraying at the exit. When manually collecting fractions according to elution profile, a delay time given by the length of the prep line should be taken in consideration.

17. To minimize coelution of several phosphopeptides, fractions should be collected according to the LC/SIM peak elution profile and not at specified time intervals.

18. When working with low-abundance phosphopeptides, it is imperative to carry out the whole phosphorylation mapping procedure in the shortest time possible, never letting the fractions sit for more than a day or two.

19. Phosphopeptide-selective detection by precursor ion scanning using nano-ES is optimized using freshly prepared stock solutions of standard phosphopeptides to produce and detect the m/z 79 (PO_3^-) marker ion. Five hundred femtomoles of phosphopeptide KRPpSQRHGSKY (University of Michigan Protein Structure Facility) is diluted just prior to analysis to a final concentration of 100 femtomole/µL using 5 µL of 50/50 methanol/water containing 10% ammonium hydroxide. The basic solution is made up fresh by diluting a 30% ammonium hydroxide stock solution with H_2O to make a 20% NH_4OH solution, which is then mixed 1:1 with methanol in a 1.5-mL Eppendorf tube. 1.5 µL of the standard solution is loaded into the nanospray capillary. The resolution of Q3 is reduced to pass a 4–5 Da window around the marker ion of interest (m/z 79) to enhance precursor scan sensitivity. A further small gain in sensitivity is obtained by decreasing the resolution of Q1. Nitrogen gas is used as the collision gas in Q2.

20. In our previously reported studies, dried down the samples prior to negative-ES precursor ion scanning for m/z 79 experiments. Samples were afterwards resuspended in 50/50 methanol/water containing 10% ammonium hydroxide. However, we found that bringing samples to dryness was the main cause of tip plugging. Therefore we changed our protocol and now just make the solution basic by adding an appropriate volume of 50/50

methanol/water containing 10% ammonium hydroxide, typically 1:1 sample to basic solution. This has reduced drastically the incidence of tip plugging. After fractions have been made basic, the samples should be analyzed as soon as possible, preferably the same day.

21. The optimal position in terms of S/N on the Sciex nano-ES source is a few mm directly in front of the orifice. To generate a stable signal and maintain a minimum flow rate (20–40 nL/min), a positive air pressure on the sample in the capillary, of slightly more than atmosphere, usually is required. Some capillaries require touching to the gate valve plate (voltages off and applying air pressure) to initiate sample flow. With sample flowing, the electronics are then turned on before the final adjustment of moving the capillary toward the orifice. This last adjustment is made using the cameras and by monitoring the changes in the mass spectra from scan to scan in real time. Tuning and calibration for full-scan data acquisition in negative ion nano-ES modes is carried out as described in **Note 12**.

22. In order to help prevent plugging of the needle and, hence, deterioration of spray stability (which can happen during negative ion nano-ES of basic solutions), a slightly higher spray air pressure is used compared to that needed to do positive ion. If plugging occurs, it can often be removed by increasing the air pressure to the capillary, and forceful touching of the tip to the entrance plate (with voltages off). When a droplet is observed, air pressure is reduced to normal operating levels.

23. When acquiring data on samples that approach the limit of detection (less than 50 fmoles total), there may be little or no obvious peptide signal in the full-scan negative ion mode. In such cases, we often add a peptide as an internal standard to the sample (50 fmole/µL). In this case, a narrow m/z window around the peptide can be used to rapidly assess S/N and spray stability while optimizing air pressure (flow rate) and sprayer position.

24. Acquisition is terminated when the overall ion statistics and appearance of the spectrum are satisfactory to the researcher. Acquisition times can range between a few minutes (3–5 min) for abundant phosphopeptides, to 30–45 min for samples that approach the limit of detection.

25. If enough material is available, it might be wise to use a small aliquot of the sample for LC/MS analysis to determine the preferential charge state for the phosphopeptide(s) under investigation. By switching from negative to positive ion and from basic to acidic condition, the most abundant charge state can not always be anticipated. LC/MS analysis will be beneficial even in these cases when more that one phosphopeptide per fraction needs to be targeted for sequencing. Knowledge of the retention time, in fact, will allow sequencing several precursors in a single-targeted LC/MS/MS experiment by changing the targeted precursor according to its elution time.

26. On a QTOF instrument (as well as on a triple quadrupole instrument) the amount of collision energy (CE) applied to induce peptide fragmentation is critical for the quality of the spectra and therefore for the completeness of the sequence. A certain correlation exists between peptide size, charge state, and optimal CE (for instance for a doubly charged peptide at m/z 650, the optimal CE should range around 26–27 eV). These value can be empirically determined. For the Micromass Q TOF, a list of collision energies for various peptide m/z and charge-state ranges is available with the instrument to assist in setting up the LC/MS/MS experiments. It should be noted that the CE values suggested by Micromass are almost invariably too high for triply charged peptides. Whenever the phosphorylation stoichiometry is low (ca. 5–10%), a viable alternative is to determine optimal conditions using the nonphosphorylated counterpart, which of course will require far less sample. The bulk of the sample can then be used for sequencing the phosphopeptide. It should be pointed out that if an ion trap instrument is used instead of a quadrupole-TOF instrument for targeted peptide LC/MS/MS, the collisional parameters are less critical.

27. In some cases, nano-ES might be a better choice for phosphopeptides sequencing over targeted LC/ES/MS/MS. For large or otherwise difficult to fragment peptides, averaging of the signal over a longer period of time can be extremely beneficial.

References

1. Hunter, T. (1987) A thousand and one protein kinases *Cell* **50,** 823–829.
2. Cohen, P. (2000) The regulation of protein function by multisite phosphorylation: a 25 year update. *Trends Biochem. Sci.* **25,** 596–601.
3. Boyle, W. J., van der Geer, P., and Hunter, T. (1991) Phosphopeptide mapping and phosphoamino acid analysis by two-dimensional separation on thin-layer cellulose plates. *Methods Enzymol.* **201,** 110–149.
4. Annan, R. S., Huddleston, M. J., Verma, R., et al. (2001) A multidimensional electrospray MS-based approach to phosphopeptide mapping. *Anal. Chem.* **73,** 393–404.
5. Zappacosta, F., Huddleston, M. J., Karcher, R. L., et al., (2002) Improved sensitivity for phosphopeptide mapping using capillary column HPLC and microionspray mass spectrometry: comparative phosphorylation site mapping from gel-derived proteins. *Anal. Chem.* **74,** 3221–32.
6. Huddleston, M. J., Annan, R. S., Bean, M. F., and Carr, S. A. (1993) Selective detection of phosphopeptides in complex mixtures by electrospray liquid chromatography/mass spectrometry. *J. Am. Soc. Mass. Spectrom.* **4,** 710–715.
7. Watts, J. D., Affolter, M., Krebs, D. L., et al. (1994) Identification by electrospray ionization mass spectrometry of the sites of tyrosine phosphorylation induced in

activated Jurkat T cells on the protein tyrosine kinase ZAP-70. *J. Biol. Chem.* **269,** 29,520–29,529.

8. Wilm, M., Neubauer, G., and Mann, M. (1996) Parent ion scans of unseparated peptide mixtures. *Anal. Chem.* **68,** 527–533.

9. Carr, S. A., Huddleston, M. J., and Annan, R. S. (1996) Selective detection and sequencing of phosphopeptides at the femtomole level by mass spectrometry. *Anal. Biochem.* **239,** 180–92.

10. Zhang, X., Herring, C. J., Romano, P. R., et al. (1998) Identification of phosphorylation sites in proteins separated by polyacrylamide gel electrophoresis. *Anal. Chem.* **70,** 2050–2059.

11. Posewitz, M. C. and Tempst, P. (1999) Immobilized gallium(III) affinity chromatography of phosphopeptides. *Anal. Chem.* **71,** 2883–2892.

12. Steen, H., Kuster, B., Fernandez, M., et al. (2001) Detection of tyrosine phosphorylated peptides by precursor ion scanning quadrupole TOF mass spectrometry in positive ion mode. *Anal. Chem.* **73,** 1440–1448.

13. Verma, R., Annan, R. S., Huddleston, M. J., et al. (1997) Phosphorylation of Sic1p by G1 cyclin/Cdk is required for its degradation and entry into S phase, *Science* **278,** 455–560.

14. Chen, S. L., Huddleston, M. J., Shou, W., et al. (2002) Mass spectrometry based methods for phosphorylation site mapping of hyperphosphorylated proteins applied to Net1, a regulator of exit from mitosis in yeast. *Mol. Cell Proteomics* **1,** 186–196.

15. Wu, X., Ranganahan, V., Weisman, D. S., et al. (2000) ATM phosphorylation of Nijmegen breakage syndrome protein is required in a DNA damage response. *Nature* **405,** 477–482.

16. Watty, A., Neubauer, G., Dreger, M., et al. (2000) The in vitro and in vivo phosphotyrosine map of activated MuSK. *Proc. Nat. Acad. Sci. USA* **97,** 4585–4590.

17. Karcher, R. L., Roland, J. T., Zappacosta, F., et al. (2001) Cell cycle regulation of myosin-V by calcium/calmodulin-dependent protein kinase II. *Science* **293,** 1317–1320.

18. Rogers, S. L., Karcher, R. L., Roland, J. T., et al. (1999) Regulation of melanosome movement in the cell cycle by reversible association with myosin V. *J. Cell Biol.* **146,** 1265–1276.

19. Biemann, K. (1990) Nomenclature for Peptide Fragment Ions, in *Methods in Enzymology,* vol. **193** (McCloskey J., ed.), Academic Press, pp. 886–887.

8

Studies of Calmodulin-Dependent Regulation

Paul C. Brandt and Thomas C. Vanaman

Summary

Methods are presented for purifying bovine testes calmodulin and the calmodulin-regulated plasma-membrane calcium ATPase from human erythrocytes by calcium dependent affinity chromatography. The assay of CaM Kinase II using a synthetic peptide substrate is also described.

Key Words: Affinity chromatography; calcium; calmodulin; plasma membrane calcium pumping ATPase; CaM-dependent protein kinase; enzyme assay; phenothiazine.

1. Introduction

Calmodulin (CaM) is a small (17 kDa) highly acidic protein first described more than 30 yr ago as an activator of cyclic nucleotide phosphodiesterase (*1,2*), later shown to be dependent on its ability to bind calcium (*3*). Subsequent work (*4*) showed that calmodulin is widely distributed throughout all eukaryotes as a highly conserved, high-affinity calcium-binding protein that serves as the archetype for the EF-hand family of calcium-regulated proteins.

To date, calmodulin has been identified as a direct calcium-dependent regulator of numerous enzymes and proteins involved in signal transduction (for review *see* **ref.** *5*) including the aforementioned phosphodiesterase, both mammalian and bacterial adenylate cyclases, numerous protein kinases, the phosphoprotein phosphatase calcineurin, three forms of mammalian nitric oxide synthase, plasma membrane calcium pumps and channels, G proteins, transcription factors, histone deacetylases, and other cellular proteins. These various

From: *Methods in Molecular Biology, vol. 284:*
Signal Transduction Protocols
Edited by: R. C. Dickson © Humana Press Inc., Totowa, NJ

proteins and their corresponding regulatory cascades are involved in many diverse aspects of cellular function, division, differentiation, and even cell death.

In almost all cases, calmodulin acts as a calcium-dependent activator of responses. This requires calcium binding to all four EF-hand calcium-binding sites that are arranged in pairs in each end of the molecule. As is now clear from three-dimensional structures of apo (*6*) and calcium replete (*7*) calmodulin, calcium binding causes conformational changes in CaM exposing hydrophobic/ aromatic binding pockets. These pockets plus acidic residues in the central helix of the molecule provide a recognition surface for proteins with complementary basic amphipathic helical segments representing calmodulin-binding domains. Conformational flexibility of these interacting surfaces (*8*) permits calmodulin to bind to a variety of sequences with moderate (μM) to high (nM) affinity.

The variety of cellular targets for calmodulin action as well as the presence of over 200 genes in mammals encoding proteins containing the EF hand structure involved in high-affinity calcium binding make assessment of a physiological role for calmodulin in regulating a biological process a challenging task. This task is made even more difficult by the fact that calmodulin is an essential protein in all eukaryotes, thus, limiting genetic approaches.

The reader is referred to the web site http://structbio.vanderbilt.edu/cabp_database/cabp.htmL for more detailed information on calmodulin structure and function and the calcium-binding protein (CaBP) super family of proteins.

This chapter will describe in detail how to isolate calmodulin and calmodulin-binding proteins by calcium-dependent affinity-binding methods and how to assay a specific calmodulin-regulated enzyme, CaM Kinase II, in either crude homogenates or following purification by immunoprecipitation.

2. Materials

2.1. Calmodulin Purification by Trifluroaminopropylphenothiazine (TAPP): Sepharose Affinity Chromatography

1. 10X Homogenization buffer: Prepare 1 L of 100 mM Tris (12.1 g Tris-free base), 10 mM EGTA (3.8 g) adjusted to pH 7.5 with 6 N HCl.
2. 10X TAPP loading buffer: Prepare 1 L of 100 mM Tris (12.1 g Tris-free base), 20 mM CaCl$_2$ (2.9 g), 2 M NaCl (116.9 g) adjusted to pH 7.5 with 6 N HCl.
3. TAPP-Sepharose elution buffer: Prepare 1 L of 10 mM Tris (1.21 g Tris-free base), 200 mM NaCl (11.7 g), 10 mM EGTA (3.9 g), 1 mM 2-mercaptoethanol (70 μL) adjusted to pH 7.5 with 6 N HCl.
4. 100 mM Phenylmethylsulfonyl fluoride (PMSF) in 95% ethanol (prepared just prior to use).

5. TAPP-Sepharose (50 mL packed resin): Prestripped with 6 M guanidinium chloride and equilibrated in 1X TAPP loading buffer containing 70 μL/L 2-mercaptoethanol.
6. Cold deionized water.
7. 6 M H_2SO_4 in saturated Ammonium sulfate.
8. 1 M Tris-free base (121.7 g/L).
9. 10 mM ammonium bicarbonate (0.79 g/L).
10. Sephadex G-50 (250 mL packed bed) in 10 mM ammonium bicarbonate.

2.2. Preparation of CaM-Sepharose

All solutions should be prepared in advance and chilled. Washing steps also require large volumes of chilled deionized water.

1. Activating buffer: 2 M sodium carbonate, pH 11.0 (212 g/L)
2. Washing buffer: 0.1 M sodium bicarbonate, pH 9.5 (8.4 g/L)
3. Coupling buffer: 0.2 M sodium bicarbonate (16.8 g/L), 0.5 M NaCl (29.3 g/L), pH 8.3.
4. CNBr stock solution: 2 g/mL in acetonitrile (prepare immediately before use).
5. CaM stock solution: 1 mg/mL in coupling buffer (prepared from dried CaM stock immediately before use).
6. Quenching solution: 1 M ethanolamine-HCl, pH 8.0.
7. Storage buffer: 40 mM Tris (4.87 g/L Tris free base), 1 mM $CaCl_2$ (0.111 g/L of anhydrous $CaCl_2$) adjusted to pH 8.0 with 1 M HCl.

2.3. CaM-Sepharose Chromatography of Human Erythrocyte Ca^{2+}-Pumping ATPase

1. RBC wash buffer: 10 mM Tris-HCl (1.58 g/L), 0.13 M KCl (9.69 g/L), adjusted to pH 7.2 with 1 M HCl or 1 M KOH.
2. Lysis buffer: 10 mM Tris-HCl (1.58 g/L), 5 mM EGTA (1.90 g/L), adjusted to pH 7.2 with 1 M HCl or 1 M KOH.
3. HEPES wash buffer: 10 mM HEPES (2.38 g/L) adjusted to pH 7.2–7.4 with 1 M HCl.
4. Ghost storage buffer: 10 mM HEPES (2.38 g/L), 0.13 M KCl (7.46 g/L), 0.5 mM $MgCl_2$ (0.05 g/L), 0.05 mM $CaCl_2$ (5.5 mg/L), adjusted to pH 7.2–7.4.
5. CaM-Sepharose loading buffer: 20 mM HEPES (4.76 g/L), 0.13 M KCl (7.46 g/L), 0.5% (w/v) Triton X-100 (S.G. = 1.07: 4.67 mL/L), 0.05% (w/v) phosphatidylcholine (20 mL of 25 mg/mL stock solution in H_2O/L), 0.1 mM $CaCl_2$ (11.1 mg/L), 20 μM PMSF (200 μL of 0.1 M stock solution in ethanol/L), pH 7.4.
6. CaM-Sepharose elution buffer: 20 mM HEPES (4.76 g/L), 0.13 M KCl (7.46 g/L), 0.5% (w/v) Triton X-100 (S.G. = 1.07: 4.67 mL/L), 0.05%

(w/v) phosphatidylcholine (20 mL of 25 mg/mL stock solution in H_2O), 5 mM EGTA (1.46 g/L), 20 μM PMSF (200 μL of 0.1 M stock solution in ethanol/L), pH 7.4.

7. ATPase assay cocktail: 30 mM HEPES (7.14 g/L), 120 mM KCl (6.89 g/L), 0.05 mM $CaCl_2$ (5.55 mg/L), 0.5 mM $MgCl_2$ (0.05 g/L), 0.5 mM K_2ATP (dihydrate: 0.31 g/L), 60 nM CaM (1 mg/L), pH 7.4.

8. Lanzetta reagent: Mix 1 volume of 4.2% (4.2 g/100 mL H_2O) ammonium molybdate in 4 N HCl with 3 volumes of 0.045% (0.045 gm/100 mL H_2O) malachite green hydrochloride in H_2O, mix for 20 min, and filter through Whatman no. 5 filter paper.

9. 34% (34 g/100 mL H_2O) sodium citrate dihydrate in H_2O.

2.4. Measurement of CaM-Kinase II Activity in Tissue Extracts

1. Calmodulin: Prepare a 1 mg/mL solution of purified calmodulin in water and store in aliquots at $-70°C$

2. Autocamtide-3: Prepare autocamtide-3 (KKALHRGETVDAL, Biosource International, Camarillo, CA or Calbiochem, San Diego, CA) at 1 mM (1.44 mg/mL) in water and store in aliquots at $-70°C$.

3. Mg-ATP solution: Mix equal volumes of 20 mM Na_2ATP and 20 mM $MgCl_2$. Adjust to pH 7.4 with either HCl or NaOH as necessary.

4. 5X CaM Kinase II assay buffer: 100 mM Tris, pH 7.4, 25 mM $MgCl_2$ and 5 mM $CaCl_2$.

5. [γ-^{32}P]-ATP working solution: Typically, [γ-^{32}P]-ATP is purchased with a specific activity of 3000 Ci/mmol and is available from a variety of sources. Prepare a working solution of [γ-^{32}P]-ATP by adding 1 μL of the 10 mM Mg-ATP solution to 89 μL water and then adding 10 μl of 10 μCi/μL [γ-^{32}P]-ATP. This will produce a final working solution of [γ-^{32}P]-ATP that is 1 μCi/μL. The new specific activity for this [γ-^{32}P]-ATP working solution is 10 Ci/mmol.

6. P-81 phosphocellulose paper: P-81 phosphocellulose paper (Whatman, Clifton, NJ) may be purchased as pre-cut 2.5 cm diameter circles or larger sheets that may be cut into 2 cm × 2 cm squares. The sample number should be marked on the filter paper in pencil because most inks dissolve in the final acetone wash.

7. Phosphoric acid: 75 mM orthophosphoric acid (4.00 mL 100% O-phosphoric acid/L).

8. Reagent grade acetone.

9. Lysis buffer: 50 mM HEPES, pH 7.4, 100 mM NaCl, 1% Triton X-100, 2 mM EDTA, 2 mM EGTA, 50 mM NaF, 5 mM β-glycerophosphate, 1 mM Na ortho-vanadate, 1 μg/mL leupeptin, 1 μg/mL pepstatin A, 1 μg/mL antipain, 1 mM PMSF.

10. Modified RIPA buffer: 50 m*M* Tris-HCl, pH 7.4, 1% NP40, 150 m*M* NaCl, 2 m*M* EDTA, 2 m*M* EGTA, 1 m*M* Na ortho-vanadate, 1 m*M* NaF, 1 μ*M* leupeptin, 1 μ*M* pepstatin A, 1 μ*M* aprotinin, 1 m*M* PMSF.
11. Protein A or G agarose: Use Protein A with rabbit polyclonal antibodies (PAbs) and Protein G with mouse monoclonal antibodies (Mabs). Wash the resin twice in phosphate-buffered saline (PBS) and prepare as a 50% slurry in PBS. Store at 4°C.
12. Anti-CaM kinase II antibody: Many commercial CaM kinase II antibodies are available for use in immunoprecipitation. However, not all will precipitate the kinase without also blocking activity. Unless, it is specified that the antibody is compatible with kinase assays, it will be necessary to determine that the antibody will precipitate an active kinase.
13. PBS for 1 L: 11.5 g Na₂HPO₄, 2 g KH₂PO₄, 80 g NaCl, 2.0 g KCl yields pH 7.4.
14. Bicinchoninic acid protein assay reagent (BCA): Pierce Chemical Company, Rockford, IL.

3. Methods

3.1. Purification of Calmodulin by Calcium-Dependent Affinity Chromatography

The ability of calmodulin to bind to hydrophobic/aromatic moieties, including small molecules, in a calcium-dependent manner provides the basis for a very simple and specific method for its isolation. The protocol set forth below was first described (*9*) for the purification of CaM using a column prepared with a reactive amine containing des-methyl derivative, chloroaminopropyl phenothiazine (CAPP), of the calmodulin antagonist, chlorpromazine. A number of other phenothiazine derivatives useful for calmodulin isolation have been described (*10*) including TAPP-Sepharose as shown below. The protocol also works well with commercially available phenyl-Sepharose with only minor alterations to the protocol (*11*).

The procedure described below is for isolating CaM from bovine testis. It involves preparing an initial extract and subjecting it to fractionation with ammonium sulfate prior to TAPP-Sepharose chromatography to remove interfering contaminants. Purification of CaM from other eukaryotic sources may require additional steps as a result of the presence of other proteins (especially related CaBPs) that also interact with the resin (*see* **Note 1**).

3.1.1. Day 1

Do all work at 4°C.

1. Starting with three adult bull testes (approx 800 g), remove the testes from the scrotal sack. (This is most easily done using frozen testes that have been

partially thawed overnight in the cold room.) Testes are chopped into small pieces, then minced by passage two times through a meat grinder and weighed.

2. Dilute homogenization buffer 1:10 and add 2-mercaptoethanol to 1 mM (70 µL/L), then add 10 mL of stock PMSF solution per liter.

3. Homogenize minced testes in 2 volumes (per weight) of homogenization buffer in a Waring blender 3 × 1 min at top speed.

4. Centrifuge at 10,400g (see http://opbs.okstate.edu/~melcher/CentHelp.html for appropriate rotor speed) for 60 min.

5. Pour supernatant through cheesecloth and save. Re-homogenize the pellet in 1 volume of homogenization buffer. Centrifuge as previously-mentioned and combine this supernatants with the first one.

6. Pass combined supernatants through cheesecloth again.

7. Sixty percent Ammonium sulfate cut: Slowly add 390 g/L of ammonium sulfate (*see* **ref.** *12*) to the combined supernatant with vigorous stirring. Maintain at pH 7.5 by titration with 1 M NH$_4$OH or 1 M H$_2$SO$_4$ as needed.

8. Stir at 4°C for at least 1 h. Recheck pH.

9. Centrifuge at 25,000 g for 60 min and collect the supernatant fraction by pouring through cheesecloth to remove residual particulates.

10. Acid Precipitation of Ammonium Sulfate Supernatant: Adjust the filtered 60% ammonium sulfate supernatant to pH 4.1 using ammonium sulfate-saturated 6 M H$_2$SO$_4$. Stir for at least 60 min. Readjust to pH 4.1.

11. Centrifuge at 25,000 g for 60 min. Decant the supernatant fraction and discard.

12. Resuspend the pH 4.1 pellet in a minimum volume of 1X TAPP loading buffer by using a dounce homogenizer. Dissolve the pellet by adjusting the suspension to pH 7.5 with 1 M Tris (free base) while stirring. After the suspension has clarified, stir for an additional 30 min while maintaining pH 7.5. Try to keep volume small (<100 mL).

13. Centrifuge the redissolved pellet fraction at 105,000g for 30 min and recover the supernatant fluid; adjust the pH to 7.5 if necessary.

14. Load the supernatant onto a 2.0 × 16 cm column of TAPP-Sepharose equilibrated with calcium-containing TAPP loading buffer, collect the flow-through, and pass over the column a second time.

15. Wash the column overnight with TAPP loading buffer (at least 2 L).

3.1.2. Day 2

1. Monitor the A$_{235nm}$ of the unbound pass through fraction against loading buffer. When it is less than 0.1, begin elution.

2. Elute the column with TAPP-Sepharose elution buffer, collecting 2.5-mL fractions in tubes containing 150 µL TAPP loading buffer that has been made 500 mM in CaCl$_2$ (*see* **Note 1**).

3. Measure the A_{235nm} or A_{280nm} and pool corresponding protein-containing fractions.

4. Desalt over a Sephadex G-50 column (2.0 × 50 cm) in 10 mM ammonium bicarbonate, collecting 2.5-mL fractions. Detect CaM by A_{280nm}, pool and lyophilize. (N.B: CaM concentration can be estimated based on extinction at 276 nm. $E_{1mg/mL, 276nm} = 0.18$).

3.2. Preparation of CaM Sepharose

The ability of CaM to interact with most target proteins only in the presence of Ca^{2+} offers an excellent method for their purification as well. The protocol below, first described (*13*) for isolating CaM-sensitive cyclic nucleotide phosphodiesterase, has been used widely for the isolation of a variety of CaM-regulated enzymes and to demonstrate their CaM dependence. It should be noted that most tissues have a large number of CaM-binding proteins. For example, we have detected over 30 proteins from synaptosomes that bind to CaM-Sepharose as judged by two-dimension gel electrophoresis. These include proteins such as plasma-membrane calcium pumps, CaM Kinase II, calcineurin, cyclic nucleotide phosphodiesterase, and other proteins known to be present at the synapse. The preparation of CaM-Sepharose is straightforward and involves coupling to CNBr-activated Sepharose 4B. This and related activated supports can be purchased commercially or prepared as described below based on the original protocol of March et al. (*14*).

3.2.1. Activation

(All operations should be performed in a chemical fume hood.)

1. Using a 600-mL coarse sintered-glass funnel, wash 50 mL (packed volume) of sepharose 4B extensively with distilled water (1–2 L), then twice with 300 mL of activating buffer.

2. Dry the resin to a moist cake (until liquid just enters resin bed) by suction and transfer to a beaker as a 1:1 slurry in activating buffer.

3. While stirring, add 0.05 volumes of the cyanogen bromide solution in acetonitrile dropwise over about 2–3 min, and continue stirring vigorously for about 5 min.

4. Transfer the activated resin to the original coarse sintered-glass funnel and wash extensively with water (2–3 L minimum) until no smell of CNBr is detected (*see* **Note 2**).

5. Wash the resin with 10 volumes of 0.1 M sodium bicarbonate, pH 9.5, followed by 10 volumes of coupling buffer (0.2 M sodium bicarbonate, 0.5 M NaCl, pH 8.3).

3.2.2. Coupling

6. Dry the activated resin again to a moist cake and transfer to a mixing flask with a suspended magnetic stirring bar (100-mL spinner culture bottles are excellent for this purpose).
7. Add 50 mL of a 1 mg/mL solution of CaM in coupling buffer and stir the reaction mixture gently overnight at 4°C.
8. Transfer the coupled resin to a sintered-glass funnel and remove the reacted coupling mixture by suction. (N.B.: This fraction may be saved to test for residual calmodulin to assess coupling efficiency [usually <80%].)

3.2.3. Quenching and Storage (see **Note 2**)

1. Wash the coupled resin with 5 volumes of coupling buffer, suction to a moist cake, and resuspend as a 1:1 slurry in quenching solution.
2. React with gentle stirring for 1 h to block any remaining activated functional groups. (It should be noted that ethanolamine is ideal for this purpose, because reaction with the amine moiety results in incorporation of an aliphatic hydroxyl moiety, preserving the properties of the Sepharose matrix.)
3. Finally, wash the resin exhaustively with storage buffer and store as a 1:1 slurry in this buffer at 4°C. The resulting CaM-Sepharose conjugate is functional for 3–6 mo.

3.3. CaM-Sepharose Chromatography of Human Erythrocyte Ca^{2+}-Pumping ATPase

CaM-Sepharose can be used for the purification of both readily soluble cytosolic proteins and membrane-localized proteins that require detergent solubilization. An excellent example of the latter is the purification of the plasma-membrane calcium-pumping ATPase (PMCA) in animal cells by the protocol below first described by Carafoli and co-workers (**15,16**) for human erythrocyte PMCA. Unless otherwise specified, all of the following operations are carried out at 4°C.

3.3.1. Day 1: Preparation of Erythrocyte Membrane (RBC Ghosts) Fraction

1. Resuspend 6 to 7 U (1.5–1.7 L) of recently outdated packed red blood cells (RBCs) from the local blood bank in 5 volumes wash buffer.
2. Centrifuge at 5,800*g* for 10 min in 1-L swinging bucket rotor bottles.
3. Gently aspirate off the supernatant, including the white blood cells (WBCs), which come off in large clumps. (Be sure to remove WBCs as completely as possible, even at the sacrifice of some RBCs, as they contain proteases as well as numerous calmodulin binding proteins.)

4. Repeat **step 1**. (N.B.: At this stage, the sedimented RBC fraction is a very loose pellet behaving like a very thick liquid rather than a hard pellet. It is easily resuspended, requiring great care in handling.)
5. Resuspend the washed RBC fraction from **step 2** by filling the 1-L centrifuge bottles with lysis buffer followed by gentle mixing with inversion and pellet the RBC ghosts by centrifuging at 24,000g for 35 min in a fixed angle rotor (routinely in 250 mL centrifuge bottles).
6. Carefully remove the supernatant containing hemoglobin and other soluble red cell constituents, taking care not to lose pellet material (very soft and difficult to see as the supernatant is dark red).
7. Resuspend the RBC ghosts in half the original volume of lysis buffer and combine the contents of pairs of centrifuge bottles to concentrate the ghost preparation, centrifuge, and remove the supernatant as above.
8. Using the same procedure as set forth in **step 5** for concentrating the ghosts, wash the resulting RBC ghost pellet two to three times further with the HEPES wash buffer by resuspension and centrifugation until the ghosts are white or light pink. During these washes, pool the washed RBC ghost (membrane) fraction with care not to disturb the red pellet (unlysed RBCs) at the bottom of the bottles. (N.B.: The final preparation should be in a single 250-mL centrifuge bottle.)
9. Perform a final wash of the recovered RBC membrane fraction by filling the centrifuge bottle with HEPES storage buffer, centrifuge, and remove the supernatant. Store the RBC ghost preparation at 4°C overnight.

3.3.2. Day 2: CaM-Sepharose Chromatography

1. Resuspend the soft RBC ghost pellet by gentle swirling. Measure the volume of the ghosts and assay for protein content. Add sufficient HEPES storage buffer to make the final protein concentration approx 5 mg/mL.
2. Solubilize RBC ghosts by the addition of 5 mL of 10% Triton X-100 (10 g/100 mL: Specific Gravity = 1.07 g/mL) per 100 mL of resuspended RBC ghosts to give a final concentration of 5 mg Triton X-100/mL (1:1 w/w ratio with protein). Stir slowly with a magnetic stirrer for 25 min at 4°C.
3. Centrifuge the solubilized RBC ghosts at 100,000g for 35 min at 4°C and collect the supernatant fractions from each centrifuge bottle and combine as necessary.
4. Measure the volume of the pooled supernatants and add, for each 100 mL of supernatant, 2 mL of stock phosphatidyl choline (25 mg/mL), 2 mL of stock dithiothreitol (DTT) (100 mM) and 100 µL of 100 mM CaCl$_2$ (final concentrations: 0.5 mg/mL phosphatidyl choline, 2 mM DTT and 0.1 mM CaCl$_2$).
5. Apply the supernatant to a 1.5 × 10 cm CaM-Sepharose column equilibrated in sample application buffer and wash with 2–3 bed volumes of this

buffer followed by at least 2–3 bed volumes of wash buffer until the $A_{280\ nm}$ reaches baseline.

6. Elute the column with elution buffer collecting 1-mL fractions into tubes containing 0.1 mL of the 50% glycerol stabilizing solution.

7. Read the $A_{280\ nm}$ and assay fractions for ATPase activity as follows:

    a. Mix 200 µL assay cocktail with 10 µL of each fraction.

    b. Incubate at 37°C for 10 min.

    c. Add 59 µL of the reaction mixture to 800 µL of Lanzetta (*17*) reagent and mix.

    d. After 1 min, add 100 µL of 34% (w/v) sodium citrate, mix, and incubate for 30 min at 25°C.

    e. Measure $A_{660\ nm}$. The color is stable for about 4 h. (N.B.: The assay must include a blank for background color correction and a sodium phosphate standard curve [1–10 nmoles/mL] for calculation of P_i release [ATP hydrolysis]).

8. Pool the active fractions and aliquot into 200–300 µL fractions and store at −70°C. (*see* **Note 3**).

3.4. Measurement of CaM-Kinase II Activity in Cell Extracts

Of the calmodulin-regulated kinases (CaM kinases), CaM kinases II and IV have among the broadest substrate specificity and consequently affect a variety of processes within the cell (*18*). The contribution of CaM kinase II to learning and memory, through its effects on long-term potentiation has been studied extensively (*19*). Both CaM kinases II and IV are involved in regulation of gene transcription (*20*).

Autophosphorylation of Thr268 and Thr305 or Thr306 in CaM kinase II can produce calmodulin-independent activation of CaM kinase II, so it is important to measure activity in the absence of calmodulin (*see* **Note 4**). Phosphorylation of CaM kinase II at Thr268 still allows further stimulation of the kinase by calmodulin, but phosphorylation at Thr305 or Thr306 does not allow binding of calmodulin and calmodulin-dependent stimulation is not observed (*21*).

3.4.1. Cell Lysis

1. Prepare $0.5–1.0 \times 10^6$ cells for assay. Either adherent or nonadherent cells may be used. For adherent cells, a semiconfluent 35-cm dish usually provides sufficient material for assay.

2. Wash cells twice in ice-cold PBS. Adherent cells may be washed directly on the plates. Nonadherent cells may be collected by centrifugation and washed.

3. Add 300 µL of ice-cold lysis buffer to each sample and let stand on ice for 10 min.

4. Collect lysate into a tube and centrifuge at 12,000 g for 10 min at 4°C.
5. Carefully remove the supernatant and assay the protein content by the BCA method relative to a protein standard.

3.4.2. Direct Assay (see **Note 4**)

1. Assay samples in at least duplicate.
2. For each sample to be assayed, mix the following components in a tube: 4 µL 5X CaM kinase II assay buffer, 2 µL 1 mM autocamide-3, 2 µL [γ-^{32}P]-ATP working solution, 2 µL 1 mg/mL calmodulin. These components may be prepared as a combined master mix that is aliquoted into individual assay tubes.
3. For each sample, also prepare a reaction without calmodulin (substitute water) to account for background phosphorylation and auto-activated CaM kinase II activity. The inclusion of samples that do not contain autocamtide-3 can be used to control for determining the background signal owing to phosphorylation of endogenous substrates that might bind to the P-81 phosphocellulose paper.
4. Add 10 µL of lysate to each tube and incubate at 30°C for 10 min. It is often a good practice to first do a dilution series with the sample with which you expect to have the greatest activity to ensure that you are working within the linear range of the assay. If 10 µL of extract will produce a signal that exceeds the capacity of the assay, it may be diluted appropriately with lysis buffer.
5. Adsorption of autocamtide-3 onto phosphocellulose paper. Apply 15 µL of each reaction onto a prelabeled piece of P-81 phosphocellulose paper. Place filters in a 500-mL beaker containing 350–400 mL 75 mM orthophosphoric acid. Stir with a glass rod for 5 min. Decant the orthophosphoric acid and repeat this wash step four additional times.
6. After decanting the last orthophosphoric acid wash, add 400 mL acetone and stir for 5 min. Remove the papers and air dry.
7. Place each dried filter in a 7 mL liquid-scintillation vial and add 4 mL of scintillation cocktail (e.g., Econofluour) and count radioactivity in a liquid-scintillation counter. Each set of analyses should include both negative control filters (no added ^{32}P) for background correction and vials with known amounts of ^{32}P to correct for counting efficiency.

3.4.3. Assay With Immunoprecipitation

If excessive nonspecific background phosphorylation or dephosphorylation of the substrate peptide or endogenous substrates is suspected, it is possible to immunoprecipitate the CaM kinase II and to a large extent separate it from these kinases and phosphatases.

1. To the cells prepared as described in **Subheading 2.4.**, add 500 µL of ice-cold modified RIPA buffer to each 35-mm dish and collect the extract by scraping.
2. Centrifuge sample at 14,000 *g* at 4°C for 15 min. Transfer supernatant to a new tube.
3. Measure the protein content as described in **Subheading 3.4.1.** and dilute to approx 1 mg/mL with modified RIPA buffer. If the concentration is below use 1 mg/mL, use the extract at that concentration.
4. Preclear the extract by adding 100 µL of protein A or G-agarose beads to 1 mg (1 mL) of extract. Incubate at 4°C with continuous inversion mixing for 1 h.
5. Remove Protein A or G agarose by centrifugation at 14,000 *g* at 4°C for 5 min. Transfer the supernatant to a fresh tube.
6. Add sufficient antibody to quantitatively precipitate the CaM kinase II in the clarified lysate and incubate at 4°C for 1 h. (N.B.: To determine if CaM kinase II is being quantitatively immunoprecipitated, a series of preliminary experiments can be undertaken in which serial immunoprecipitations of the same batch of extract are prepared and assayed for CaM kinase II activity. When no more CaM kinase II activity is precipitated, this is the minimum amount of antibody needed for quantitative recovery of enzyme from the extract.)
7. Add sufficient Protein A or G agarose to the sample to precipitate all of the added anti-CaM kinase II antibody. Incubate at 4°C with inversion mixing for 1 h. The IgG binding capacity of Protein A or G agarose is usually supplied by the manufacturer.
8. Collect immune complex by centrifugation at 14,000 *g* at 4°C for 5 min. Remove the supernatant, add 1 mL modified lysis buffer to the agarose pellet, mix, and collect by centrifugation. Repeat three times.
9. Wash one time with 1X kinase assay buffer and collect the immune complex by centrifugation.
10. This immobilized complex may be used directly in the CaM kinase II assay described earlier. Substitute water for the 10 µL of extract that would normally be added.

3.4.4. Calculations

1. Determination of counting efficiency. For the control sample containing a known amount of ^{32}P, determine the theoretical dpm in the vial. 1 µCi = 2.2 × 10^6 dpm, so a counting efficiency of 100% would produce 2.2 × 10^6 cpm for every µCi in the vial. Although the counting efficiency for ^{32}P is high, it is never 100%. Counting efficiency (%) = (cpm measured) / (theoretical dpm for µCi ^{32}P in vial) × 100.

2. Average the replicates of each sample.
3. For each sample, convert cpm to dpm using the counting efficiency determined in **step 23**.
4. For each sample, convert dpm to µCi using the conversion factor: 1 µCi = 2.2×10^6 dpm.
5. Convert the µCi of ^{32}P present on the P-81 phosphocellulose paper to mole P_i using the specific activity of the ^{32}P in the reaction. In this case, the final specific activity of $[\gamma\text{-}^{32}P]$-ATP calculated in **Subheading 2.4.** is 10 Ci/mmol = 10 µCi/nmol ATP.
6. The incorporation of P_i into autocamtide-3 is expressed as nmol P_i /µmol peptide. In this case, the final amount of peptide in the kinase reaction was 2 µmol. The enzyme-specific activity is typically expressed as nmol P_i/µmol peptide/min/µg total protein. The time would be 10 min and the total protein would be the amount of protein contained in the 10 µL of extract added to the reaction.

4. Notes

1. Calmodulin purification by TAPP-Sepharose Affinity Chromatography. As shown in **Fig. 1**, this procedure yields essentially homogeneous calmodulin. We routinely isolate 80–100 mg of CaM/800 g testis. Testis is favored over other mammalian tissues as it has the highest quantity of calmodulin and lacks most other related calcium binding proteins (especially S100 found in brain), which also show calcium-dependent binding to TAPP-Sepharose and related supports. This method is excellent for the purification of recombinant CaM expressed in bacteria that lack endogenous proteins having this property.
2. Preparation of CaM sepharose. The preparation of CaM-sepharose is relatively straightforward. However, it is important to note that CaM has a high content of methionines that are susceptible to cleavage with CNBr. It is therefore essential to make sure that the activated resin is washed free of this reagent prior to adding CaM for coupling. Quenching reactive functional groups following coupling is an equally important step. One simple test to determine that complete quenching has been achieved is to add ^{3}H-leucine (or an equivalent amino acid) to a small fixed amount of resin in buffer at pH 8.5, incubate for 30 min, wash the resin, and count in a scintillation counter to verify that no radioactive amino acid has been coupled to the column. The functionality of bound CaM can be determined by testing its ability to act as a Ca^{2+}-dependent activator of brain 3,5-cyclic nucleotide phosphodiesterase (*13*).
3. CaM-Sepharose Chromatography of Human Erythrocyte Ca^{2+}-pumping ATPase. **Figure 2** shows the results from a typical preparation of human

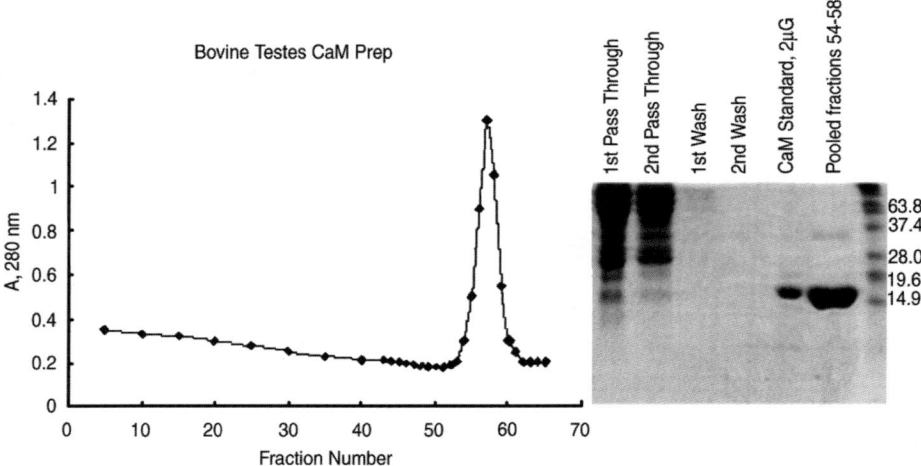

Fig. 1. TAPP-Sepharose purification of bovine testes calmodulin. The panel at the left shows the elution profile for the TAPP-Sepharose column after the addition of EGTA-containing elution buffer at fraction 0. The right panel shows Coomasie Blue stained SDS-PAGE (**24**) of the indicated pooled fractions from the column. Samples were loaded to be equivalent fractions of the total.

erythrocyte Ca^{2+}-ATPase prepared as described above. The bulk of the activity eluted in fractions 13, 14, and 15 coincident with the major peak of $A_{280\,nm}$ absorbing material. The shoulder of $A_{280\,nm}$ material eluting just ahead of the activity does not appear to contain substantial amounts of protein. Fractions 13–15 from this separation were pooled yielding 3.2 mL of purified PMCA (MW approx138 kDa) at a concentration of 0.31 mg/mL or a total of approx 1 mg of total protein. The purified enzyme was stimulated three-fold by CaM with a specific activity + CaM of approx 22.5 μmoles ATP hydrolyzed/h/mg protein. The enzyme is stable for up to 6 mo at −70°C.

4. Measurement of CaM-kinase II activity in tissue extracts. One of the major problems in obtaining accurate measurements of kinase activity in crude extracts or partially purified samples is the presence of multiple protein phosphatases. The lysis buffer described in materials usually contains sufficient phosphatase inhibitory capacity from the NaF, β-glycerophosphate, and ortho-vanadate to block dephosphorylation of autophosphorylated CaM kinase II and the substrate during assay. However, it is sometimes necessary to add additional phosphatase inhibitors such as bromotetramisole, microcystin LR, Na molybdate, Na tartrate, and imidazole. Premade phosphatase inhibitor cocktails (Sigma) may also be added to the lysis buffer.

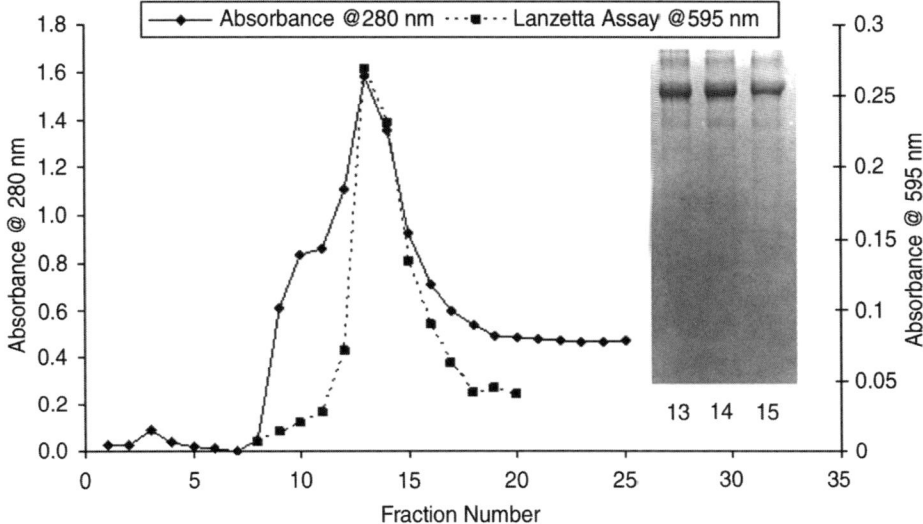

Fig. 2. Calcium-dependent chromatography of detergent solubilized erythrocyte Ca^{2+}-pumping ATPase on CaM-Sepharose. The graph shows both the A280 nm and activity (A595 nm) elution profile for fractions from the CaM-Sepharose column operated as described in **Subheading 3.3.2.** The inset silver stained 7% Sodium dodecyl sulfate polymerase gel electrophoresis (SDS-PAGE) gel **(24)** shows 10 µL aliquots of fractions 13, 14, and 15 as indicated.

CaM kinase II activity determined as described in Methods can be presented as two activities: the CaM-stimulated activity and the endogenous activity owing to autophosphorylation at Thr286 **(22)**. Also, if phosphorylation of autocamtide-3 is calmodulin-independent, this could be a measure of the Thr305 autophosphorylated CaM kinase II. However, this is less likely owing to the expected rapid turnover of this form of the enzyme. Thus, calmodulin-independent phosphorylation of autocamtide-3 would more likely represent a kinase not previously known to phosphorylate this substrate or endogenous phosphorylated material binding to the P-81 paper. Alternatively, calmodulin-stimulation of CaM kinase II activity can be presented as the fold induction of activity by taking the ratio of the kinase activity when calmodulin is present to the activity without added calmodulin.

The assays described for CaM kinase II also can be used for CaM kinase IV by simply replacing the substrate peptide with one that is specifically recognized by CaM kinase IV. The γ-peptide (KSDGGVKKRKSSSS) derived from CaM kinase II γ is a highly specific substrate for CaM kinase IV **(23)**. It does not appear to be recognized by other calmodulin-dependent kinases. The γ-peptide is commercially available from Biosource International or Biomol.

It is also possible to discriminate between these and other kinases by using specific inhibitors that are commercially available. KN-93 will inhibit calmodulin activation of CaM kinase II and IV by binding to the calmodulin-binding domain of each enzyme and preventing calmodulin binding. An inactive isomer of KN-93, KN-92, can be used as a control for possible nonspecific effects (e.g., owing to the need to solubilize the compounds in dimethyl sulfoxide [DMSO]). KN-92 and KN-93 are available from a variety of sources, including Calbiochem and Alexis Biochemicals. They are added to cellular extracts to a final concentration of 10 μ*M* for 30 min before adding the extracts to the kinase reaction mixture. Numerous other inhibitors of CaM and its regulated enzymes are also available commercially and can be used both in vitro and in vivo.

References

1. Cheung, W. Y. (1970) Cyclic 3', 5'-nucleotide phosphodiesterase. Demonstration of an activator. *Bioche. Biophys. Res. Commun.* **38,** 533–538.
2. Kakiuchi, S. and Yamazaki, R. (1970) Calcium dependent phosphodiesterase activity and its activating factor (PAF) from brain studies on cyclic 3', 5'-nucleotide phosphodiesterase (3). *Biochem. Biophys. Res. Commun.* **41,** 1104–1110.
3. Teo, T. S. and Wang, J. H. (1973) Mechanism of activation of a cyclic adenosine 3', 5'-monophosphate phosphodiesterase from bovine heart by calcium ions. Identification of the protein activator as a Ca^{2+} binding protein. *J. Biol. Chem.* **248,** 5950–5955.
4. Vanaman, T. C. (1980) The structure, function and evolution of calmodulin, in *Calcium Binding Proteins as Cellular Regulators* (Cheung, W. Y., ed.), Academic Press, New York, NY, pp. 41–58.
5. Van Eldik, L. J. and Watterson, D. M. (1998) Calmoduin and calcium signal transduction: an introduction, in *Calmodulin and Signal Transduction* (Van Eldik, L. J. and Watterson, D. M., eds.), Academic Press, New York, NY, pp. 1–14.
6. Kuboniwa, H., Tjandra, N., Grzesiek, S., et al. (1995). Solution structure of calcium-free calmodulin. *Nat. Struct. Biol.* **2,** 768–776.
7. Babu, Y. S., Sack, J. S., Greenhough, T. J., et al. (1985). Three-dimensional structure of calmodulin. *Nature* **315,** 37–40.
8. Vetter, S. W. and Leclerc, E. (2003) Novel aspects of calmodulin target recognition and activation. *Eur. J. Biochem.* **270,** 404–414.
9. Jamieson, G. A., Jr. and Vanaman, T. C. (1979) Calcium-dependent affinity chromatography of calmodulin on an immobilized phenothiazine. *Biochem. Biophys. Res. Commun.* **90,** 1048–1056.
10. Hart, R. C., Hice, R. E., Charbonneau, H., et al. (1983) Preparation and properties of calcium-dependent resins with increased selectivity for calmodulin. *Anal. Biochem.* **135,** 208–220.
11. Gopalakrishna, R. and Anderson, W. B. (1982) Ca^{2+}-induced hydrophobic site on

calmodulin: application for purification of calmodulin by phenyl-sepharose affinity chromatography. *Biochem. Biophys. Res. Commun.* **104**, 830–836.

12. England, S. and Seifter, S. (1990) Precipitation techniques. *Methods Enzymol.* **182**, 285–300.
13. Watterson, D. M. and Vanaman, T. C. (1976) Affinity chromatography purification of a cyclic nucleotide phosphodiesterase using immobilized modulator protein, a troponin C-like protein from brain. *Biochem. Biophys. Res. Commun.* **73**, 40–46.
14. March, S. C., Parikh, I,. and Cuatrecasas, P. (1974) A simplified method for cyanogen bromide activation of agarose for affinity chromatography. *Anal. Biochem.* **60**, 149–152.
15. Niggli, V., Penniston, J. T., and Carafoli, E. (1979) Purification of the $(Ca^{2+}-Mg^{2+})$-ATPase from human erythrocyte membranes using a calmodulin affinity column. *J. Biol. Chem.* **254**, 9955–9958.
16. Niggli, V., Zurini, M., and Carafoli, E. (1987) Purification, reconstitution, and molecular characterization of the Ca^{2+} pump of plasma membranes. *Methods Enzymol.* **139**, 791–808.
17. Lanzetta, P. A., Alvarez, L. J., Reinach, P. S., and Candia, O. A. (1979) An improved assay for nanomole amounts of inorganic phosphate. *Anal. Biochem.* **100**, 95–97.
18. Hanson, P. I. and Schulman, H. (1992) Neuronal Ca^{2+}/calmodulin-dependent protein kinases. *Annu. Rev. Biochem.* **61**, 559–601.
19. Lisman, J., Schulman, H., and Cline, H. (2002) The molecular basis of CaMKII function in synaptic and behavioural memory. *Nat. Rev. Neurosci.* **3**, 175–190.
20. Means, A. R. (2000) Regulatory cascades involving calmodulin-dependent protein kinases. *Mol. Endocrinol.* **14**, 4–13.
21. Fujisawa, H. (2001) Regulation of the activities of multifunctional Ca^{2+}/calmodulin-dependent protein kinases. *J. Biochem. (Tokyo)* **129**, 193–199.
22. Hudmon, A. and Schulman, H. (2002) Structure-function of the multifunctional Ca^{2+}/calmodulin-dependent protein kinase II. *Biochem. J.* **364**, 593–611.
23. Miyano, O., Kameshita, I., and Fujisawa, H. (1992). Purification and characterization of a brain-specific multifunctional calmodulin-dependent protein kinase from rat cerebellum. *J. Biol. Chem.* **267**, 1198–1203.
24. Laemmli, U. K. (1970) Cleavage of structural proteins during the assembly of the head of bacteriophage T4. *Nature* **227**, 680–685.

9

Measurement of Protein–DNA Interactions In Vivo by Chromatin Immunoprecipitation

Hogune Im, Jeffrey A. Grass, Kirby D. Johnson, Meghan E. Boyer, Jing Wu, and Emery H. Bresnick

Summary

Elucidating mechanisms controlling nuclear processes requires an understanding of the nucleo-protein structure of genes at endogenous chromosomal loci. Traditional approaches to measuring protein–DNA interactions in vitro have often failed to provide insights into physiological mechanisms. Given that most transcription factors interact with simple DNA sequence motifs, which are abundantly distributed throughout a genome, it is essential to pinpoint the small subset of sites bound by factors in vivo. Signaling mechanisms induce the assembly and modulation of complex patterns of histone acetylation, methylation, phosphorylation, and ubiquitination, which are crucial determinants of chromatin accessibility. These seemingly complex issues can be directly addressed by a powerful methodology termed the chromatin immunoprecipitation (ChIP) assay. ChIP analysis involves covalently trapping endogenous proteins at chromatin sites, thereby yielding snapshots of protein–DNA interactions and histone modifications within living cells. The chromatin is sonicated to generate small fragments, and an immunoprecipitation is conducted with an antibody against the desired factor or histone modification. Crosslinks are reversed, and polymerase chain reaction (PCR) is used to assess whether DNA sequences are recovered immune-specifically. Chromatin-domain scanning coupled with quantitative analysis is a powerful means of dissecting mechanisms by which signaling pathways target genes within a complex genome.

Key Words: Chromatin; histone; immunoprecipitation; DNA; transcription; crosslink.

1. Introduction

A critical mechanism that creates diversity from cells to organisms involves the regulated activation and repression of genes, thereby establishing the

From: *Methods in Molecular Biology, vol. 284:*
Signal Transduction Protocols
Edited by: R. C. Dickson © Humana Press Inc., Totowa, NJ

complement of proteins characteristic of a specific cell type. Great efforts have been made to elucidate mechanisms that confer cell type-specific patterns of gene expression. Unraveling the underlying mechanisms has enormous potential for revealing the molecular basis of diseases and for the development of novel therapeutics through the specific modulation of gene expression.

Perhaps the most fundamental step in transcriptional control involves the sequence-specific recognition of DNA by transcription factors known as *trans*-acting factors. The development of a grab bag of experimental approaches has ushered in major progress in understanding how proteins bind their cognate DNA recognition sequences specifically within the complex milieu of cellular DNA. Filter-binding assays were initially used to measure protein–DNA interactions in vitro *(1,2)*. This assay involves a short incubation to allow protein binding to DNA, after which protein–DNA complexes are trapped on a filter. Through the use of radiolabeled DNA, quantitative measurements of equilibrium-binding constants and rate constants can be obtained. Although filter-binding assays have considerable utility for analyzing binding with purified or semipurified components, this method is not ideal for analyzing binding with complex mixtures of proteins. By contrast, the electrophoretic mobility shift assay (EMSA) *(3)* allows one to identify proteins with sequence-specific DNA binding activity in crude extracts containing hundreds or even thousands of proteins. Protein–DNA complexes are rapidly resolved from unbound radiolabeled DNA on a nondenaturing gel. Similarly, DNaseI footprinting *(4)* allows for the detection of sequence-specific DNA binding with crude protein extracts. DNaseI footprinting involves binding an extract or purified protein to a DNA fragment of approx 100–500 base pairs uniquely labeled on the five or three end. Complexes are incubated with DNaseI, which readily cleaves free DNA. Sequence-specific protein–DNA interactions protect the bound sequence from cleavage, revealing a "footprint" (the protected region) upon analysis of DNA fragments on a denaturing urea-polyacrylamide gel. These methods are used routinely to rapidly define the DNA sequence preferences for sequence-specific DNA binding proteins. Thus, it can be straightforward to identify proteins that interact with conserved *cis*-acting elements on regulatory regions (e.g., promoters, enhancers, silencers, and locus-controls regions) of genes in vitro, an important first step in analyzing transcriptional-control mechanisms.

The physiological template for transcription and other nuclear processes, such as DNA replication, recombination, and repair, is chromatin rather than naked DNA *(5–7)*. Although EMSA and footprinting assays are powerful approaches for analyzing transcriptional mechanisms, studying the interaction of *trans*-acting factors with naked DNA has intrinsic limitations. A reoccurring problem is that multiple, highly related transcription factors often bind an identical DNA sequence in vitro, whereas these factors can have different functions in vivo *(8)*. For example, multiple E-proteins exist that bind E-box consensus sequences (CANNTG). Certain proteins, such as upstream stimulatory factor, are readily detected using in vitro DNA binding assays, whereas the binding of other factors such as c-myc is difficult to measure, in part owing to the low abundance of c-myc in nuclear extracts *(9,10)*. In vitro binding assays commonly identify the most abundant, highest-affinity interactor in the extract, which is often not the functional factor in vivo. Furthermore, transcription factors can reside in the cell nucleus in an "active" state, and therefore such factors in cell-free extracts readily bind cognate recognition sites. However, chromatin structure can exclude factors from constitutively binding recognition motifs in vivo *(6,11,12)*, thus establishing a chromatin-remodeling requirement for factor access to the template. Given these serious issues, it is imperative that protein–DNA interactions detected in vitro be validated via in vivo analysis.

Despite the seemingly formidable problem of how to detect binding of a protein to a target DNA sequence in a living cell, major progress has been made in developing the requisite technology. Although it is not the intent of this chapter to comprehensively review such technologies, it is important to recognize the advance afforded by knowing what factors occupy specific *cis*-elements in a living cell. Initial studies utilized in vivo footprinting, which often involves treating cells with dimethyl sulfate, which methylates the N7 position of guanosine *(13)*. Guanosine methylation is efficient and relatively uniform in vitro. However, if a high percentage of the templates in the cell population are occupied by a DNA-binding protein, the protein can severely inhibit or enhance methylation, thus providing the basis for the footprinting assay. Genomic DNA is purified from cells and subjected to Maxam-Gilbert chemistry to cleave the DNA at the methylated sites. By incorporating an elegant ligation-mediated polymerase chain reaction (PCR) step *(14)*, fragments can be displayed on a denaturing urea-containing polyacrylamide gel at base-pair resolution. Through comparisons of cleavage patterns of naked DNA and DNA from dimethyl sulfate-treated cells, conclusions can be reached regarding whether a particular site is occupied by protein(s) in vivo. A modification of the assay has been reported in which protein contacts with both adenosine and guanosine residues can be measured *(15)*, which expands the repertoire of *cis*-elements that can be analyzed. In addition, another variation involves adding DNaseI to permeabilized cells *(16)* or to isolated nuclei *(17)*. The differential cleavage of chromatinized free DNA reflects occupancy of specific sequences by proteins.

In vivo footprinting assays have utility for scanning a defined region, such as a promoter, to delineate sequences bound by proteins in living cells. In certain cases, the enhancement or diminishment of cleavage can be correlated with the pattern obtained by in vitro footprinting, which can provide important clues as to the identity of the protein bound. Importantly, the footprinting assay cannot identify the bound protein, but can only provide a clue, based on the DNA sequence of the respective region. Thus, this highlights a major limitation of in vivo footprinting assays. A goal of such studies is to pinpoint *cis*-elements bound by factors in vivo, which should provide important clues regarding the identity of the cognate-binding protein. However, the identity of binding proteins is based upon inference, because related factors often bind identical DNA sequences.

Studies utilizing ultraviolet (UV) light and lasers to crosslink histones to chromatin provided a foundation for the development of methodology to overcome the major limitations of in vivo footprinting *(17–20)*. The chromatin immunoprecipitation or ChIP assay has evolved to use a relatively nonspecific crosslinking reagent formaldehyde to rapidly crosslink proteins to cognate-binding sequences in cells or tissues *(21,22)*. Following the covalent trapping of proteins at their site of interaction in native chromatin, the chromatin is fragmented into small pieces (less than approx 1 kb), and an antibody against the protein is used to co-immunoprecipitate the protein and the bound chromatin *(23)*. Antibodies can be directed against any protein of interest—*trans*-acting factors that bind DNA sequence-specifically, as well as proteins that bind frequently throughout genomes, such as acetylated or methylated core histones. Upon deproteinization of the immunoprecipitate, PCR is used to determine if a sequence is recovered in an immune-specific manner, which would indicate that the protein associated with the sequence in living cells.

The simplest scenario resulting from ChIP analysis is that the protein binds directly to a known *cis*-element, thereby revealing the endogenous interactor with the site in living cells. However, given the extensive network of protein–DNA and protein–protein crosslinks generated by formaldehyde, the recovery of a specific DNA fragment does not necessarily mean that the protein associates directly with the fragment. This is an intrinsic limitation of ChIP analysis vis-à-vis identifying interactors with defined sequences in chromatin. It is possible, however, to assign the contribution of specific *cis*-elements to binding. A DNA fragment containing or lacking a site suspected of mediating binding in vivo can be stably integrated into chromosomal DNA, and ChIP can be used to determine if the site is required for binding *(24)*.

ChIP analysis can also yield functionally important information on higher-order nucleoprotein complexes. Sequences directly and indirectly bound to the protein can be immunoprecipitated. Complex scenarios arise upon consideration of indirect crosslinking, an example being the tethering of coregulator complexes to a DNA-bound protein via protein–protein interactions. Alternatively, a DNA-bound protein might be crosslinked to a distant site owing to long-range physical interactions between distinct chromatin regions. As long-range chromatin interactions commonly control nuclear processes *(23)*, measurements of such interactions can be highly significant.

Beyond defining protein interactions with known *cis*-elements, entire chromatin domains can be scanned by ChIP, revealing the native nucleoprotein structure of the domain *(25–28)*. As certain *cis*-elements are scattered throughout genomes, based on the simplicity of the specific recognition motif, domain scanning provides a powerful means of distinguishing endogenous, physiologically relevant sites of interaction from the numerous sites that would be detected by in vitro binding analyses with naked DNA. This situation is exemplified by analysis of the binding of GATA family transcription factors to GATA consensus motifs (WGATAR) *(29,30)*. Although the murine μ-globin and GATA-2 loci contain 286 and 88 consensus GATA motifs, respectively, only a small subset of these sites are occupied in vivo, based upon ChIP analysis *(26,31)*. Domain scanning coupled with the use of quantitative real-time PCR is a powerful way to define the biological consequences of protein–DNA interactions in vivo.

ChIP methodology continues to be modified to increase the utility of the assay. A genomic microarray step has been incorporated so that preimmune and immune reactions are labeled with fluorophores and hybridized to genomic microarrays *(32,33)*. This approach allows one to define a large complement of genes bound by a factor of interest in a single experiment, without having to analyze data from PCR reactions with hundreds of site-specific primer sets. It should be obvious that methodologies to define protein–DNA interactions in vivo have advanced considerably from the initial approaches in which major effort was required to determine whether specific base pairs at a single locus are protected from modification. This chapter describes in detail the core ChIP methodology used in our laboratory, which can be modified based on the specific application desired.

2. Materials

2.1. Buffers

The following buffers, should be prepared before beginning the protocol:

1. 1 *M* Tris-HCl, pH 8.0, 0.5 *M* EDTA, pH 8.0, 5 *M* NaCl, 10% Nonidet P40, 20% sodium dodecyl sulfate (SDS), and 20% Triton X-100.
2. 1X phosphate-buffered saline (PBS), 2.5 *M* glycine, 3 *M* sodium acetate, pH 5.2, cell lysis buffer, nuclei lysis buffer, IP dilution buffer, IP wash buffer 1, IP wash buffer 2, 1X TE, and elution buffer. All buffers are stored at room temperature, and when necessary, an aliquot is chilled on ice before use.

2.2. Crosslinking

1. Platform shaker (Hoefer, San Francisco, CA, cat. no. PR70/75-115V).
2. 37% Formaldehyde (Sigma, St. Louis, MO, cat. no. F-1268).
3. 2.5 *M* glycine (Promega, Madison, WI, cat. no. H-5073).

4. Deionized-distilled water (ddH$_2$O) (Millipore, Billeria, MA, Milli-Q system).
5. 1X PBS, pH 7.4: 8.0 g NaCl, 0.2 g KCl, 2.71 g Na$_2$HPO$_4$ · 7H$_2$O, 0.24 g KH$_2$PO$_4$, ddH$_2$O to 1 L.

2.3. Cell Lysis

1. Histone deacetylase and protease inhibitors (10 m*M* *n*-butyric acid, sodium salt (Sigma, cat. no. B-5887) (−20°C), 1 µg/mL leupeptin (Roche Applied Science, cat. no. 1-017-128) (−20°C), 50 µg/mL phenylmethylsulfonyl fluoride (PMSF) in 100% ethanol (Boehringer, Danbury, CT, cat. no. 837-091) (−20°C), add to an aliquot of the ChIP buffer immediately before use (*see* **Note 1**).
2. 1 *M* Tris-HCl, pH 8.0.
3. 5 *M* sodium chloride.
4. 10% Nonidet P40 (Igepal CA-630, Sigma).
5. Cell lysis buffer: 10 m*M* Tris-HCl, pH 8.0, 10 m*M* NaCl, 0.2% NP40. Add histone deacetylase and protease inhibitors to the buffer immediately before use.

2.4. Nuclei Lysis

1. 0.5 *M* ethylenediaminetetraacetic acid (EDTA), pH 8.0.
2. Nuclei lysis buffer: 50 m*M* Tris-HCl, pH 8.0, 10 m*M* EDTA, 1% SDS. Add histone deacetylase and protease inhibitors to the buffer immediately before use.

2.5. Sonication and Preclearing

1. 20% Triton X-100.
2. Immunoprecipitation (IP) dilution buffer: 20 m*M* Tris-HCl, pH 8.0, 2 m*M* EDTA, 150 m*M* NaCl, 1% Triton X-100, 0.01% SDS. Add histone deacetylase and protease inhibitors to the buffer immediately before use.
3. Normal rabbit serum (−20°C) (Covance Research Products, Princeton, NJ).
4. Branson sonifier 250 (Branson Ultrasonics, Danbury, CT, cat. no. 101-063-196) with microtip horn-1/8 in solid-tapered (cat. no. 101-148-062).
5. Protein A-sepharose, 20% slurry (Sigma, cat. no. P-3391) (4°C) (*see* **Note 2**).
6. Hematology/chemistry mixer (Fisher, Pittsburg, PA, cat. no. 14-059-346).
7. Falcon polypropylene tubes (no. 2096).

2.6. Immunoprecipitation

1. IP dilution buffer (*see* **Subheading 2.5.**).
2. Nuclei lysis buffer (*see* **Subheading 2.4.**).
3. Protein A-sepharose, 20% slurry (4°C).

2.7. Washing Protein A-Sepharose/Immune Complexes

1. IP wash buffer 1: 20 m*M* Tris-HCl, pH 8.0, 2 m*M* EDTA, 50 m*M* NaCl, 1% Triton X-100, 0.1% SDS.
2. Lithium chloride.
3. Deoxycholic acid.
4. IP wash buffer 2: 10 m*M* Tris-HCl, pH 8.0, 1 m*M* EDTA, 250 m*M* LiCl, 1% NP40, 1% deoxycholic acid.
5. 1X Tris-EDTA (TE) buffer: 10 m*M* Tris-HCl, pH 8.0, 1 m*M* EDTA.

2.8. Immune Complex Elution

1. Sodium bicarbonate ($NaHCO_3$).
2. Elution Buffer (100 m*M* $NaHCO_3$, 1% SDS).

2.9. Reverse Crosslinking

1. 1 mg/mL RNase A (Sigma, cat. no. R-4875) (−20°C) (*see* **Note 3**).
2. 5 *M* NaCl.
3. 20 mg/mL Proteinase K (Promega, cat. no. V-3021) (−20°C).

2.10. DNA Purification

1. 1X TE buffer.
2. 10 mg/mL tRNA (Sigma, cat. no. R-7876) (−20°C).
3. Phenol.
4. Phenol saturated with 0.1 *M* Tris-HCl, pH 8.0 (*see* **Note 4**).
5. Chloroform.
6. Sodium acetate.
7. Glacial acetic acid.
8. 3 *M* sodium acetate (NaOAc), pH 5.2.
9. 10 mg/mL Glycogen (−20°C).
10. 100% and 70% ethanol (EtOH).
11. 10 m*M* Tris-HCl, pH 8.0.
12. Agarose.

2.11. Quantitative Real-Time Polymerase Chain Reaction

1. ABI 7000 sequence detection system (Applied Biosystems, Foster City, CA).
2. 96-well plates (Eppendorf, cat. no. 954-02-030-3).
3. SYBR Green PCR master mix (Applied Biosystems, cat. no. 4309155).
4. ABI PRISM optical adhesive covers (Applied Biosystems, cat. no. 4311971) (*see* **Note 5**).
5. ABI PRISM optical caps (8 caps/strip) (Applied Biosystems, cat. no. 4323032).

3. Methods

The following steps apply to the analysis of a single-cell line or a primary cell sample, in which five conditions will be used for the immunoprecipitation. Although most of our studies have utilized blood cell lines and primary blood cells, which are largely non-adherent, only minor modifications are required for analysis of adherent cells (*see* **Note 6**). Up to four ChIP experiments routinely are conducted in parallel. The protocol requires at least 3 d to complete, with the final steps in **Subheadings 3.4.** and **3.8.** requiring overnight incubations.

3.1. Crosslinking

1. Use 1×10^7 cells per IP condition. Transfer the cells to a 500-mL Erlenmeyer flask (*see* **Note 7**).
2. Dilute the cells to 90 mL (5×10^5 cells/mL, final) with sterile tissue-culture medium at room temperature (*see* **Note 8**).
3. Crosslink by adding 1 mL 37% formaldehyde (0.4% final). Mix for 10 min at room temperature (setting 4–5, platform shaker).
4. Stop the crosslinking reaction by adding 5 mL of 2.5 M glycine (0.125 M final). Mix for 5 min at room temperature (setting 4–5, platform shaker) (*see* **Note 9**).
5. Transfer the cells to 50-mL conical centrifuge tubes, and centrifuge at 240g for 7 min at 4°C in a swinging bucket rotor. Keep the sample on ice after this step.
6. Resuspend the cells with 0.5 mL ice-cold PBS and divide equally into two 1.5-mL microfuge tubes.
7. Wash the 50-mL tubes with an additional 0.5 mL of ice-cold PBS and transfer to the 1.5-ml microfuge tubes from **step 6**.
8. Pellet the cells at 600 g for 5 min at 4°C (*see* **Note 10**).

3.2. Cell Lysis

1. Chill 1 mL cell lysis buffer on ice for 5 min.
2. Carefully remove the supernatant from the cell pellet with a pipet (*see* **Note 11**).
3. Estimate the total cell pellet volume.
4. Resuspend the cells in 1.5 × pellet volumes of ice-cold cell lysis buffer by gently pipetting up and down (*see* **Note 12**).
5. Combine the resuspended cells into one tube and then incubate on ice for 10 min.
6. Centrifuge at 600 g for 5 min at 4°C.
7. Remove the supernatant with a pipet, being careful not to disturb the nuclei pellet (*see* **Note 13**).

3.3. Nuclei Lysis

1. Chill 1.5 mL nuclei lysis buffer on ice for 3 min (*see* **Note 14**).
2. Resuspend the nuclei pellet in 1 mL nuclei lysis buffer by repetitive pipetting.
3. Incubate the sample for 10 min at 4°C (*see* **Note 15**).

3.4. Sonication and Preclearing

1. Chill 1 mL of IP Dilution Buffer on ice for 5 min. Transfer the lysed nuclei to a 15 mL Falcon 2096 conical tube, and add 0.8 mL of ice-cold IP Dilution buffer.
2. Sonicate the chromatin to reduce the average size to approx 500 bp (*see* **Note 16**):
   a. Number of bursts: 8.
   b. Length of bursts: 30–40 s.
   c. Output control setting: 20%.
   d. Duty cycle: constant.
3. Transfer the sonicated chromatin to a 2 mL microfuge tube and centrifuge at 16,000 g for 10 min at 4°C to pellet any insoluble material.
4. Transfer the supernatant to a 15-mL conical polypropylene Falcon 2096 tube.
5. Add 2 mL IP dilution buffer to the sample to bring the total ratio of nuclei lysis buffer: IP dilution buffer to 1:4 (*see* **Note 17**).
6. Preclear the chromatin by adding 50 μL of preimmune rabbit serum. Mix at 4°C for at least 1 h.
7. In a 1.5-ml microfuge tube, pellet 500 μL of (20% slurry) by centrifuging at 4000g for 2 min at 4°C.
8. Use 500 μL of the chromatin solution to resuspend the Protein A-sepharose pellet and transfer back to the 15-mL tube. Mix at 4°C overnight.
9. End of d 1.

3.5. Immunoprecipitation

1. Pellet the from the preclearing reaction by centrifuging in a swinging bucket rotor at 1500g for 2 min at 4°C.
2. Aliquot the chromatin into 1.5-ml microfuge tubes: 180 μL for the input sample and 900 μL for each IP sample (*see* **Note 18**).
3. Set up a "no chromatin" sample, as a negative control, by mixing 180 μL nuclei lysis buffer and 720 μL IP Dilution Buffer in a 1.5-mL microfuge tube. Treat the sample identical to samples incubated with antibodies (*see* **Note 19**).
4. Add antibodies: Usually 0.8% to 1.6% (i.e., 7.5 μL anti-acetyl-histone H3, 1 mg/mL; Upstate Biotechnology, cat. no. 06-599). Add 10 μL of normal rabbit serum to one aliquot of chromatin as a negative control. In addition,

set up a "no chromatin" sample lacking antibody. Mix for 2–3 h at 4°C (*see* **Note 20** and **21**).

5. Centrifuge the IP samples at 16,000 g for 5 min at 4°C to remove any insoluble material.
6. For each IP sample, prepare one 1.5-mL microfuge tube containing 150 μL of Protein A-sepharose (20% slurry) centrifuge at 4000 g for 2 min at 4°C, and discard the supernatant.
7. Transfer the samples to the Protein A-sepharose pellets. Mix for 1–2 h at 4°C.

3.6. Washing Protein A-Sepharose/Immune Complexes

1. Pellet the Protein A-sepharose-bound immune complexes by centrifuging at 4000 g for 2 min at 4°C.
2. Wash the Protein A-sepharose-bound immune complexes with 500 μL of each solution: twice with ice-cold IP wash buffer 1, once with ice-cold IP wash buffer 2, and twice with ice-cold 1X TE buffer.
3. Transfer the Protein A-sepharose/immune complex pellets to new 1.5-mL microfuge tubes with first wash of IP wash buffer 1. This step is critical for reducing background.
4. Resuspend the immunoadsorbed immune complexes by vortexing for 1–2 s at low speed after each buffer addition.
5. Pellet the Protein A-sepharose at 4000 g for 2 min at 4°C, and discard the supernatant (*see* **Note 22**).

3.7. Immune Complex Elution

1. Elute the immune complexes from the Protein A-sepharose with 150 μL of elution buffer at room temperature (*see* **Note 23**).
2. Vortex for 2–3 s at medium speed, centrifuge at 4000 g for 2 min at room temperature, and transfer the supernatants to new 1.5-mL microfuge tubes.
3. Elute the immune complexes with an additional 150 μL of elution buffer, and combine both eluted samples in the same tube.

3.8. Reverse Crosslinking

1. To the IP samples add 1 μL of 1 mg/mL RNase A and 18 μL of 5 M NaCl.
2. To the input sample add 1 μL of 1 mg/mL RNase A and 10.8 μL of 5 M NaCl.
3. Incubate the samples in a heating block at 67°C for a minimum of 4 h (*see* **Note 24**).
4. Add 3 μL of 20 mg/mL Proteinase K to each sample, and then vortex for 1–2 s at low setting. Reduce the temperature to 45°C and incubate overnight (*see* **Note 25**).
5. End of d 2.

3.9. Nucleic Acid Purification

1. Add 120 μL of 1X TE buffer to the input sample.
2. Add 1 μL of 10 mg/mL tRNA to the input and IP samples.
3. Extract the input and IP samples with 300 μL of 1:1 phenol/chloroform by vortexing well (10–15 s at highest speed) and centrifuging at 16,000 *g* for 5 min at room temperature. Transfer the top aqueous phase to a new 1.5-mL microfuge tube after each extraction.
4. Extract the input sample with 1:1 phenol/chloroform an additional time, vortex well, and centrifuge at 16,000 *g* for 5 min at room temperature (*see* **Note 26**).
5. Extract the input and IP samples once with 300 μL of chloroform, vortex well, and centrifuge at 16,000 *g* for 5 min at room temperature.
6. Add 30 μL of 3 *M* NaOAc, pH 5.2, 0.5 μL of 10 mg/mL tRNA and 0.5 μL of 10 mg/mL glycogen to the input and IP samples.
7. Add 750 μL of 100% EtOH (2.5 sample volumes) to each sample, and mix by inverting tubes several times. Incubate at least 30 min at −80°C (*see* **Note 27**).
8. Centrifuge the samples at 16,000 *g* for 20 min at 4°C. Discard the supernatant (*see* **Note 28**).
9. Wash the pellets with 800 μL of ice cold 70% EtOH, and centrifuge at 16,000 *g* for 10 min at 4°C. Discard the supernatant (*see* **Note 29**).
10. Air-dry the pellets for 5–10 min. Do not overdry the pellets, as they may not completely dissolve.
11. Dissolve the IP samples and the input samples in 30 μL and 66.7 μL of autoclaved 10 m*M* Tris-HCl, pH 8.0, respectively (*see* **Note 30**).
12. Incubate samples for 10 min at room temperature. Mix with moderate vortexing for 2–3 s.
13. Run 4 μL of the input sample on a 1.6% agarose gel to check the chromatin size (*see* **Note 31**).
14. Store samples at −20°C.

3.10. Quantitative Real-Time Polymerase Chain Reaction

1. Set up a real-time PCR plate document on the instrument. For each primer set/sample set combination, use a unique detector.
2. Prepare the following serial dilution of the input sample in 10 m*M* Tris-HCl, pH 8.0: 1/5, 1/25, 1/125, and 1/625. Use the diluted input samples to generate a standard curve for each primer set/sample set combination.
3. Load each well to be used in the plate with assay mix containing; 12.5 μL SYBR green master mix, 100 n*M* forward and 100 n*M* reverse primer, and

bring each sample to a final volume of 23.5 µL with ddH₂O. A mix of the three components can be made and then pipetted into each well. Most of our real-time PCR primer sets work best at 100 nM or 220 nM final concentrations. However, some primer sets work considerably better at higher concentrations (up to 900 nM).

4. Assay 1.5 µL of diluted input or IP sample per well.
5. Cover the wells with either Optical Adhesive Covers or strip caps (*see* **Note 5**).
6. Centrifuge the plate at 200 g for 1 min at 4°C to sediment the reaction mixture to the bottom of the plate.
7. Load the plate into the instrument and start the thermal cycle protocol, based on the manufacturers instructions.

4. Notes

1. The inhibitors have relatively short half-lives in the ChIP buffers and therefore are added to an aliquot of the ChIP buffer immediately before use.
2. Preparation of a 20% slurry of Protein A-sepharose: Transfer 2.0 g protein A-sepharose (1.0 g dry = 4 mL hydrated) to a 50-ml conical centrifuge tube. Add ice-cold 1X TE to a final volume of 40 mL. Mix gently at 4°C overnight. Pellet the Protein A-sepharose at 1500 g for 2 min at 4°C in a swinging bucket rotor. Wash three times with 40 mL ice-cold 1X TE, gently invert the tube until the pellet is resuspended; pellet the Protein A-sepharose after each wash. Resuspend to a 20% slurry (v/v) in cold 1X TE containing 2.0% BSA + 0.02% sodium azide. Store at 4°C.
3. Dissolve the RNase A in ddH₂O to 10 mg/mL, boil for 15 min, and store at −20°C.
4. Thaw phenol in a 70°C water bath. Transfer 100 mL of melted phenol to a 500 ml bottle, and add an equal volume of 0.5 M Tris-HCl, pH 8.0, at room temperature. Stir the mixture by inverting the bottle multiple times for approx 5–10 min. When the two phases have separated, aspirate as much of the upper aqueous phase as possible using a glass pipet attached to a vacuum line equipped with appropriate traps. Add an equal volume of 0.1 M Tris-HCl, pH 8.0, and mix. Remove the upper aqueous phase, and repeat the extraction until the pH of the lower phase is >7.8. After the phenol is equilibrated and the final aqueous phase has been removed, add 0.1 volume of 0.1 M Tris-HCl, pH 8.0. Store at −20°C in 50-mL conical tubes wrapped in aluminum foil. Phenol is highly caustic, necessitating great care in handling even the smallest volumes of phenol.
5. Use the optical adhesive covers if more than half a plate is used; otherwise use the optical caps.

6. For analysis of adherent cells such as mouse embryonic stem (ES) cells, ES cells are grown to approx 50–75% confluence. One 10 cm plate (approx 2–4 $\times$ 10^7 cells) is sufficient for a single ChIP experiment with five immunoprecipitation conditions. Culture media is changed prior to crosslinking, and formaldehyde is added directly to the culture plate at a final concentration of 0.4%. Cells are then incubated on a shaking platform for 10 min at room temperature. The crosslinking reaction is terminated by adding glycine to a final concentration of 0.125 M and incubating with constant shaking for 5 min. Cells are harvested by discarding media, rinsing with 1 mL of ice-cold PBS, and scraping into 1 ml of PBS. Plates are washed twice with 1 mL of PBS to collect the remaining cells, which are pooled with the rest of the cells. Cells are isolated by centrifugation at 400 g for 8 min at 4°C, and the cell pellet is washed once with ice-cold PBS containing PMSF at a final concentration of 50 µg/mL. A typical cell pellet size from a single plate is approx 100 µL. Typsinization can be used to isolate highly adherent cells such as primary human umbilical vein endothelial cells (HUVEC). As usual, trypsin is inactivated with 10% fetal calf or calf serum. In the case of HUVEC, the cell pellet from five dishes in which cells are 50–70% confluent is approx 100 µL.
7. The amount of cells used per IP can be varied from 5 $\times$ 10^6 to 2 $\times$ 10^7 depending on the crosslinking efficiency and the quality of the antibody.
8. The volume of the crosslinking reaction can be varied from 30 mL to 180 mL as long as the final concentration of formaldehyde and glycine is 0.4% and 0.125 M, respectively.
9. After adding glycine to the cells in media containing phenol red, the color of the media will change from red to orange.
10. At this point the cell pellet can be frozen in a dry ice/EtOH bath, and then stored at -80°C for processing later. A typical cell pellet size from 2 $\times$ 10^7 suspension cells is 100 µL.
11. The nuclei can be easily lysed during cell lysis, resulting in the loss of chromatin. Thus, the sample should be handled gently prior to the sonication step. However, the cell lysis step can be omitted if cell pellets were frozen or if smaller numbers of cells are used.
12. An accurate determination of the pellet volume can be made with a pipet after the pellets are suspended in cell lysis buffer. If the pellet volume is greater than 500 µL, use a Dounce B homogenizer to lyse the cells.
13. The supernatant should be removed immediately after centrifugation, as the nuclei pellet is soft and can be accidentally removed with the supernatant if the pellet sits in the supernatant too long.
14. If the nuclei lysis buffer is placed on ice for more than 5 min, the SDS in the buffer will precipitate. If this happens, warm the buffer to room temperature, vortex or mix gently, and chill on ice again.

15. The sample can be incubated on ice for up to 2 h.

16. Place the Falcon tube in between the inner wall of a small beaker (150 mL) and the ice. Keep the bottom of the tube approx 1 cm up from the bottom of the beaker. Begin sonicating with the tip near the bottom of the tube. Turn on the sonicator and adjust the settings. Gradually raise the tip up from the bottom of the tube. Raise the tip until it is 3–5 mm below the surface. Incubate the tube on ice for at least 1 min between each sonication burst. After each sonication burst, repack the ice around the tube. If foaming occurs, incubate the tube on ice for at least a minute. However, if the foam persists for more than a minute, centrifuge the tube at 100 g for 1 min, resuspend any insoluble material, and recommence sonication.

17. The final ratio of nuclei lysis buffer to IP dilution buffer must be greater than or equal to 1:4 to reduce the SDS concentration so that it does not exceed 0.2% to minimize interference with the immunoprecipitation in d 2 (total volume of chromatin needed = 0.9 mL × the number of IP conditions + 0.3 mL [extra volume for the input]); for example, for 6 conditions use 0.9 mL × 6 + 0.3 mL = 5.7 mL. Add 4.1 mL IP dilution buffer rather than 3.4 mL.

18. Incubate the input at 4°C on ice until the reverse crosslinking step is conducted.

19. This is strongly recommended for initial attempts, because it is an excellent negative control. However, this step can be eliminated once the technique has been optimized.

20. The amount of antibody used will differ for different antibodies. When using an antibody that has never been used for ChIP, perform a titration with the antibody to determine the minimum concentration of antibody necessary to saturate the signal.

21. For antibodies that do not bind Protein A-sepharose (for example, antibodies raised in rats), incubate 30 µL of Protein A-sepharose with 25 µL of AffiniPure Rabbit anti-Rat IgG (H+L) (2.4 mg/mL) (Jackson Immuno-Research, 312-005-003) in 900 µL of IP-dilution buffer for 2–3 h with mixing alongside the IP samples. We normally include a chromatin sample with only the rabbit antirat antibody to measure the background from the rabbit antirat IgG. Alternatively, Protein G sepharose can be used. Centrifuge at 4000 g for 2 min at 4°C, and discard the supernatant.

22. When removing the supernatant, draw it off from the meniscus by following the meniscus down with the pipet tip. Be **very careful** not to remove the Protein A-sepharose at the bottom of the tube. Leaving 25–50 µL of buffer above the Protein A-sepharose during the washes will not negatively affect the results. After the last TE wash, carefully remove as much TE buffer as possible before proceeding to the elution step.

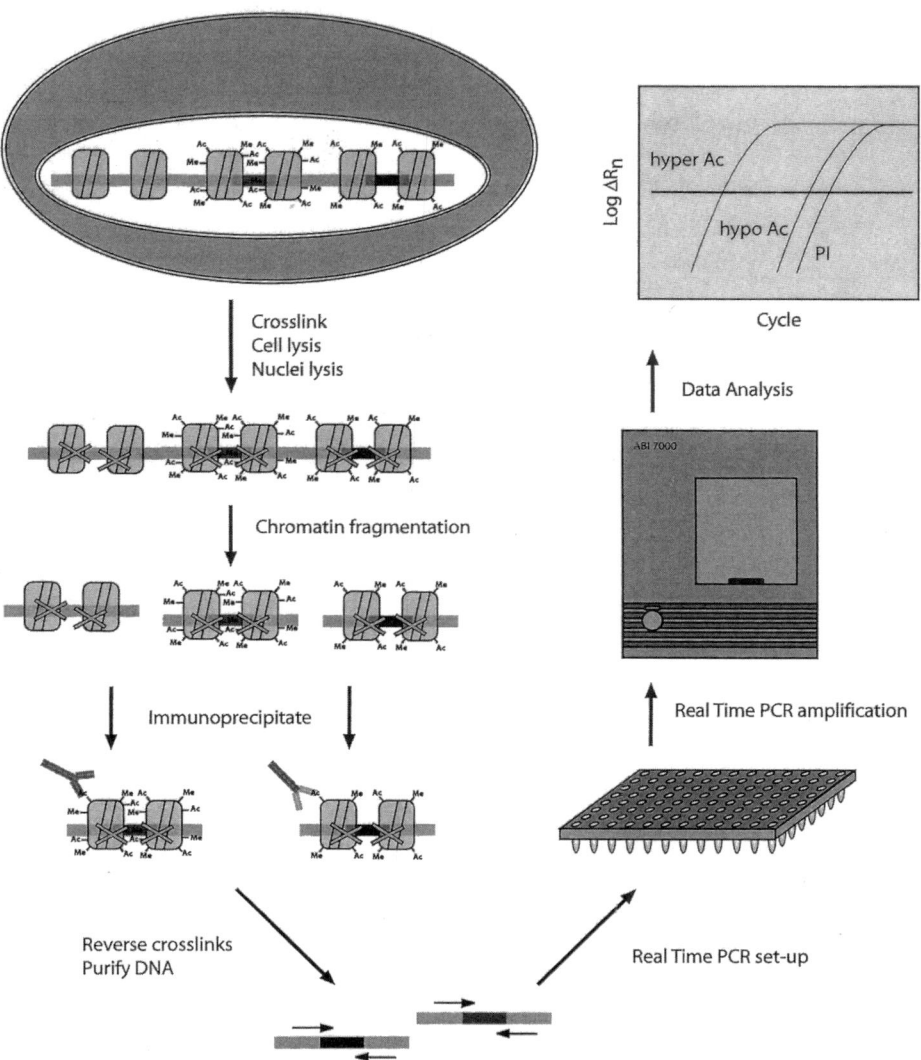

Fig. 1. Chromatin immunoprecipitation (ChIP) assay to measure histone modifications in living cells. Cells or tissue are treated briefly with formaldehyde to crosslink histones to DNA. "X" depicts protein–DNA crosslinks. Posttranslational modifications of histone amino-terminal tails are shown: Ac, acetylation; Me, methylation. Specific antibodies are used to immunoprecipitate acetylated and methylated histones. DNA fragments recovered from the washed immune complex are used as a template for quantitative real-time PCR. When measuring histone modifications, hyperacetylated chromatin yields a signal that appears early in the reaction cycle. Chromatin containing low levels of acetylation or hypoacetylated chromatin yields signals that appear later in the reaction cycle.

23. Do not chill the elution buffer before or during use, because the SDS will precipitate, and the buffer will need to be warmed at 37°C for 1–2 min.

24. A water bath may be used rather than a heating block. This step reverses the crosslinking and digests any RNA that is bound to the immune pellet.

25. The Proteinase K treatment can be stopped after 2 h, following either a 4 h or overnight RNase A treatment. After this step is completed, the samples can be stored at −20°C for processing at a later time.

26. Certain input samples may require more than two extractions with 1:1 phenol/chloroform.

27. The samples can be stored at −80°C for as long as necessary.

28. The IP pellets should appear as small translucent or white ovals of about 1–2 µL, and the input pellet should be white, about 5–10 µL.

29. After discarding the 70% EtOH wash, centrifuge for 1–2 s to isolate the remaining EtOH at the bottom. Using a pipet, remove as much of the EtOH as possible.

30. Autoclaved ddH$_2$O can be used in place of 10 mM Tris-HCl, pH 8.0. Inputs are dissolved in a volume to achieve 0.3% total chromatin immunoprecipitated per µL and used to generate a 125-fold range for the standard curve.

31. If agarose gel analysis reveals that the chromatin is too large, the length of bursts and/or the output setting can be increased to reduce chromatin length in subsequent experiments. However, increasing the sonication time and/or output may elevate the temperature of the sample and may decrease the yield of immunoprecipitated chromatin. Adjusting the number of bursts to reduce the chromatin size is usually not as effective as adjusting the burst duration or the output setting.

References

1. von Hippel, P. H. and McGhee, J. D. (1972) DNA-protein interactions. *Ann. Rev. Biochem.* **41,** 231–300.

2. Woodbury, C. P. and von Hippel, P. H. (1983) On the determination of deoxyribonucleic acid-protein interaction parameters using the nitrocellulose filter-binding assay. *Biochem.* **22,** 4730–4737.

3. Garner, M. M. and Revzin, A. (1981) A gel electrophoresis method for quantifying the binding of proteins to specific DNA regions: application to components of the *Escherichia coli* lactose operon regulatory system. *Nucleic Acids Res.* **9,** 3047–3060.

4. Galas, D. J. and Schmitz, A. (1978) DNaseI footprinting: a simple method for the detection of protein–DNA binding specificity. *Nucleic Acids Res.* **5,** 3157–3170.

5. Felsenfeld, G. (1992) Chromatin as an essential part of the transcriptional mechanism. *Nature* **355,** 219–224.

6. Hager, G. L., Archer, T. K., Fragoso, G., et al. (1993) Influence of chromatin structure on the binding of transcription factors to DNA. *Cold Spring Harb. Symp. Quant. Biol.* **58**, 63–71.

7. Wolffe, A. P. (1994) Transcription: in tune with the histones. *Cell* **77**, 13–16.

8. Weinmann, A. S., Bartley, S. M., Zhang, T., et al. (2000) Use of chromatin immunoprecipitation to clone novel e2f target promoters. *Mol. Cell Biol.* **21**, 6820–6832.

9. Boyd, K. E. and Farnham, P. J. (1997) Myc versus USF: discrimination at the cad gene is determined by core promoter elements. *Mol. Cell Biol.* **17**, 2529–2537.

10. Boyd, K. E., Wells, J., Gutman, J., et al. (1998) c-Myc target gene specificity is determined by a post-DNA binding mechanism. *Proc. Natl. Acad. Sci. USA.* **95**, 13,887–13,892.

11. Cordingley, M. G., Riegel, A. T., and Hager, G. L. (1987) Steroid-dependent interaction of transcription factors with the inducible promoter of mouse mammary tumor virus in vivo. *Cell* **48**, 261–270.

12. Cordingley, M. G. and Hager, G. L. (1988) Binding of multiple factors to the MMTV promoter in crude and fractionated nuclear extracts. *Nucleic Acids Res.* **16**, 609–628.

13. Nick, H. and Gilbert, W. (1985) Detection in vivo of protein–DNA interactions within the lac operon of Escherichia coli. *Nature* **313**, 795–798.

14. Mueller, P. R. and Wold, B. (1989) In vivo footprinting of a muscle specific enhancer by ligation mediated PCR. *Science* **246**, 780–786.

15. Strauss, E. C. and Orkin, S. H. (1992) In vivo protein–DNA interactions at hypersensitive site 3 of the human beta-globin locus control region. *Proc. Natl. Acad. Sci. USA.* **89**, 5809–5813.

16. Tanguay, R. L., Preifer, G. P., and Riggs, A. D. (1990) PCR-aided DNaseI footprinting of single copy gene sequences in permeabilized cells. *Nucleic Acids Res.* **18**, 5902.

17. Bresnick, E. H., Bustin, M., Marsaud, V., et al. (1992) The transcriptionally-active MMTV promoter is depleted of histone H1. *Nucleic Acids Res.* **20**, 273–278.

18. Stefanovsky, V., Dimitrov, S. I., Angelov, D., and Pashev, I. G. (1989) Interactions of acetylated histones with DNA as revealed by UV laser induced histone-DNA crosslinking. *Biochem. Biophys. Res. Commun.* **164**, 304–310.

19. Nacheva, G. A., Guschin, D. Y., Preobrazhenskaya, O. V., et al. (1989) Change in the pattern of histone binding to DNA upon transcriptional activation. *Cell* **58**, 27–36.

20. Postnikov, Y. V., Shick, V. V., Belyavsky, A. V., et al. (1991) Distribution of high mobility group proteins 1/2, E and 14/17 and linker histones H1 and H5 on transcribed and non-transcribed regions of chicken erythrocyte chromatin. *Nucleic Acids Res.* **19**, 717–725.

21. Solomon, M. J., Larsen, P. L., and Varshavsky, A. (1988). Mapping protein–DNA interactions in vivo with formaldehyde: evidence that histone H4 is retained on a highly transcribed gene. *Cell* **53**, 937–947.

22. Dedon, P. C., Soults, J. A., Allis, C. D., and Gorovsky, M. A. (1991) Formaldehyde cross-linking and immunoprecipitation demonstrate developmental changes in H1 association with transcriptionally active genes. *Mol. Cell Biol.* **11**, 1729–1733.

23. Johnson, K. D. and Bresnick, E. H. (2002) Dissecting long-range transcriptional mechanisms by chromatin immunoprecipitation. *Methods* **26,** 27–36.
24. Forsberg, E. C., Downs, K. M., and Bresnick, E. H. (2000) Direct interaction of NF-E2 with hypersensitive site 2 of the beta-globin locus control region in living cells. *Blood* **96,** 334–339.
25. Forsberg, E. C., Downs, K. M., Christensen, H. M., et al. (2000) Developmentally dynamic histone acetylation pattern of a tissue-specific chromatin domain. *Proc. Natl. Acad. Sci. USA.* **97,** 14,494–14,499.
26. Johnson, K. D., Grass, J. D., Boyer, M. E., et al. (2002) Cooperative activities of hematopoietic regulators recruit RNA polymerase II to a tissue-specific chromatin domain. *Proc. Natl. Acad. Sci. USA* **99,** 11,760–11,765.
27. Kiekhaefer, C. M., Grass, J. A., Johnson, K. D., et al. (2002) Hematopoietic activators establish an overlapping pattern of histone acetylation and methylation within a tissue-specific chromatin domain. *Proc. Natl. Acad. Sci. USA* **99,** 14,309–14,314.
28. Grass, J. A., Boyer, M. E., Paul, S., et al. (2003) GATA-1-dependent transcriptional repression of GATA-2 via disruption of positive autoregulation and domain-wide chromatin remodeling. *Proc. Natl. Acad. Sci. USA* **100,** 8811–8816.
29. Merika, M. and Orkin, S. H. (1993) DNA-binding specificity of GATA family transcription factors. *Mol. Cell Biol.* **13,** 3999–4010.
30. Ko, L. J. and Engel, J. D. (1993) DNA-binding specificities of the GATA transcription factor family. *Mol. Cell Biol.* **13,** 4011–4022.
31. Grass, J. A., Boyer, M. E., Paul, S., et al. et al. (2003) GATA-1-dependent transcriptional repression of GATA-2 via disruption of positive autoregulation and domain-wide chromatin remodeling. *Proc. Natl. Acad. Sci. USA* **100,** 8811–8816.
32. Lee, T. I., Rinaldi, N. J., Robert, F., et al. (2002) Transcriptional regulatory networks in Saccharomyces cerevisiae. *Science* **298,** 799–804.
33. Weinmann, A. S., Yan, P. S., Oberley, M. J., et al. (2002) Isolating human transcription factor targets by coupling chromatin immunoprecipitation and CpG island microarray analysis. *Genes Dev.* **16,** 235–244.

10

Characterization of Protein–DNA Association In Vivo by Chromatin Immunoprecipitation

Laurent Kuras

Summary

Chromatin immunoprecipitation (ChIP) is one of the most powerful methods to identify and characterize the association of proteins with specific genomic regions in the context of intact cells. In this method, cells are first treated with formaldehyde to crosslink protein–protein and protein–DNA complexes *in situ*. Next, the crosslinked chromatin is sheared by sonication to generate small chromatin fragments, and the fragments associated with the protein of interest are immunoprecipitated using antibodies to the protein. Finally, protein–DNA crosslinks are reversed and the DNA is examined for the presence of particular sequences by quantitative polymerase chain reaction (PCR). Enrichment of specific sequences in the precipitate indicates that the sequences are associated with the protein of interest in vivo. The ChIP method described here is intended for studying protein–DNA association in the budding yeast *Saccharomyces cerevisiae,* but it can be easily implemented in other cell types, including fly, mammalian, and plant cells.

Key Words: Chromatin; crosslinking reagents; DNA; DNA-binding proteins; formaldehyde; immunoprecipitation; polymerase chain reaction; protein–DNA interactions; yeast.

1. Introduction

Chromatin immunoprecipitation (ChIP) is a highly powerful method to identify and characterize protein–DNA interactions in vivo within a natural chromatin environment. The basic ChIP method is remarkably versatile and has been used in a wide range of cell types, including yeast, fly, mammalian and plant cells (for examples, *see* **refs. *1–20***). The ChIP method uses formaldehyde

From: *Methods in Molecular Biology, vol. 284:*
Signal Transduction Protocols
Edited by: R. C. Dickson © Humana Press Inc., Totowa, NJ

(molecular formula: HCHO) as a crosslinking agent to "freeze" protein–DNA interactions directly *in situ*, within intact cells, and to prevent dissociation and redistribution of proteins during chromatin extraction and immunoprecipitation *(9,21)*. Formaldehyde is a membrane soluble and dipolar compound that reacts with amino and imino group of amino acids (side-chain of lysines, arginines, and histidines) and of nucleic acids (e.g., adenines and cytosines). Formaldehyde produces both protein–protein, protein–DNA and protein–RNA crosslinks. Therefore, it can be used to study proteins that are not directly crosslinkable to DNA but that interact with proteins that are themselves directly crosslinkable to DNA *(22)*. Another key advantage of using formaldehyde is that the crosslinks are fully reversible *(9)*, which simplifies subsequent characterization of the interacting molecules. Following fixation with formaldehyde, cells are lysed and the crosslinked chromatin is isolated and sonicated to produce short DNA fragments (400 bp average). Then the DNA fragments associated with the protein of interest are purified by selective immunoprecipitation with an antibody specific for the protein (either antibodies to the native protein or antibodies to the tag in case the protein was tagged). Finally, following reversal of crosslinks and purification of DNA, the pool of immunoprecipitated DNA fragments is examined for the presence of specific regions by quantitative polymerase chain reaction (PCR).

2. Materials

2.1. Cell Growth and Formaldehyde Crosslinking

1. Liquid media for yeast *(23)*.
2. Formaldehyde, 37% aqueous solution.
3. 2.5 *M* glycine, autoclaved and stored at room temperature.
4. 20 m*M* Tris-HCl, pH 8.0, ice cold.
5. FA lysis buffer: 50 m*M* HEPES-KOH, pH 7.5, 150 m*M* NaCl, 1 m*M* EDTA, 1% (v/v) Triton X-100, 0.1% (w/v) deoxycholic acid sodium salt, 0.1% (w/v) sodium dodecyl sulfate SDS, filter-sterilized through 0.22 μm filter and stored at 4°C (*see* **Note 1**).
6. Falcon 14-mL polypropylene round-bottom tubes (Becton Dickinson).
7. 100 m*M* phenylmethylsulfonyl fluoride (PMSF) in isopropanol, stored up to 9 mo at room temperature. PMSF is a strong neurotoxin and should be used with extreme caution.

2.2. Preparation of Crosslinked Chromatin Extracts

1. Ice-cold FA lysis buffer.
2. 100 m*M* PMSF.

3. Protease inhibitor cocktail tablets (Roche).
4. 0.5-mm Zirconia/silica beads (BioSpec Products).
5. Vortexer (e.g., Vortex-Genie 2, Scientific Industries).
6. Kimwipes.
7. Black (22 G) hypodermic needle.
8. 50-mL NALGENE Oak Ridge centrifuge tubes (Nalge Nunc International).
9. 15-mL COREX centrifuge tubes.
10. 2-mL microcentrifuge tube (Eppendorf).
11. Sonication device fitted with a microtip (e.g., Bronson sonifier 250, Branson Ultrasonices).
12. 1.5 mL Safe-Lock tubes (Eppendorf).
13. 5X Elution buffer: 125 mM Tris-HCl, pH 7.5, 25 mM EDTA, 2.5% (w/v) SDS. Filter through 0.22-µm filter and store at room temperature.
14. 20 mg/mL Pronase in water (Roche).
15. TE buffer: 10 mM Tris-HCl pH 8.0, 1 mM EDTA. Autoclave and store at room temperature.
16. Agarose and DNA gel electrophoresis equipment.

2.3. Immunoprecipitation

1. 1.5-mL Costar low-binding microcentrifuge tubes (Corning Life Sciences).
2. Protein A sepharose CL-4B gel (Amersham Biosciences), 25% slurry in PBS prepared from dry powder as recommended by the manufacturer. Store at 4°C.
3. PBS: 20 mM Na phosphate, pH 7.0, 150 mM NaCl, 10 mM EDTA. Filter through 0.45-µm filter, autoclave and store at room temperature.
4. 20 mg/mL Acetylated bovine serum albumin (Ac-BSA) (Sigma).
5. 5 mg/mL Sonicated salmon sperm DNA (400 bp average size) in TE.
6. Rotating wheel.
7. Drawn-out Pasteur pipets.
8. FA-lysis/0.5 M NaCl buffer: 50 mM HEPES-KOH pH 7.5, 500 mM NaCl, 1 mM EDTA, 1% (v/v) Triton X-100, 0.1% (w/v) deoxycholic acid sodium salt, 0.1% (w/v) SDS. Filter-sterilize through 0.22-µm filter. Store at room temperature.
9. LiCl/detergent buffer: 10 mM Tris-HCl, pH 8.0, 0.25 M LiCl, 1 mM EDTA, 0.5% (v/v) Nonidet P40, 0.5% (w/v) deoxycholic acid sodium salt. Filter-sterilize through 0.22-µm filter and store at room temperature.
10. Elution buffer: 25 mM Tris-HCl, pH 7.5, 5 mM EDTA, 0.5% (w/v) SDS. Filter-sterilize through 0.22-µm filter and store at room temperature.
11. 65°C Water bath or heat block.
12. 1.5 mL Safe-Lock tubes (Eppendorf).

2.4. Reversal of the Crosslinks and Purification of DNA

1. 5X Elution buffer.
2. 65°C Air incubator or oven.
3. 10 mg/mL DNase-free RNase (Roche).
4. 25:24:1 (v/v/v) phenol/chloroforme/isoamyl alcohol.
5. Chloroform.
6. 20 mg/mL Glycogen, molecular biology-grade (Roche).
7. 4 M LiCl, filter through 0.45-μm filter and autoclave.
8. 100% Ethanol.

2.5. Analysis of Immunoprecipitated DNA by Quantitative PCR

1. Oligonucleotides: 24- to 26-mers with approx 40% GC content and similar melting temperature (around 60°C), designed to produce 150–300 bp fragments. Prepare 20 μM working dilutions in water and store at −20°C.
2. HotStarTaq DNA polymerase with 10X reaction buffer (Qiagen).
3. High-purity dNTP mix containing 2 mM dATP, dCTP, dGTP, and dTTP each (Amersham Biosciences).
4. 10 mCi/mL [α-^{32}P]dATP (specific activity: 3000 Ci/mmol).
5. Thin-walled 0.2-mL PCR tubes, DNase- and RNase-free.
6. Aerosol filter pipet tips (optional).
7. Thermal cycler with a heated cover.
8. Gel loading buffer: 50% glycerol (w/v), 10 mM EDTA, 0.1% (w/v) bromophenol blue, 0.1% (w/v) xylene cyanol. Store at 4°C.
9. Apparatus and accessories for polyacrylamide gel electrophoresis.
10. 40% Acrylamide-bisacrylamide solution, 37.5:1.
11. 10% (w/v) Ammonium persulfate in water, store up to 2 wk at 4°C.
12. TEMED (N,N,N',N'-tetramethylethylenediamine).
13. 10X TBE (0.89 M Tris-HCl base, 0.89 M boric acid, 0.025 M EDTA, pH 8.0).
14. Whatman 3MM chromatography paper.
15. Plastic wrap.
16. Gel-drying system.
17. Autoradiography films, intensifying screen, and cassette.
18. Phosphorimaging equipment, e.g., PhosphorImager (Molecular Dynamics) or Storm System (Amersham Biosciences).

3. Method

The method described here is based on that developed by Strahl-Bolsinger et al. *(5)* and focuses more specifically on the budding yeast *Saccharomyces cerevisiae* (*see* **Note 2**). This section outlines: (1) the growth and fixation with formaldehyde of yeast cells, (2) the extraction of the crosslinked chromatin, (3) the immunoprecipitation of the crosslinked chromatin, (4) the reversal of the crosslinks and purification of DNA, and (5) the analysis of DNA by quantitative PCR.

3.1. Cell Growth and Formaldehyde Fixation

3.1.1. Cell Growth

1. Inoculate 5 mL of appropriate liquid growth medium with a single colony picked from a fresh plate and allow to grow overnight at 30°C.
2. Dilute an aliquot into 100 mL of fresh liquid-growth medium in a 500-mL flask so that the culture will have reached an optical density (OD) at 600 nm of 1–2 (ca. 2–3 10^7 cells/ mL) the next morning (*see* **Note 3**).

3.1.2. Formaldehyde Fixation

1. When the culture has reached the desired density, add 2.8 mL of 37% formaldehyde (*see* **Note 4**) directly to the medium (final concentration = 1%). After adding formaldehyde, mix rapidly and leave the suspension at room temperature for 15 min with occasional shaking (every 5 min) (*see* **Note 5**).
2. Add 20 mL of 2.5 M glycine (final concentration = 0.4 M) to stop the crosslinking reaction, mix, and incubate for 5 min at room temperature.
3. Transfer the suspension into a 250-mL centrifuge bottle and pellet cells by centrifugation for 8 min at 10,000 g at 4°C.
4. Discard the supernatant (*see* **Note 6**) and resuspend cells in 250 mL of cold 20 mM Tris-HCl, pH 8.0.
5. Centrifuge for 8 min at 10,000 g at 4°C to pellet cells and discard the supernatant.
6. Resuspend cells in 5 mL of cold FA-lysis buffer and add 50 µL of 100 mM PMSF (final concentration = 1 mM) (*see* **Note 7**). Transfer the suspension into a 14-mL round-bottom polypropylene tube and pellet cells by centrifugation for 5 min at 2500 g at 4°C.
7. Discard the supernatant by aspiration with a Pasteur pipet connected to a vacuum aspirator and keep on ice. At this point cells can either be directly extracted or frozen and stored in a −80°C freezer for several weeks.

3.2. Extraction of the Crosslinked Chromatin

Tubes should be kept on ice between the successive manipulations during extraction of crosslinked chromatin.

1. If cells were frozen, thaw on ice first.
2. Resuspend cells in 1 mL of ice-cold FA-lysis buffer containing 1 mM PMSF and 1X protease inhibitor cocktail.

3. Add 1.5 mL of 0.5 mm zirconia/silica beads (*see* **Note 8**).
4. To lyse cells, vortex vigorously 12 times for 1 min and place on wet ice for 1 min in between each time (*see* **Note 9**).
5. Add 4 mL of FA-lysis buffer containing 1 m*M* PMSF and 1X protease inhibitor cocktail.
6. To collect the suspension without the beads, wipe off ice and water from the outside of the tube with a Kimwipe, invert the tube, puncture the bottom with a hot black (22 G) needle and insert the tube in a 50-mL Oak Ridge centrifuge tube. Centrifuge for 2 min at 1000 *g* at 4°C to bring the suspension down. Repeat centrifugation if part of the suspension has not been transferred to the 50 mL tube.
7. Transfer the whole extract in a 15 mL COREX tube and centrifuge for 20 min at 2000 *g* at 4°C to pellet the crosslinked chromatin (*see* **Note 10**).
8. Discard the supernatant and add 1.6 mL of cold FA-lysis buffer containing 1 m*M* PMSF and 1X protease inhibitor cocktail to wash the pellet. Break up the pellet by gently pipeting up and down with a 1-mL micropipetor.
9. Transfer the whole suspension into a 2-mL tube taking care that no material is left behind in the COREX tube. Spin at maximum speed in a microcentrifuge for 20 min at 4°C.
10. Discard the supernatant and add 1.6 mL of cold FA-lysis buffer containing 1 m*M* PMSF and 1X protease inhibitor cocktail. Break up the pellet as in **step 8** and place the tube on a holder in an ice-water bath.
11. Sonicate the suspension to yield DNA fragments in a size range between 100 and 1000 bp with an average of 400 bp. Use a sonication device fitted with a microtip. Place the microtip at the 0.5 mL graduation mark on the 2 mL tube and hold the tube firmly to prevent movement during sonication. Sonicate for short cycles (20 maximum) at the maximum microtip power setting. Place the tube in ice-water at least 5 min in between two sonication cycles to avoid excessive heating in the sample. If several samples are processed in turns, wash the microtip thoroughly with water and ethanol in between samples to avoid cross-contamination. Every sonication device is different and the number of cycles required with a particular device to achieve the desired DNA fragment size should be determined in a pilot experiment. Four to six cycles are generally sufficient to get DNA fragments with an average size of 400 bp (*see* **Note 11**).
12. Transfer the suspension to a 15-mL COREX tube and add 4 mL of cold FA-lysis buffer containing 1 m*M* PMSF and 1X protease inhibitor cocktail. Let stand on ice for 30–60 min.
13. Centrifuge for 20 min at 12,000 *g* at 4°C to remove cell debris and insoluble components.

14. Transfer the supernatant (which contains the fragmented crosslinked chromatin) into a fresh 14-mL round-bottom polypropylene tube and discard the pellet. Save 100 µL to check the size of the DNA fragments and divide the rest into 0.5 mL aliquots. Freeze the aliquots in liquid nitrogen and store at −80°C (*see* **Note 12**)

15. To check the size of the DNA fragments, mix 100 µL of extract with 25 µL of 5X elution buffer in a 1.5-mL Safe-Lock tube, add 6 µL of 20 mg/mL Pronase, incubate 1 h at 37°C, and proceed as described in **Subheading 3.4., Steps 2–8** to reverse crosslinks and purify DNA. Resuspend purified DNA in 50 µL of TE and analyze 20 µL by electrophoresis on a 1.5% agarose gel and staining with ethidium bromide (*see* **Note 13**).

3.3. Immunoprecipitation

A control immunoprecipitation with either pre-immune serum, immune serum depleted of the antibodies, peptide-block antibodies, or no antibody at all should be performed to determine the specificity of the immunoprecipitation (*see* **Note 14**).

3.3.1. Coupling of Antibodies to Protein A Beads

1. In a 1.5-mL low-binding microcentrifuge tube, combine the appropriate amount of antibody (amount sufficient to deplete at least 90% of the antigen from 500 µL of chromatin extract, *see* **Note 15**) with 60 µL of 25% Protein A sepharose bead slurry (*see* **Note 16**), 25 µL of 20 mg/mL Acetylated-Bovine Serum Albumin (Ac-BSA), and 4 µL of 5 mg/mL sonicated salmon sperm DNA. Bring the volume to 1 mL with ice-cold PBS.
2. Incubate for 4 h to overnight at 4°C on a rotating wheel.
3. Microcentrifuge 5 s at maximum speed at room temperature.
4. Aspirate the supernatant (which contains unbound antibodies) using a drawn out Pasteur pipet connected to a vacuum aspirator.
5. Add 1 mL of ice-cold PBS and resuspend the beads by inverting the tube.
6. Repeat **steps 3** and **4** to collect beads.
7. Add 25 µL of 20 mg/mL Ac-BSA, 4 µL of 5 mg/mL sonicated salmon sperm DNA, and mix by gently flicking the bottom of the tube. Keep on ice.

3.3.2. Immunoprecipitation

1. If chromatin extracts were frozen, thaw on ice first and spin in a microcentrifuge for 20 min at 16,000 g at 4°C to clarify the extract from protein aggregates that may have formed during freezing and thawing.
2. Add 500 µL of chromatin extract to the antibody-conjugated Protein A beads prepared before.
3. Incubate for 1–2 h at room temperature on a rotating wheel (*see* **Note 17**).

4. Microcentrifuge 1 min at 1000 *g* at room temperature to collect the beads and remove the supernatant by aspiration with a drawn out Pasteur pipet connected to a vacuum aspirator.
5. Add 1.4 mL of FA-lysis/0.5 *M* NaCl buffer and incubate for 5 min at room temperature on a rotating wheel (*see* **Note 18**).
6. Microcentrifuge 1 min at 1000*g* at room temperature and remove the supernatant as in **step 4** (*see* **Note 19**).
7. Repeat **steps 5** and **6** twice.
8. Repeat **steps 5** and **6** with 1.4 mL of LiCl/detergent buffer.
9. Repeat **steps 5** and **6** with 1.4 mL of TE buffer.
10. Add 125 µL of elution buffer, mix by vortexing, and incubate 20 min in a 65°C water bath or heat block to elute the immunoprecipitate from protein A beads.
11. Microcentrifuge 1 min at maximum speed at room temperature and transfer eluate to a 1.5 mL Safe-Lock tube.

3.4. Reversal of the Crosslinks and Purification of DNA

It is important to treat at the same time as the immunoprecipitate an aliquot of the chromatin extract (routinely 100 µL, i.e., 20% of the volume used for the immunoprecipitation, adjusted to 25 m*M* Tris-HCl, pH 7.5, 5 m*M* EDTA, 0.5% (w/v) SDS with 25 µL of 5X elution buffer). This will serve as the total (or input) control for later analysis.

1. Add 6 µL of 20 mg/mL Pronase to the immunoprecipitate and the total chromatin aliquot, and incubate 1 h at 37°C to digest proteins.
2. Place samples in a air incubator or an oven at 70°C and incubate for at least 6 h up to overnight to reverse protein-DNA crosslinks (*see* **Note 20**).
3. Add 5 µg of DNase-free RNase and incubate 30 min at 37°C.
4. Extract samples once with an equal volume of 25:24:1 phenol/chloroforme/ isoamyl alcohol and once with an equal volume of chloroform.
5. Add 20 µg of glycogen and 1/10 volume of 4 *M* LiCl, and vortex. Add 2 vol of 100% ethanol and vortex again. Incubate at least 2 h up to overnight at −20°C to precipitate DNA (*see* **Note 21**).
6. Microcentrifuge at maximum speed for 20 min at 4°C.
7. Remove the supernatant, wash the pellet with 1 mL of 100% ethanol, and microcentrifuge at maximum speed for 5 min at 4°C (*see* **Note 22**).
8. Remove the supernatant and air dry the pellet for 10–15 min at room temperature.
9. Resuspend immunoprecipitate and total DNA in 200 µL of TE buffer (*see* **Note 23**).

3.5. Analysis of DNA by Quantitative PCR

Our method of choice to detect the presence of specific genomic fragments in the immunoprecipitate DNA is PCR (*see* **Notes 24** and **25**). The relative enrichment of a particular fragment in the immunoprecipitate DNA is calculated by dividing the amount of PCR product obtained from the immunoprecipitate by that obtained from the total input DNA. For accurate quantification, it is essential to make sure that calculation is based on PCR signals that are in the linear range of the amplification reaction. Indeed, the amount of product generated during amplification is proportional to the amount of template only during the logarithmic-linear phase of the reaction. Therefore, it is essential to process several dilutions of the immunoprecipitate and input DNA (two- or threefold serial dilutions in TE buffer) and check that the amount of PCR product decreases proportionally with the amount of template.

It is equally important to perform appropriate PCR controls (i.e., with oligonucleotide primers specific for genomic regions that are not expected to associate, and therefore coimmunoprecipitate, with the protein of interest). These controls will serve to determine the background level of the whole ChIP experiment and to appreciate the significance of the enrichment obtained for the other regions.

3.5.1. PCR Analysis

PCR amplifications are carried out in 15 µL reaction volumes containing 1 µM of each oligonucleotide primer, 0.2 mM of each dNTP, 0.1 mCi/mL of [α-^{32}P]dATP (specific activity, 3000 Ci/mmol) and 0.75 U of HotStarTaq DNA polymerase (Qiagen) or equivalent Taq polymerase (*see* **Note 26**). To micropipet reagents and DNA samples, use aerosol filter tips to prevent PCR contaminations.

1. Prepare a common PCR premix containing the appropriate amount of H$_2$O, 10X PCR buffer, primers, dNTPs, [α-^{32}P]dATP, and Taq polymerase (*see* **Note 27**). The minimal number of samples includes two or three different dilutions (two-or threefold serial dilutions) of the immunoprecipitate and the input DNA. Adjust the final premix volume assuming that 5 µL of DNA template will be added next.
2. Distribute 10 µL of the premix into 0.2-mL PCR tubes at room temperature.
3. Add the DNA template. The amount of template added is typically 5 µL of 1:1 (undiluted), 1:2, and 1:4 dilutions of the immunoprecipitate DNA (which corresponds to 1/40, 1/80, and 1/160 of the immunoprecipitate assuming a sample volume of 200 µL) and 5 µL of 1:25, 1:50, and 1:100 dilutions of the input DNA sample (which corresponds to 1/5000, 1/10,000, and 1/20,000 of the total input DNA assuming a sample volume of 200 µL and reversal of 100 µL of chromatin extract).

4. Place the PCR tubes in a thermal cycler with a heated cover and start amplification using a cycling program consisting of a 2-min initial denaturation at 95°C (15 min if Qiagen HotStarTaq DNA polymerase is used) followed by 25 cycles with 30 in. at 94°C (denaturation), 30 in. at 55°C (annealing), and 60 in. at 72°C (elongation), and by a final extention step at 72°C for 5 min.

3.5.2. Gel Electrophoresis and Quantitation

PCR products are separated by vertical electrophoresis on nondenaturing polyacrylamide gels and visualized by film autoradiography. Intensity of radioactive PCR signals is quantitated by phosphorimager analysis.

1. After PCR is completed, microcentrifuge tubes for a few seconds to bring condensation down.
2. Add 2 μL of loading buffer and mix by vortexing.
3. Load 4 μL on a 8% polyacrylamide/1X TBE gel (20cm × 20cm × 1mm, 25 wells) and run the gel at 10 volt/cm, taking care to avoid excessive heating.
4. Stop migration when the Bromophenol Blue dye has reached two-thirds of the plate length. Discard the 1X TBE buffer present in the lower reservoir of the electrophoresis tank in a radioactive waste (this buffer may contain free [α-^{32}P]dATP).
5. Detach the plates from the electrophoresis apparatus and carefully pry them apart so that the gel is still attached to one plate.
6. Transfer the gel onto a piece of Whatman 3MM paper, place on a preheated gel dryer connected to a vacuum pump, cover with plastic wrap, and dry for 20 min at 80°C.
7. Expose the dried gel to an autoradiography film or to a storage phosphor screen for visualization and quantification of PCR products (*see* **Notes 28** and **29**).

4. Notes

1. Solutions used for chromatin extraction and immunoprecipitation should be prepared from Molecular Biology grade reagents using distilled and deionized water. It is convenient to prepare solutions from individual concentrated stock solutions (e.g., prepare FA lysis buffer from 0.5 M HEPES-KOH, pH 7.5, 5 M NaCl, 0.5 M EDTA, pH 8.0, 20% [v/v] Triton X-100, 10% [w/v] deoxycholic acid sodium salt, and 10% [w/v] SDS solutions). Solutions must be filtered through 0.22- or 0.45-μm filter (respectively, for sterilization and/or removal of particles) and, when appropriate, autoclaved. Do not autoclave solutions with HEPES or detergent.

2. Implementation of the ChIP method to cell types other than yeast will require to adapt the steps of cell fixation and chromatin extraction to the particular characteristics of the cells used. For examples, *Drosophila* and human cells do not have a cell wall as yeast cells do and therefore can be lysed without vortexing in the presence of glass beads (*see* **refs. *10,12,16***). Other steps (immunoprecipitation, reversal, and DNA analysis) can be implemented without modifications.

3. Growth properties of strains depend on their genetic background. Important differences exist among strains. In addition, the growth rate of a strain depends on the composition of the growth medium. Therefore the size of the inoculate that will lead to the desired cell density the next morning should be determined in advance in pilot experiments.

4. Formaldehyde is toxic and should be manipulated with gloves and in a fume hood to prevent contact with skin and inhalation of noxious fumes. In addition, flasks should be kept covered (for example with aluminium foil) after addition of formaldehyde.

5. The optimal concentration of formaldehyde, incubation time, and temperature at which the reaction is performed should be determined experimentally. Crosslinking times reported previously range from 10 min at room temperature to overnight at 4°C. The use of 1% formaldehyde at room temperature for 15 min represents standard conditions that were shown to work well with a number of proteins. However, these conditions should not be considered to be suitable for all proteins. The extent of crosslinking is one of the most important parameters for the success of the ChIP experiment. Excessive crosslinking may result in poor cell breakage and poor chromatin fragmentation during extraction and sonication. Excessive crosslinking may also alter reactivity of the epitopes recognized by the antibodies, resulting in poor immunoprecipitation. On the other hand, suboptimal crosslinking may result in fewer crosslinks between the protein of interest and DNA, leading to weak enrichment of the genomic targets of the protein.

6. Formaldehyde-containing wastes must be disposed of as hazardous chemical waste. Do not discard them down the sink but keep them in appropriate bottles and consult your institution's safety office for the rules regarding storage and disposal.

7. PMSF is unstable in aqueous solution (half-life of 35 min at pH 8.0) and should be added to buffers immediately before use.

8. 0.5-mm acid-washed glass beads (Sigma) can be used instead of 0.5-mm zirconia/silica beads.

9. Cell-breakage efficiency should be greater than 90% to achieve maximum extraction. It is important to make sure that bead movement is not constricted during vortexing, which could otherwise result in poor and unequal

cell lysis among samples. Eight samples can be processed in no longer than 48 min if two vortexers are used at the same time. It may be convenient to use a multivortexer or a BeadBeater (BioSpec Products) if one wants to process more samples.

10. After centrifugation of the whole-cell lysate at 20,000 g, the crosslinked chromatin is found in the pellet with cell debris and potential unbroken cells. This characteristic of the crosslinked chromatin allows to separate it from soluble proteins that might interfere during immunoprecipitation.

11. Sonication is a crucial step because it promotes solubilization of the crosslinked chromatin and allows it to separate from cell debris and insoluble components. In addition, shorter DNA fragments allows for mapping DNA-protein interactions with higher resolution, and to define more accurately regions where proteins are recruited.

12. The amount of chromatin extract prepared from 100 mL of culture grown to an OD_{600} of 1–2 is sufficient to carry out up to 10 immunoprecipitations. Chromatin extracts can be stored at −80°C for several months without damage. However, it is recommended to use fresh extracts in experiments studying proteins never assayed before.

13. It is important to check the size of the DNA fragments to confirm that they are in the expected range.

14. An alternative control in experiments using cells expressing epitope-tagged proteins (e.g., HA- or Myc-tagged proteins) is to perform a parallel immunoprecipitation (including antibodies) with chromatin extract prepared from untagged, isogenic cells.

15. Differences in the amount of protein–DNA complexes between samples will be accurately measured only if almost all complexes are immunoprecipitate in each sample. It is recommended to perform preliminary experiments to determine the appropriate amount of antibodies to use. Chromatin extract should be subjected to immunoprecipitation with varying amounts of antibodies, and the supernatant after immunoprecipitation examined by immunoblotting for the presence of the antigen. It is important to remember that immunoprecipitation of proteins is less efficient in crosslinked extracts compared to noncrosslinked ones, probably owing to modification or masking of the epitopes, and it has been reported that epitopes of some proteins seem to become inaccessible when the protein gets crosslinked to the DNA (either directly or indirectly) (*see* **ref. 3**). This potential problem can be circumvented by using polyclonal antibodies (p*Abs*) rather than monclonal ones.

16. Substitute Protein G for Protein A if the antibody is from a species or subclass that does not bind to Protein A (*see* **ref. 24** for Protein A/G affinities for antibodies from various species and subclasses).

17. Prolonged incubations (e.g., overnight in the cold room) tend to result in increased background (i.e., nonspecific precipitation of chromatin fragments) and should be avoided.

18. It is not recommended to complete all the washes too quickly (i.e., without incubation time) because this may not allow enough time for proteins included within the antigen-antibody-protein A bead latticework to diffuse out of it. Instead, beads should be washed over approx 30 min (all washes included).

19. Remove supernatant as completely as possible during washes to achieve the lowest background as possible. The Pasteur pipet should be moved progressively toward the bottom of the tube as the wash buffer is withdrawn until the very surface of the beads. Use gentle vacuum to avoid removing beads.

20. It is preferable to use an air incubator or oven instead of a heat-block or water bath because it prevents concentration of the sample by condensation.

21. Glycogen serves as a carrier to maximize DNA precipitation and helps to visualize the pellet after centrifugation.

22. The DNA pellet from the immunoprecipate sample will be very small and nearly invisible. Therefore, the supernatant should be removed with extreme caution to avoid any loss of material.

23. Alternatively, the DNA can be purified using the QIAquick PCR purification kit (Qiagen). Follow the protocol described in the QIAquick spin handbook supplied with the kit. To elute DNA from the QIAquick column, add 200 µL of TE buffer, let the column stand for 10 min at room temperature, and microcentrifuge at maximum speed for 1 min.

24. Alternatively, coimmunoprecipitated DNA can be analyzed by slot blot or Southern hybridization (for examples, *see* **refs. *1,4,10***). For slot-blot analysis, immunoprecipitated and total DNA samples are immobilized on a nylon membrane by slot blot, and the membrane is hybridized with ^{32}P-labeled probes for specific genes. For Southern analysis, immunoprecipitated DNA is radiolabeled and used as a probe against genomic DNA fragments separated by agarose gel electrophoresis and immobilized onto a hybridization membrane. Both methods have been shown to give good results but we favor PCR analysis because of high sensitivity and specificity of the PCR method.

25. PCR, slot blot or Southern hybridization analyses can be carried out in any laboratory provided with standard equipment. Unfortunately, these analyses are restricted to studying interaction of proteins with a limited number of target fragments. A method of global analysis using DNA microarrays have recently been developed to monitor protein–DNA interactions across the entire yeast genome *(25–27)*. In these methods, immunoprecipitate and input DNA fragments are amplified and differentially labeled using

different fluorescent dyes, and then hybridized to microarray plates that contain the whole genome (or only intergenic regions in case of transcription factors that specifically target promoter region). This approach allows one to localize DNA-bound proteins at the whole genome level.

26. It is highly convenient to use a Taq DNA polymerase that is inactive until PCR cycles start. This permits set-up at ambient temperature and reduces formation of nonspecific products and primer-dimers.

27. To minimize exposure to radiation, PCR reactions should be set up behind an appropriate shield and in a place designated for radioactive work. PCR samples must be in a shielded container during storage and transportation.

28. Exposition for 1 h at −80°C should be long enough to detect a signal when using an intensifying screen. After exposition, the film can be scanned and signals quantitated using an imaging software if the laboratory is not equipped for phosphorimager autoradiography.

29. A faster and more convenient way to quantitate the immunoprecipitate DNA is using real-time PCR if one has the necessary instrument (e.g., LightCycler from Roche or ABI PRISM 7700 from Applied Biosystems).

Acknowledgments

I thank Nadia Benaroudj for helpful comments on the manuscript. L.K. is supported by the CNRS and an ACI grant from the French Ministry of Research.

References

1. Braunstein, M., Rose, A. B., Holmes, S. G., et al. (1993) Transcriptional silencing in yeast is associated with reduced nucleosome acetylation. *Genes Dev.* **7,** 592–604.

2. Kuo, M. H., Brownell, J. E., Sobel, R. E., et al. (1996) Transcription-linked acetylation by Gcn5p of histones H3 and H4 at specific lysines. *Nature* **383,** 269–272.

3. Aparicio, O. M., Weinstein, D. M., and Bell, S. P. (1997) Components and dynamics of DNA replication complexes in *S. cerevisiae*: redistribution of MCM proteins and Cdc45p during S phase. *Cell* **91,** 59–69.

4. Meluh, P. B. and Koshland, D. (1997) Budding yeast centromere composition and assembly as revealed by in vivo cross-linking. *Genes Dev.* **11,** 3401–3412.

5. Strahl-Bolsinger, S., Hecht, A., Luo, K., and Grunstein, M. (1997) SIR2 and SIR4 interactions differ in core and extended telomeric heterochromatin in yeast. *Genes Dev.* **11,** 83–93.

6. Tanaka, T., Knapp, D., and Nasmyth, K. (1997) Loading of an Mcm protein onto DNA replication origins is regulated by Cdc6p and CDKs. *Cell* **90,** 649–660.

7. Kuras, L. and Struhl, K. (1999) Binding of TBP to promoters in vivo is stimulated by activators and requires Pol II holoenzyme. *Nature* **399,** 609–613.

8. Kuras, L., Kosa, P., Mencia, M., and Struhl, K. (2000) TAF-Containing and TAF-independent forms of transcriptionally active TBP in vivo. *Science* **288,** 1244–1248.

9. Solomon, M. J. and Varshavsky, A. (1985) Formaldehyde-mediated DNA-protein crosslinking: a probe for in vivo chromatin structures. *Proc. Natl. Acad. Sci U S A* **82,** 6470–6474.

10. Orlando, V. and Paro, R. (1993) Mapping Polycomb-repressed domains in the bithorax complex using in vivo formaldehyde cross-linked chromatin. *Cell* **75,** 1187–1198.

11. Park, J. M., Werner, J., Kim, J. M., et al. (2001) Mediator, not holoenzyme, is directly recruited to the heat shock promoter by HSF upon heat shock. *Mol. Cell* **8,** 9–19.

12. Boyd, K. E., Wells, J. Gutman, J., et al. (1998) c-Myc target gene specificity is determined by a post-DNAbinding mechanism. *Proc. Natl. Acad. Sci. USA* **95,** 13,887–13,892.

13. Chen, H., Lin, R. J., Xie, W., et al. (1999) Regulation of hormone-induced histone hyperacetylation and gene activation via acetylation of an acetylase. *Cell* **98,** 675–686.

14. Parekh, B. S. and Maniatis, T. (1999) Virus infection leads to localized hyperacetylation of histones H3 and H4 at the IFN-beta promoter. *Mol. Cell* **3,** 125–129.

15. Agalioti, T., Lomvardas, S., Parekh, B., et al. (2000) Ordered recruitment of chromatin modifying and general transcription factors to the IFN-beta promoter. *Cell* **103,** 667–678.

16. Shang, Y., Hu, X., DiRenzo, J., et al. (2000) Cofactor dynamics and sufficiency in estrogen receptor-regulated transcription. *Cell* **103,** 843–852.

17. Frank, S. R. Schroeder, M., Fernandez, P., et al. (2001) Binding of c-Myc to chromatin mediates mitogen-induced acetylation of histone H4 and gene activation. *Genes Dev.* **15,** 2069–2082.

18. Christova, R. and Oelgeschlager, T. (2002) Association of human TFIID-promoter complexes with silenced mitotic chromatin in vivo. *Nat. Cell Biol.* **4,** 79–82.

19. Johnson, C., Boden, E., Desai, M., et al. (2001) In vivo target promoter-binding activities of a xenobiotic stress-activated TGA factor. *Plant J.* **28,** 237–243.

20. Wang, H., Tang, W., Zhu, C., and Perry, S. E. (2002) A chromatin immunoprecipitation (ChIP) approach to isolate genes regulated by AGL15, a MADS domain protein that preferentially accumulates in embryos. *Plant J.* **32,** 831–843.

21. Solomon, M. J., Larsen, P. L., and Varshavsky, A., et al. (1988) Mapping protein-DNA interactions in vivo with formaldehyde: evidence that histone H4 is retained on a highly transcribed gene. *Cell* **53,** 937–947.

22. Orlando, V., Strutt, H., and Paro, R. (1997) Analysis of chromatin structure by in vivo formaldehyde cross-linking. *Methods* **11,** 205–214.

23. Sherman, F. (2002) in Methods in Enzymology, "Guide to yeast genetics and molecular and all biology" (Guthrie, C. and Fink, J. R., eds.), **350,** pp. 3–41, Academic Press, Amsterdam.

24. Harlow, E. and Lane, D. (1988) *Antibodies: A Laboratory Manual.* Cold Spring Harbor Laboratory Press, Cold Spring Harbor, NY.

25. Ren, B., Robert, F., and Wyrick, J. J., et al. (2000) Genome-wide location and function of DNA binding proteins. *Science* **290,** 2306–2309.

26. Iyer, V. R., Horak, C. E., and Scafe, C. S., et al. (2001) Genomic binding sites of the yeast cell-cycle transcription factors SBF and MBF. *Nature* **409,** 533–538.

27. Lieb, J. D., Liu, X., Botstein, D., and Brown, P. O. (2001) Promoter-specific binding of Rap1 revealed by genome-wide maps of protein–DNA association. *Nat. Genet.* **28,** 327–334.

11

Nonradioactive Methods for Detecting Activation of *Ras*-Related Small G Proteins

Douglas A. Andres

Summary

Ras-related small GTPases serve as critical regulators for a wide range of cellular signaling pathways and are activated by the conversion of the GDP-bound state to the GTP-bound conformation. Until recently, measurement of the GTP-bound active form of *Ras*-related G proteins involved immunoprecipitation of ^{32}P-labeled protein followed by separation of the labeled GTP/GDP bound to GTPase. A new method based on the large affinity difference of the GTP- and GDP-bound form of *Ras* proteins for specific binding domains of effector proteins in vitro has been developed. By using glutathione *S*-transferase (GST) fusion proteins containing these binding domains, the GTP-bound form of the GTPase can be precipitated from cell lysates. In principle, this method can be used for all members of the *Ras* superfamily. Here we describe a general procedure to monitor the GTP-bound form of *Ras*-related GTPases.

Key Words: GTP-bound; *Ras*; GTPase; activation probe.

1. Introduction

Small *Ras*-like GTPases are activated by the conversion of the GDP-bound conformation into the GTP-bound conformation and inactivated by GTP hydrolysis. In their activated GTP-bound forms, the *Ras* proteins interact with a variety of intracellular effector proteins to initiate signaling pathways that contribute to cell proliferation, differentiation, or cell survival, depending on the cellular context and individual G protein *(1–4)*. Therefore the ability to measure

From: *Methods in Molecular Biology, vol. 284:*
Signal Transduction Protocols
Edited by: R. C. Dickson © Humana Press Inc., Totowa, NJ

the activation state of *Ras* family proteins provides a useful tool in the examination of a number of biological systems.

The level of *Ras*-related protein activation (the fraction of the total cellular protein bound to GTP) and the duration of this activated state are tightly controlled by the opposing actions of guanine-nucleotide exchange factors (GEFs), which stimulate GDP for GTP exchange, and GTPase-activating proteins (GAPs), which stimulate the intrinsic GTPase activity of *Ras*-related proteins. Both the level and duration of *Ras* family activation appear to contribute to both downstream signal strength and the resulting biological effects. Therefore, to fully understand the role that *Ras* GTPases play in regulating cellular signaling pathways, it is essential to be able to monitor the level, duration, and timing of cellular *Ras* family protein activation.

A variety of methods for studying the activation of *Ras*-like small G proteins have been described. The earliest of these methods directly measured the ratio of GTP to GDP bound to *Ras*-like GTPases extracted from cells *(5,6)*. These approaches rely on the use of radiolabeled guanine nucleotides and purified proteins to measure exchange activity in vitro or radiolabeling cells with orthophosphate and specifically immunoprecipitating the G protein of interest to measure the ratio of bound radiolabeled GTP to GDP by thin-layer chromatography (TLC). Although these assays have been useful in studies of *Ras* function, the approach suffers from several technical drawbacks that have limited its use for many members of the *Ras* superfamily. First, the success of this method is based in large part on the ability to find an effective immunoprecipitating antibody that also blocks GAP interaction and therefore acts to inhibit GTP hydrolysis on the G protein during the experimental procedure. Because antibodies for many *Ras*-related GTPases with appropriate immunprecipitating/inhibiting characteristics are not available, it is often necessary to introduce epitope-tagged G protein by transient overexpression or to analyze purified proteins to study activation. Second, these methods require large amounts of radioactivity to metabolically label cells. This can be costly, requires strict radiation safety procedures, and the radiolabeling procedure may itself influence cellular pathways and effect the experimental outcome.

Nonisotopic methods have recently been devised that enable the detection of the activation status of numerous *Ras*-like small G proteins in treated cell lysates (**Fig. 1**). The method is based on the large affinity difference of the GTP- vs GDP-bound form for specific binding domains of downstream effector proteins in vitro. By using glutathione *S*-transferase (GST) fusion proteins containing these binding domains, the GTP-bound form of the GTPase can be precipitated from cell lysates *(7–10)*. In principle, this method can be used for all small GTPases and has proved to function for the *Ras (11)*, Rap1 *(12,13)*, Rap2 *(14,15)*, R-*Ras (10)*, ARF *(16,17)*, Ral *(18,19)*, and Rin GTPases *(20)* (**Table 1**).

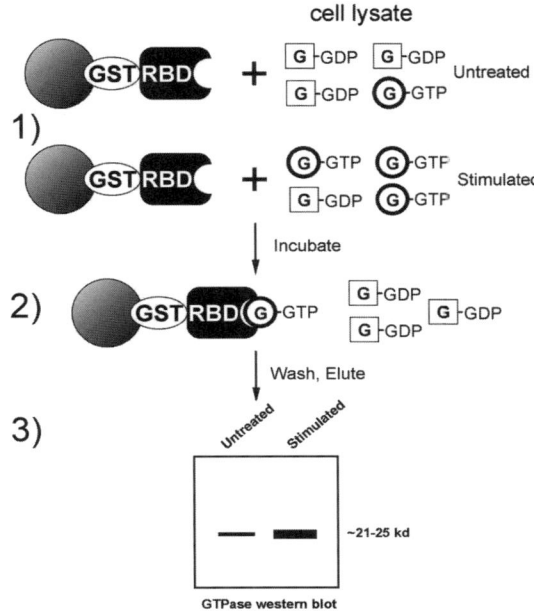

Fig. 1. Schematic of activation probe assay. Small *Ras*-related GTPases exist in an equilibrium between GTP-bound (active) and GDP-bound (inactive) states. For many members of the *Ras* family, effector proteins (*see* **Table 1**) have been isolated that bind with much higher affinity to the GTP-bound (active) state of the GTPase. Chimeric proteins combining glutathione *S*-transferase (GST) with the *Ras*-like GTPase-binding domain (RBD) can be used to detect activated GTPases with high affinity and can be used in vitro to select GTP-bound molecules. (1) Cells are treated with potential GTPase activators and lysed. (2) Lysates are mixed with bacterially purified GST-RBD precoupled to glutathione-coupled beads and incubated at 4°C to allow binding of activated GTP-bound GTPase. (3) SDS-PAGE separates protein within the GST-RBD complex and specific antibodies can be used in Western blotting to identify the relative amount of activated G protein. Because relatively little GDP-bound GTPase can bind GST-RBD, the level of GTPase recovered provides a useful index of G protein activation.

2. Materials

1. pGEX Expression system (Amersham Pharmacia, Piscataway, NJ).
2. *Escherichia coli* strain BL21.
3. Ampicillin.
4. IPTG (isopropyl-β-D-thio-galactopyranoside).
5. GST-RBD lysis buffer: 50 mM Tris-HCl, pH 7.5, 150 mM KCl, 10% (v/v) glycerol, 20% (w/v) sucrose, 5 mM MgCl$_2$, 1 mM EDTA, 2 mM dithiothreitol

Table 1 Detection of Ras-Related Small G Proteins Using Activation Probes

G protein	Recognition domain	References
Rin	N-terminus of Raf-1 (aa 1–140)[a]	*(20)*
R-*Ras*	N-terminus of Raf-1 (aa 1–147)	*(10)*
Rap1a, Rap1b	*Ras*-binding domain of RalGDS (aa 726–828)	*(12,13)*
H-*Ras*	N-terminus of Raf-1 (aa 1–147)	*(11)*
Rap2	*Ras*-binding domain of RalGDS (aa 726–828)	*(14,15)*
Ral	Ral-binding domain of RLIP76 (aa 397–518)	*(18,19)*
Rac1	p21-binding domain of PAK1 (aa 67–150)	*(23,24)*
Cdc42	p21-binding domain of PAK1 (aa 67–150)	*(23,24)*
RhoA	Rho-binding domain of Rhotekin (aa 7–89)	*(25)*
ARF1, ARF6	GAT domain of GGA3	*(16,17)*

[a]aa, amino acid.

    (DTT), leupeptin (10 µg/mL), aprotinin (10 µg/mL), 1 mM phenylmethyl-sulfonyl fluoride (PMSF), and lysozyme (1 mg/mL) at 4°C.

6. French press.
7. Sodium dodecyl sulfate-polyacrylamide gel electrophoresis (SDS-PAGE) equipment.
8. Glutathione-sepharose 4B resin (Amersham Pharmacia).
9. Washing buffer: 50 mM Tris-HCl, pH 7.5, 150 mM NaCl, 10 mM MgCl$_2$, 10% (v/v) glycerol, 1% Nonidet P-40, 2 mM DTT, 1 mM PMSF, aprotinin (1 µg/mL) at 4°C.
10. Cell lysis buffer: 50 mM Tris-HCl, pH 7.5, 150 mM NaCl, 20 mM MgCl$_2$, 10% (v/v) glycerol, 1% (v/v) Nonidet P-40, 2 mM DTT, 1 mM PMSF, aprotinin (1 µg/mL), and leupeptin (1 µg/mL) at 4°C.
11. Polyvinylidine difluoride (PVDF) membrane.
12. Ponceau S stain.
13. Chemiluminescence detection kit (New England Nuclear, Boston, MA).
14. Rubber policeman.
15. 2X SDS sample buffer: 250 mM Tris-HCl, pH 6.8, 2% SDS, 0.572 M 2-mercaptoethanol (ME), 10% glycerol, and 0.2% Bromophenol Blue.

3. Methods

3.1. In Vitro Assay of Small G Proteins: Precipitating Activated GTP-Bound GTPases With Activation-Specific Probes

    This method is an adaptation of the method first described for examining the activation of *Ras* and Rap1 *(11,13)*. Because this method relies on purified proteins that are not always commercially available, we will first discuss the bacterial purification of GST-fusion activity probes as well as their utilization in mammalian cell lysates for monitoring *Ras*-like GTPase activity.

3.1.1. Expression and Isolation of Activation Probes

A series of activation-specific probes have been identified and successfully used for a range of small *Ras*-related GTPases (7–10). These include, for *Ras*, Rin, and R-*Ras*, the *Ras*-binding domain (RBD) of c-Raf1; for Rap1 and Rap2, the RBD of RalGDS; and for Ral, the Ral-binding domain of RLIP76 (RalBP) (*see* **Table 1**). Each of the domains was cloned in pGEX vectors (Pharmacia) to obtain bacterial expression vectors that produce GST fusion proteins with the GST portion at the NH$_2$-terminus. Cloning and characterization of each of these probes have been previously described (*see* **Table 1**).

1. Transform *E. coli* BL21 with GST fusion expression constructs using standard molecular techniques. Incubate the bacteria in 50 mL of LB medium, containing ampicillin (50 µg/mL), overnight at 30°C with shaking.
2. Dilute the bacterial culture 1:50 in 1 l of LB medium containing ampicillin (50 µg/mL) and 0.4% (w/v) D-glucose and allow the bacteria to grow to an OD$_{600}$ of 0.6–0.7. Add IPTG to a final concentration of 0.8 mM and incubate the bacteria for an additional 4–6 h at 30°C. Centrifuge the bacteria at 7000g at 4°C for 15 min (*see* **Note 1**). The harvested cell pellet can be stored at −80°C for months before extraction without significant breakdown of the GST fusion protein.
3. Resuspend the pellet in 25 mL of GST-RBD lysis buffer, incubate on ice for 30 min, and pass three times through a French press to achieve bacterial lysis (*see* **Note 2**). Centrifuge the lysates for 1 h at 12,000 g at 4°C to remove bacterial debris. Collect the supernatant fluid and store aliquots at 80°C. The presence of GST fusion protein in the cleared lysate can be checked by SDS-PAGE. The bacterial lysates are stable for several months at −80°C (*see* **Note 3**). However, multiple freeze-thaw cycles should be avoided.
4. The fusion proteins are purified from the supernatant with glutathione-sepharose 4B beads (Amersham Pharmacia, Piscataway, NJ) previously washed five times in RBD lysis buffer without lysozyme (approx 200–400 µL of packed beads). Glutathione-sepharose beads are incubated with the supernatant for 2 h at 4°C with constant rotation, the beads centrifuged at 2000 rpm at 4°C, and washed five times with ice cold washing buffer. Beads can be aliquoted at this point in washing buffer and stored at −80°C (*see* **Note 4**). The integrity of the GST-fusion protein is checked by SDS-PAGE and should represent ≥80% of the total purified protein. Contaminating proteins or the presence of a minor fraction of GST-RBD breakdown

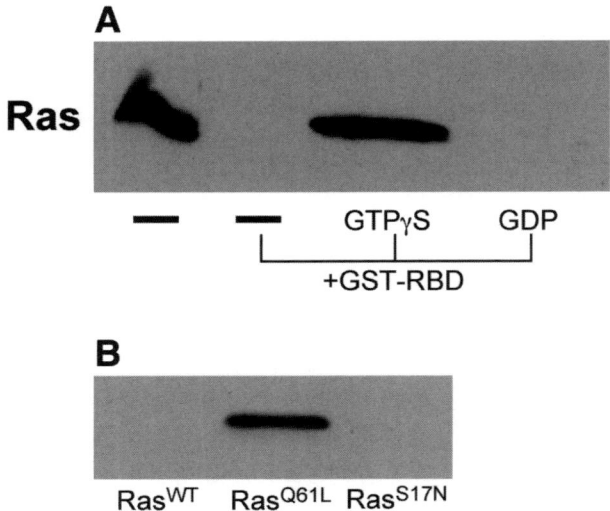

Fig. 2. Detection of in vitro loaded GTPases. (**A**) Precipitation with GST-RafBD, using recombinant H-*Ras* either EDTA treated (unloaded -), loaded with GDP, or loaded with GTPγS. The lane at the left represents the total H-*Ras* in the sample prior to GST-RafBD precipitation. (**B**) Precipitation with GST-RafBD, using HEK293 cell lysates expressing various HA-tagged H-Ras mutants and immunoblotted with anti-HA antibody to detect expressed H-*Ras*. H-*Ras*^{Q61L} is constitutively GTP bound, whereas H-*Ras*^{S17N} is found predominantly in the GDP-bound state.

products is typical in these preparations and should not interfere with the assay. The final protein concentration is determined and the beads can be used directly for the assay.

5. The utility of the GST-RBD activation probe should be verified after each GST-RBD purification by using recombinant GTPases loaded with either GDP as a negative control or with GTPγS as a positive control (**Fig. 2A**). Simple control experiments can also be performed by generating GDP-bound GTPase and GTP-bound GTPase in vitro from the cell lysates to be used for study. This is also an effective means to assess the total amount of activatable GTPase present in the sample. Either overexpressed recombinant GTPases or endogenous GTPases can be loaded with nucleotides in this manner. Briefly, the cell lysate should be incubated with an excess of guanine nucleotide ([100–200 μM] GTPγS, a nonhydrolyzable GTP analog, or 1 m*M* GDP), in a high concentration of EDTA (10–20 m*M*) and a low concentration of Mg^{2+} (<5 m*M*) at 30°C to promote nucleotide exchange and induce defined nucleotide binding (*21*). Subsequently increas-

ing the Mg^{2+} concentration (to approx 75 mM) and returning the treated lysate to 4°C will stop nucleotide exchange. Loaded GTPases should be used immediately in the affinity precipitation assay. Alternatively, GTPase mutants that have demonstrated effects on the activation status of the *Ras* protein can be used (**Fig. 2B**).

3.2. Preparation of Cell Lysate

Preparation of a cell extract represents a critical step in the assay and determination of an optimal lysis buffer must be established empirically for each individual GTPase before any affinity precipitation. (1) In general, cells are pretreated or stimulated to generate the active GTP-bound form of the GTPase. For cells in suspension, the stimulation is stopped by addition of an equal volume of cold 2X lysis buffer. For adherent cells, the stimulation medium is removed, the plates transferred to ice and washed twice with ice-cold ice PBS, cold 1X lysis buffer is added and allowed to sit for 5–10 min, and the cells are scraped from the plates while still on ice. (2) The resulting cell lysates are subjected to centrifugation at 4°C for 10 min at 10,000 rpm in a refrigerated microfuge. (3) Cell lysates can be used directly for the affinity-precipitation assay (*see* **Note 5**), or can be immediately frozen in liquid nitrogen and stored at −80°C until use. (4) For each affinity precipitation sample, an equal number of cells or the same amount of total cell lysate should be mixed with the GST-RBD resin. Activation can also be monitored for transiently overexpressed GTPases. This is particularly useful when examining the activation state of a GTPase for which specific antibodies are lacking *(20)*. In this case, it is important that care is taken to assure equal recombinant protein expression in each plate of transfected cells.

Additional issues must be considered when preparing cell lysates:

1. The number of cells required to obtain a readily detectable GTPase signal will depend on many factors, including the expression level of the GTPase under study and the level of activation of this G protein to the stimuli under investigation. If using a cell type not previously assayed for activation by this method, we suggest starting with one or two 100-mm plates, 70–80% confluent (*see* **Note 6**), with or without a known or expected activator of the GTPase, preferably alongside positive controls (as described in **Subheading 3.1.**) (**Fig. 3**).
2. Detergents are critical components of the lysis buffer. Each cell type has a different detergent requirement to obtain optimal protein solubilization without disruption of the nucleus. In a variety of neuronal cell lines, including PC12 cells and HEK293 cells we have developed a cell lysis buffer that appears to present a good starting point *(19,20)*. Detergents are also present in the binding buffer for the affinity precipitation to prevent nonspecific

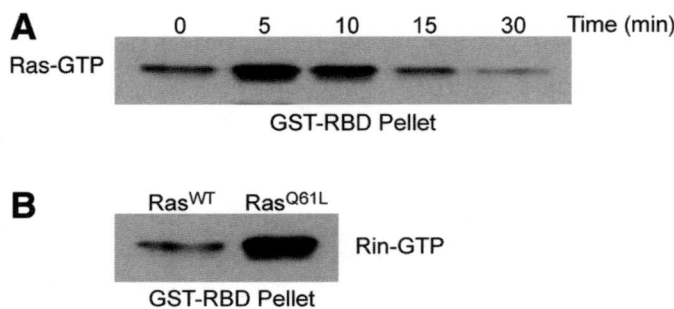

Fig. 3. Detection of GTPase activation. **(A)** PC6 cells were transiently transfected with an expression vector for HA-tagged H-*Ras*. Prior to the preparation of whole cell lysates, cells were serum-starved for 12 h and then stimulated with nerve growth factor (NGF) (100 ng/mL) for the indicated periods of time. GTP-bound H-*Ras* was precipitated with activation-specific probe (GST-RafBD) and analyzed by SDS-PAGE followed by Western blotting using anti-HA antibody. **(B)** In vitro activation of Rin by activated H-*Ras* (H-Ras^{Q61L}). PC6 cells were transiently co-transfected with a vector expressing HA-tagged Rin and either an expression vector for wild-type or H-*Ras*^{Q61L}. Cells were serum-starved for 12 h prior to recovery of GTP-bound Rin by GST-RBD precipitation analysis.

binding to GST-RBD. However, we and others have found that some detergents can also disrupt specific binding. Therefore, analysis of the effect of detergents and detergent concentration on specific vs nonspecific binding is recommended when establishing an activation probe assay.

3.3. Identification of GTP-Bound GTPases With Activation Probes

The GTP-bound GTPases are precipitated from cleared total cell lysate, using activation-specific GST-RBD fusion proteins that are precoupled to glutathione-Sepharose beads. Detection is achieved using gel electrophoresis and Western blotting using antibodies specific for the small GTPases (*see* **Fig. 1** for an overview).

1. Transfer the culture dishes (usually one to two 100-mm dishes per treatment) containing the cells for analysis to ice and wash two times with ice-cold PBS, and ice-cold 1X cell lysis buffer. Scrape the cells with a rubber policeman and transfer the cell lysate to a 1.5-mL microfuge tube. Centrifuge the lysate for 10 min at 14,000 rpm in a refrigerated microfuge at 4°C.
2. Add the cleared supernatant to the activation-specific probe (GST-RBD) precoupled to glutathione-sepharose beads (*see* **Subheading 3.1.**) and incubate for 45 min on a rotary tumbler at 4°C (*see* **Note 7**). The GST-RBD/bead complex is pelleted by centrifugation at 2000 rpm for 2 min at

4°C. The beads are washed four times with 1X lysis buffer (*see* **Note 8**). After the final wash remove the remaining liquid with an insulin syringe and resuspend the beads in 20 μL of Laemmli buffer. Heat the samples at 95°C for 5 min to elute the bound proteins.

3. Separate the proteins on a 12.5% (w/v) sodium dodecyl sulfate-polyacrylamide gel, and transfer to PVDF membrane. Stain the membrane with Ponceau S to check for equal transfer of GST-probe in all samples.

4. Probe the immunoblot with GTPase-specific primary antibody (directed against the GTPase itself, or toward an epitope tag if overexpression of recombinant G protein is being analyzed) and visualize the bound secondary antibody by enhanced chemiluminescence (New England Nuclear, Boston, MA) according to the manufacturer's protocol.

4. Notes

1. GST-activation probe proteins are stable under these production conditions, although slightly higher yields of protein can be obtained by growing cells at lower temperatures (20–25°C) for longer periods (6–12 h) following induction with IPTG.

2. Alternatively, lysis can be achieved by sonication (microtip, 10 times, 30 s each). Oversonication can adversely affect performance of the GST-fusion protein: if this appears to be a problem it may be worthwhile trying different output settings to determine the lowest necessary for efficient GST-RBD recovery.

3. For most purposes, 10% of the lysate from a 1 l bacterial preparation will provide more than enough fusion protein for an experiment. Alternatively, multiple unlysed bacterial pellets can be stored and processed as needed.

4. We have found that the stability of the GST-RBD beads varies between different preparations. There is some loss in G protein-binding capacity after storage for 1–2 d at 4°C. It is therefore recommended that new batches of beads be either stored at −80°C (and the efficiency of thawed beads be determined as described above) or that new beads be made on the day of the experiment. Because the GST-RBD beads can be prepared rapidly, we have found this to be convenient. However, if performing this assay infrequently, or with few samples per experiment, it is advisable to use fresh thawed beads or scale down the resin preparation to conserve both recombinant GST-RBD protein and glutathione beads.

5. It is important to work quickly and keep lysates on ice, because the GTP-bound form of *Ras*-like GTPases will be susceptible to GAP activity until incubation with GST-RBD.

6. It may be important to maintain cells at subconfluency because certain *Ras*-like GTPases may be regulated by cell density *(22)*.

7. A general guideline for each sample is to add approx 40 µg of purified GST-RBD protein precoupled to glutathione-sepharose beads to a 0.5–1 mL total reaction volume containing 0.5–1 mg of total cellular protein. An important variable to consider when designing experiments examining small G protein activation is that the time-course for activation can vary greatly depending on the *Ras*-like protein being examined, cell type, and the stimulus. Therefore it is important to examine a wide range of time-points (from 30 s to 1–2 h) following exposure to stimuli.
8. Nonspecific precipitation of GDP-bound GTPase from cell lysates is generally not a problem. However, if signal to background is low it may be advisable to perform a parallel affinity precipitation with GST alone or with an irrelevant GST fusion protein.

References

1. Macara, I. G., Lounsbury, K. M., Richards, S. A., et al. (1996) The *Ras* superfamily of GTPases. *FASEB J.* **10,** 625–630.
2. Campbell, S. L., Khosravi-Far, R., Rossman, K. L., et al. (1998) Increasing complexity of *Ras* signaling. *Oncogene* **17,** 1395–1413.
3. Hall, A. (1998) Rho GTPases and the actin cytoskeleton. *Science* **279,** 509–514.
4. Malumbres, M. and Pellicer, A. (1998) *RAS* pathways to cell cycle control and cell transformation. *Front Biosci.* **3,** d887–d912.
5. Satoh, T. and Kaziro, Y. (1995) Measurement of *Ras*-bound guanine nucleotide in stimulated hematopoietic cells. *Methods Enzymol.* **255,** 149–155.
6. Gibbs, J. B. (1995) Determination of guanine nucleotides bound to *Ras* in mammalian cells. *Methods Enzymol.* **255,** 118–125.
7. Carey, K. D. and Stork, P. J. (2002) Nonisotopic methods for detecting activation of small G proteins. *Methods Enzymol.* **345,** 383–397.
8. Benard, V. and Bokoch, G. M. (2002) Assay of Cdc42, Rac, and Rho GTPase activation by affinity methods. *Methods Enzymol.* **345,** 349–359.
9. Taylor, S. J., Resnick, R. J., and Shalloway, D. (2001) Nonradioactive determination of *Ras*-GTP levels using activated ras interaction assay. *Methods Enzymol.* **333,** 333–342.
10. van Triest, M., de Rooij, J., and Bos, J. L. (2001) Measurement of GTP-bound *Ras*-like GTPases by activation-specific probes. *Methods Enzymol.* **333,** 343–348.
11. Taylor, S. J. and Shalloway, D. (1996) Cell cycle-dependent activation of *Ras Curr. Biol.* **6,** 1621–1627.
12. Spaargaren, M. and Bischoff, J. R. (1994) Identification of the guanine nucleotide dissociation stimulator for Ral as a putative effector molecule of R-*ras*, H-*ras*, K-*ras*, and Rap. *Proc. Natl. Acad. Sci. USA* **91,** 12,609–12,613.
13. Franke, B., Akkerman, J. W., and Bos, J. L. (1997) Rapid Ca_{2+}-mediated activation of Rap1 in human platelets. *EMBO J.* **16,** 252–259.
14. Ohba, Y., Mochizuki, N., Matsuo, K., et al. (2000) Rap2 as a slowly responding molecular switch in the Rap1 signaling cascade. *Mol. Cell Biol.* **20,** 6074–6083.

15. Reedquist, K. A. and Bos, J. L. (1998) Costimulation through CD28 suppresses T cell receptor-dependent activation of the *Ras*-like small GTPase Rap1 in human T lymphocytes. *J. Biol. Chem.* **273,** 4944–4949.
16. Dell'Angelica, E. C., Puertollano, R., Mullins, C., et al. (2000) GGAs: A family of ADP ribosylation factor-binding proteins related to adaptors and associated with the Golgi complex. *J. Cell Biol.* **149,** 81–94.
17. Santy, L. C. and Casanova, J. E. (2001) Activation of ARF6 by ARNO stimulates epithelial cell migration through downstream activation of both Rac1 and phospholipase D. *J. Cell Biol.* **154,** 599–610.
18. Goi, T., Rusanescu, G., Urano, T., and Feig, L. A. (1999) Ral-specific guanine nucleotide exchange factor activity opposes other *Ras* effectors in PC12 cells by inhibiting neurite outgrowth. *Mol. Cell Biol.* **19,** 1731–1741.
19. Shao, H. and Andres, D. A. (2000) A novel RalGEF-like protein, RGL3, as a candidate effector for rit and *Ras* [In Process Citation]. *J. Biol. Chem.* **275,** 26,914–26,924.
20. Spencer, M. L., Shao, H., Tucker, H. M., and Andres, D. A. (2002) Nerve Growth Factor-dependent Activation of the Small GTPase Rin. *J. Biol. Chem.* **277,** 17,605–17,615.
21. Self, A. J. and Hall, A. (1995) Measurement of intrinsic nucleotide exchange and GTP hydrolysis rates. *Methods Enzymol.* **256,** 67–76.
22. Posern, G., Weber, C. K., Rapp, U. R., and Feller, S. M. (1998) Activity of Rap1 is regulated by bombesin, cell adhesion, and cell density in NIH3T3 fibroblasts. *J. Biol. Chem.* **273,** 24,297–24,300.
23. Benard, V., Bohl, B. P., and Bokoch, G. M. (1999) Characterization of rac and cdc42 activation in chemoattractant-stimulated human neutrophils using a novel assay for active GTPases. *J. Biol. Chem.* **274,** 13,198–13,204.
24. Bagrodia, S., Taylor, S. J., Jordon, K. A., et al. (1998) A novel regulator of p21-activated kinases. *J. Biol. Chem.* **273,** 23,633–23,636
25. Ren, X. D., Kiosses, W. B., and Schwartz, M. A. (1999) Regulation of the small GTP-binding protein Rho by cell adhesion and the cytoskeleton. *EMBO J.* **18,** 578–585.

12

Nucleocytoplasmic Glycosylation, O-GlcNAc

Identification and Site Mapping

Natasha Elizabeth Zachara, Win Den Cheung, and Gerald Warren Hart

Summary

β-*O*-linked *N*-acetylglucosamine (O-GlcNAc) is posttranslationally added to serine and threonine residues of many nuclear and cytoplasmic proteins found in metazoans. This modification is dynamic and responsive to numerous stimuli and conditions, suggesting an important role in many regulatory pathways. Moreover, the O-GlcNAc modification seems to compete with phosphorylation for sites of attachment, indicating a reciprocal relationship with phosphorylation. This chapter includes protocols for: (1) identifying the O-GlcNAc modification on proteins through immunoblotting, lectin affinity chromatography, and galactosyltransferase labeling; and (2) identifying and enriching for the sites of attachment using the mass spectrometry-based β-elimination followed by <u>M</u>ichael <u>a</u>ddition with <u>d</u>ithiothreitol (BEMAD) technique.

Key Words: β-*O*-linked *N*-acetylglucosamine; posttranslational modification; glycosylation; site-mapping; detection.

1. Introduction

Hundreds, if not thousands, of nuclear and cytoplasmic proteins in metazoans are modified by monosaccharides of β-O-linked N-acetylglucosamine, also known as O-GlcNAc. Notably, O-GlcNAc is added and removed to proteins in the cytoplasm and nucleus, on serine and threonine residues. O-GlcNAc levels respond to extracellular glucose concentrations, morphogens, and the cell cycle, suggesting that O-GlcNAc plays an important role in different signal-transduction pathways. Moreover, aberrations in the regulation of O-GlcNAc

From: *Methods in Molecular Biology, vol. 284:*
Signal Transduction Protocols
Edited by: R. C. Dickson © Humana Press Inc., Totowa, NJ

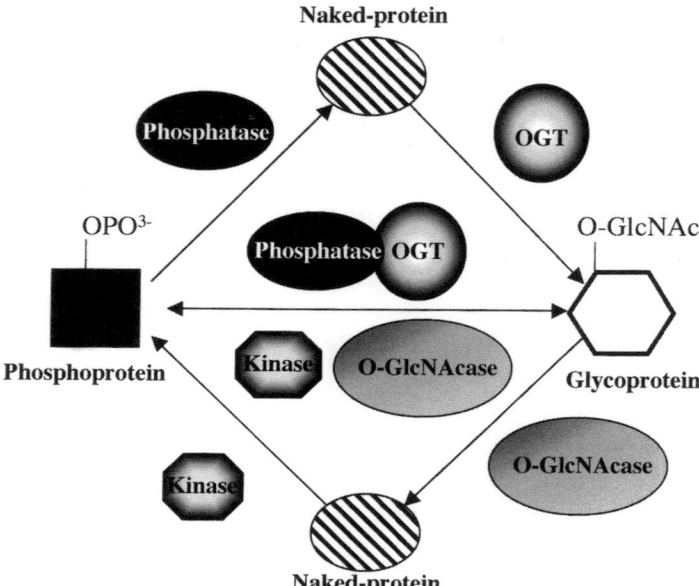

Fig. 1. A model demonstrating how O-GlcNAc and O-phosphate may provide the cell with different levels of regulation of proteins. Here, glycosylation is shown to block phosphorylation and vice versa, giving the cell at least three populations of any given protein.

have been implicated in the etiology of cancer, insulin resistance, and several neurodegenerative diseases *(1,2)*.

Interestingly, in many cases, the residues modified by O-GlcNAc are known phosphorylation sites or are adjacent to phosphorylation sites. Several groups have shown that increasing O-GlcNAc levels negatively affects phosphorylation levels; and that protein phosphatase inhibitors increase O-GlcNAc levels *(3,4)*. These data support a model where a complex relationship exists between these two posttranslational modifications, providing the cell an extra layer of control, rather than just a simple dephosphorylation and phosphorylation reaction (*see* **Fig. 1**). The exact nature of how O-GlcNAc affects the regulation of proteins remains to be elucidated *(1,2)*. This article describes techniques currently being used to determine if proteins are modified by O-GlcNAc and where the attachment of this saccharide occurs.

2. Materials

2.1. Control Proteins

1. Ovalbumin (*see* **Note 1**).
2. Bovine serum albumin (BSA) (*see* **Note 2**).
3. BSA-GlcNAc (*see* **Note 3**).

2.2. Immunoblotting With CTD 110.6

1. Purified or crude protein separated by sodium dodecyl sulfate polyacrylamide gel electrophoresis (SDS-PAGE) and electroblotted to polyvinylidene difluoride (PVDF) or nitrocellulose.
2. Tris-HCl-buffered saline (TBS): 10 mM Tris-HCl, pH 7.5, 150 mM NaCl.
3. TBS-High Tween (TBS-HT): 10 mM Tris-HCl, pH 7.5, 150 mM NaCl, 0.3% (v/v) Tween-20.
4. CTD 110.6 (Covance, Richmond, CA) ascites diluted 1/2500 in TBS-HT.
5. Anti-Mouse IgM-horseradish peroxidase (HRP) (Sigma-Aldrich, St Louis, MO) diluted 1/5000 in TBS-HT.
6. Enhanced chemiluminescence (ECL) (Amersham Biosciences, Piscataway, NJ).
7. 10 mM GlcNAc in TBS-HT.

2.3. Immunoblotting With Succinylated Wheat-Germ Agglutinin (sWGA)

1. Purified or crude protein separated by SDS-PAGE and blotted to PVDF or nitrocellulose.
2. Tris-HCl-buffered saline Tween (TBST): 10 mM Tris-HCl, pH 7.5, 150 mM NaCl, 0.05% (v/v) Tween-20.
3. Blocking agent: 3% (w/v) BSA in TBST.
4. TBS.
5. High-salt TBST: (HS-TBST): 10 mM Tris-HCl, pH 7.5, 1M NaCl, 0.05% (v/v) Tween-20.
6. 0.1 µg/mL sWGA-HRP (EY Labs, San Mateo, CA) in TBST (*see* **Note 4**).
7. 1 M GlcNAc in HS-TBST.
8. ECL reagent (Amersham Biosciences).

2.4. In Vitro Transcription Translation

1. cDNA subcloned into an expression vector with an SP6 or T7 promoter (approx 0.5–1 µg/µL).
2. Rabbit reticulocyte lysate in vitro transcription translation kit (Promega, Madison, WI).
3. Label, ^{35}S-Met, or ^{35}S-Cys, or ^{14}C-Leu.
4. sWGA-agarose (Vector Labs, Burlingame, CA).
5. 1-mL tuberculin syringe with a glass wool frit or a Bio-Spin disposable chromatography column (Bio-Rad, Hercules, CA).
6. Sephadex G-50.
7. Phosphate-buffered saline (PBS): 10 mM phosphate buffer, pH 7.5, 150 mM NaCl.
8. sWGA wash buffer: PBS, 0.2% (v/v) NP-40.
9. sWGA Gal elution buffer: PBS, 0.2% (v/v) NP-40, 1 M D-(+)-Gal.

10. sWGA GlcNAc elution buffer: PBS, 0.2% (v/v) NP-40, 1 M GlcNAc.
11. SDS-PAGE equipment and buffers.
12. Gel dryer.
13. Liquid-scintillation counter.

2.5. Galactosyltransferase Labeling

1. Protein sample(s).
2. Buffer H: 50 mM HEPES, pH 6.8, 50 mM NaCl, 2% (v/v) Triton-X100.
3. 10X Labeling buffer: 100 mM HEPES, pH 7.5, 100 mM Gal, 50 mM MnCl$_2$.
4. 25 mM 5'-adenosine monophosphate (5'-AMP), in Milli-Q water, pH 7.0.
5. UDP-[^{3}H]Gal, 1.0mCi/mL (specific activity 17.6 Ci/mmol) in 70% (v/v) ethanol.
6. UDP-Gal (not radioactive).
7. Stop solution: 10% (w/v) SDS, 0.1 M EDTA.
8. Desalting column, Sephadex G-50 (30 × 1cm) equilibrated in 50 mM ammonium formate, 0.1% (w/v) SDS.

2.6. Autogalactosylated Galactosyltransferase

1. 10X galactosyltransferase buffer: 100 mM HEPES, pH 7.4, 100 mM Gal and 50 mM MnCl$_2$.
2. Galactosyltransferase storage buffer: 2.5 mM HEPES, pH 7.4, 2.5 mM MnCl$_2$, 50% (v/v) glycerol.
3. Saturated ammonium sulfate, ≥7.4 g (NH$_4$)$_2$SO$_4$ in 25 mL Milli-Q water.
4. 85% ammonium sulfate 14 g (NH$_4$)$_2$SO$_4$ in 25 mL Milli-Q water.
5. 25 mM 5'-AMP, in Milli-Q water, pH 7.0.
6. Aprotinin.
7. β-Mercaptoethanol.
8. UDP-Gal.
9. 30–50mL centrifuge tubes.

2.7. PNGase F

1. Peptide N: glycosidase F (PNGase F).
2. 10X PNGase F denaturing buffer: 5% (w/v) SDS, 10% (v/v) β-mercaptoethanol in 50 mM sodium phosphate buffer, pH 7.5.
3. 10X PNGase F reaction buffer: 500 mM sodium phosphate buffer, pH 7.5.
4. 10% (v/v) NP-40.
5. PIC 1: Leupeptin, 1 mg/mL, antipain, 2 mg/mL, benzamide, 10 mg/mL, dissolved in aprotinin, 10,000 U/mL.
6. PIC 2: Chemostatin, 1 mg/mL, Papstatin, 2 mg/mL, dissolved in dimethyl sulfoxide (DMSO).

2.8. Hexosaminidase Digestion

1. N-acetyl-β-D-glucosaminidase, from jack bean (Sigma-Aldrich, V-Labs, Covington, LA).
2. 2% (w/v) SDS.
3. 2X reaction mixture: 80 mM citrate-phosphate buffer, pH 4.0, 1 U N-acetyl-β-D-glucosaminidase (V-labs), 8% (v/v) Triton X-100 or (v/v) NP-40, 0.01 U aprotinin, 1 μg of leupeptin, 1 μg α$_2$-macroglobulin.

2.9. Increasing Levels of O-GlcNAc With PUGNAc

1. PUGNAc (Carbogen; Switzerland) 20 mM stock in Milli-Q water.

2.10. Mapping Sites of O-GlcNAc Attachment

1. Trypsin, sequencing-grade modified (Promega).
2. 40 mM Ammonium bicarbonate, pH 8.0.
3. Trifluoroacetic acid (TFA).
4. Performic acid oxidation buffer (made fresh): 45% (v/v) formic acid, 5% (v/v) hydrogen peroxide, in Milli-Q water.
5. MgCl$_2$.
6. Alkaline phosphatase (Promega).
7. Dithiothreitol (DTT), high purity (Amersham Biosciences).
8. BEMAD solution (made fresh): 1% (v/v) Triethylamine, 0.1% (v/v) NaOH, 10 mM DTT.
9. C$_{18}$ Reversed-phase macro-spin columns (The Nest Group, Southborough, MA).
10. Buffer A: 1% (v/v) TFA.
11. Buffer B: 1% (v/v) TFA, 75% (v/v) Acetonitrile.
12. Thiol column buffer (made fresh), degassed: PBS, 1 mM EDTA.
13. Thiol column elution buffer (made fresh), degassed: PBS, 1 mM EDTA, 20 mM DTT.
14. Thiopropyl sepharose™ 6B (Amersham Biosciences).
15. 1% (v/v) acetic acid.
16. Savant Speed-Vac concentrator.
17. Finnigan LCQ with nanospray source.
18. Control peptides (*see* **Note 5**)
19. Approximately 1–100 pmol of protein sample in 40 mM ammonium bicarbonate, pH 8.0 (*see* **Note 6**).
20. Seal-Rite™ Natural microcentrifuge tubes (USA Scientific, Ocala, FL) (*see* **Note 7**).

3. Methods

The methods presented are broken down into two categories (**Fig. 2**), those that can be used to detect if proteins are modified by O-GlcNAc (**Subheading 3.1.**) and those that can be used to map glycosylation sites (**Subheading 3.2.**).

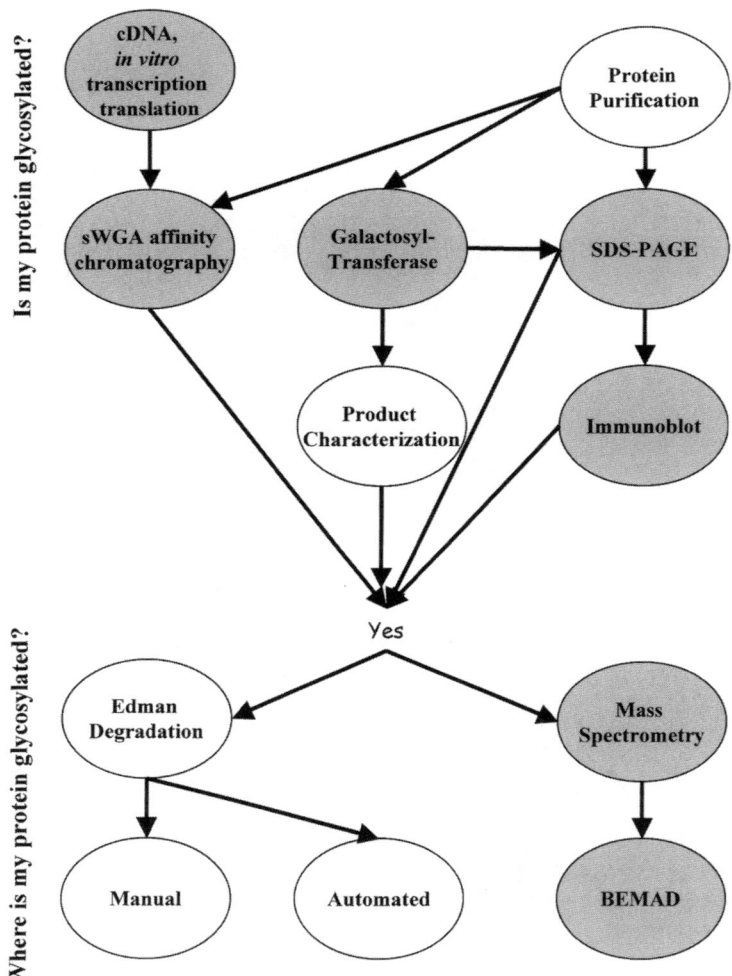

Fig. 2. A schematic demonstrating possible approaches to characterizing O-GlcNAc and its sites of attachment. Methods highlighted in gray are discussed at length in this chapter.

3.1. Detection of O-GlcNAc

Several methods are commonly used to detect O-GlcNAc on proteins *(5–7)* which have been purified from whole-cell extract either by conventional techniques or immunoprecipitated and separated from immunoglobulin using SDS-PAGE *(see Note 8)*. O-GlcNAc was originally detected in lymphocytes using β-D-1-4-galactosyltransferase from bovine milk *(8)*. The enzyme is used either in vitro or in vivo to label terminal GlcNAc residues on proteins with [6³H]Gal, forming a [³H]-βGal1-4βGlcNAc disaccharide *(5–8)*. Labeling of proteins and subsequent product analysis to determine the size and chemical composition of the released carbohydrate remains the gold standard for the detection and characterization of O-GlcNAc. In addition, labeling O-GlcNAc allows for the subsequent detection of the proteins and peptides of interest during SDS-PAGE, high-performance liquid chromatography (HPLC), protease digestion, and Edman degradation steps *(5,6)*. Although this technique is used for comprehensive studies, several other techniques have been developed to screen proteins for the O-GlcNAc modification; these include: immunoblotting with O-GlcNAc-specific antibodies (**Subheading 3.1.1.**) *(9–13)* and lectins (**Subheading 3.1.2.**), and screening low-abundance proteins for O-GlcNAc using sWGA-affinity chromatography of in vitro transcribed and translated proteins (**Subheading 3.1.3.**) *(5)*.

3.1.1. Detection of O-GlcNAc Using the Monoclonal Antibody CTD 110.6

Several monoclonal antibodies (MAbs) have been developed that included O-GlcNAc as part of their epitope, specifically RL2 *(9)*, HGAC85 *(10)* and MY95 *(13)*. However, many of these antibodies also have some peptide specificity, and as a result only recognize a subset of O-GlcNAc-modified proteins *(11)*. Recently, Comer and co-workers *(12)* showed that an antibody raised against the glycosylated C-terminal domain of the RNA Polymerase II large subunit was a general O-GlcNAc antibody. It should be noted that CTD 110.6 does not recognize all proteins modified by O-GlcNAc.

To confirm that the antibody is working in a specific manner, a number of controls should be included in this experiment. Controls include: treating purified proteins with hexosaminidase (**Subheading 3.1.5.2.**) to remove O-GlcNAc, treating cell extract with PUGNAc (**Subheading 3.1.5.3.**) to increase O-GlcNAc levels, competing the signal away with free GlcNAc, and running appropriate positive and negative controls on gels (*see* **Note 1**) such as ovalbumin (negative; Sigma-Aldrich) and BSA-GlcNAc (positive; Sigma-Aldrich).

1. Proteins of interest are blotted onto either a PVDF or a nitrocellulose membrane (*see* **Note 9**).
2. Block blots with TBS-HT for 60 min at room temperature (*see* **Note 10**).
3. Incubate blots with CTD 110.6 (1/2500) in TBS-HT with and without 10 m*M* GlcNAc overnight at 4°C.
4. Wash blots in TBS-HT 3 × 10 min.
5. Incubate blots with anti mouse IgM (Sigma), at approx 1/5000 in TBS-HT, with shaking at room temperature for 50 min.

6. Wash blots in TBS-HT 5 × 10 min.
7. Wash blots in TBS 1 × 10 min.
8. Develop the HRP-reaction (*see* **Note 11**).

3.1.2. Immunological Detection of O-GlcNAc Using sWGA

Typically sWGA has been used to detect O-GlcNAc (*see* **Note 12**). However, because sWGA will recognize any terminal β-GlcNAc residue, samples should be treated with PNGase F to remove N-linked sugars (**Subheading 3.1.5.1.**), and further characterization would be required to determine if the activity was associated with a monosaccharide or a longer complex O-glycan. The controls for this method are similar to those used for CTD 110.6.

1. Wash duplicate blots for 10 min in 3% (w/v) BSA in TBST (*see* **Note 13**).
2. Incubate blots in 3% (w/v) BSA in TBST for 60 min at room temperature (*see* **Note 14**).
3. Wash blots 3 × 10 min in TBST.
4. Incubate blots in 0.1 µg/mL sWGA-HRP in TBST with and without 1 *M* GlcNAc overnight at 4°C.
5. Wash blots in HS-TBST 6 × 10 min.
6. Wash blots in TBS 1 × 10 min.
7. Develop the HRP-reaction.

3.1.3. In Vitro Transcription Translation

When initially characterizing the O-GlcNAc transferase, Haltiwanger and colleagues observed activity of this enzyme in rabbit reticulocyte lysate *(14)*. Later, it was determined that proteins expressed in a rabbit reticulocyte lysate were modified by O-GlcNAc. This has been utilized as a screening technique to determine if low copy number proteins, and proteins that are challenging to purify, are glycosylated. Proteins, labeled with ^{35}S-Met, ^{35}S-Cys, or ^{14}C-Leu in an in vitro transcription translation (ITT) system (Promega), are tested for their ability to bind sWGA immobilized on an agarose column.

3.1.3.1. Synthesis of Proteins in Rabbit Reticulocyte Lysate

1. Synthesize proteins using the rabbit reticulocyte lysate ITT system according to the manufacturer's instructions. Include the protein of interest, a positive control for sWGA binding (for example, nuclear pore protein p62), and a negative control (luciferase, supplied with kit).
2. Treat half of each sample with hexosaminidase (**Subheading 3.1.5.2.**).
3. Desalt samples using spin filtration, Amersham Biosciences Microspin™ G50 columns (*see* manufacturer's instructions), or a 1 mL G50 desalting column.

3.1.3.2. DESALTING

1. Pour exactly 1 mL Sephadex G-50 in a tuberculin syringe that has been packed with a small glass wool frit (approx 200 μL).
2. Wash column with 5 mL sWGA wash buffer.
3. Load protein sample onto column. The volume of sample can be up to 200 μL.
4. Wash the column with sWGA wash buffer so that the total volume of this wash and the protein sample is 350 μL. For example, if sample volume is 150 μL, add 200 μL desalting buffer.
5. Transfer the syringe column to a clean prechilled tube. Elute protein with 200 μL desalting buffer. This is the desalted sample.

3.1.3.3. sWGA CHROMATOGRAPHY

The following steps are carried out at 4°C.

1. Equilibrate approx 150 μL of sWGA-agarose in sWGA wash buffer.
2. Apply sample to the column and stand at 4°C for 30 min, or seal and incubate for 30 min with rotating or rocking.
3. Let the unbound material run though the column.
4. Wash the column with 15 mL of wash buffer at approx 10 mL/h, collecting 0.5 mL fractions.
5. Load the column with 500 μL of Gal elution buffer and stand at 4°C for 20 min.
6. Wash column with 5 mL of Gal elution buffer, collecting 0.5 mL fractions.
7. Repeat **steps 5** and **6** using the GlcNAc elution buffer.
8. Count 25 μL of each fraction using a liquid scintillation counter.
9. Pool the positive fractions that elute in the presence of GlcNAc and precipitate using TCA or methanol (*see* **Note 15**).
10. Analyze the pellet by SDS-PAGE and autoradiography to confirm that the label has been incorporated into a protein of an appropriate size.

3.1.4. Galactosyltransferase Labeling

Samples should be denatured prior to labeling with galactosyltransferase, for example by boiling in the presence of 10 mM DTT and 0.5% (w/v) SDS. Up to 0.5% (w/v) SDS can be used (*see* **Note 16**). However, it should be titrated out with 10 times more NP-40 (v/v), so the solution should be brought to 5% (v/v) NP-40. Note that the solution should then be diluted to reduce the total NP-40 concentration to less than 2% (*see* **Note 16**). Galactosyltransferase requires 1–5 mM Mn^{2+} for activity, but is inhibited by higher concentrations (>20 mM) and is inhibited by Mg^{2+}.

To control for specificity, samples should be treated with PNGase F (**Subheading 3.1.5.1.**) to remove any contaminating *N*-linked sugars. Once labeled, proteins can be detected by autoradiography after separation by SDS-PAGE. To confirm that label was incorporated onto a single GlcNAc residue, "product analysis" should be performed. Product analysis entails: (1) the release of carbohydrates as sugar alditols by reductive β-elimination *(15)*; (2) confirmation that a disaccharide has been released by size-exclusion chromatography *(15,16)*; (3) confirmation that the product is [^{3}H]βGal1-4βGlc-NAcol (from galactosyltransferase labeling), which is typically performed by high-performance anion-exchange chromatography on a DIONEX CarboPac PA100 column *(17,18)*.

1. Remove solvent from label in a Speed-Vac or under a stream of nitrogen, dry approx 1–2 μCi/reaction (*see* **Note 17**). Resuspend label in 25 m*M* 5'-AMP, 50 μL per reaction (*see* **Note 18**).
2. Set reactions up as follows:

Sample, final concentration 0.5–5 mg/mL	up to 50 μL
Buffer H	350 μL
10X labeling buffer	50 μL
UDP-[^{3}H]Gal/5'-AMP	50 μL
Galactosyltransferase 30–50 U/Ml	2–5 μL
Calf intestinal alkaline phosphatase (*see* **Note 19**)	1–4 U
Milli Q water, to a final volume of	500 μL

3. Labeling is typically done at 37°C for 2 h or at 4°C overnight.
4. Add cold UDP-Gal to a final concentration of 0.5–1.0 m*M* and another 2–5 μL of galactosyltransferase (*see* **Note 20**).
5. Add 50 μL of stop solution to each sample and heat to 100°C for 5 min.
6. Resolve the protein from unincorporated label using a Sephadex G-50 column (30 × 1cm, equilibrated in 50 m*M* ammonium formate, 0.1% (w/v) SDS), collect 1 mL fractions (*see* **Note 21**).
7. Count an aliquot (50 μL) of each fraction using a liquid-scintillation counter.
8. Approximately 2 × 10^6 DPM of ^{3}H-Gal should be incorporated into 2 μg of Ovalbumin.
9. Combine the fractions containing the void volume and lyophilize to dryness.
10. Resuspend the sample in (100–1000 μL) Milli-Q water and acetone precipitate.
11. Add 3–5 volumes of cold acetone (−20°C) to the sample.
12. Incubate for 2–18 h at −20°C.
13. Pellet protein at 4°C for 10 min at 3000–16,000 *g*.

3.1.4.1. AUTOGALACTOSYLATION OF GALACTOSYLTRANSFERASE

Because galactosyltransferase contains *N*-linked carbohydrates, it is necessary to block these before using this enzyme to probe other proteins for terminal GlcNAc.

1. Resuspend 25 U of galactosyltransferase (Sigma) in 1 mL of 1X galactosyltransferase buffer.
2. Transfer sample to 30–50 mL centrifuge tube.
3. Remove a 5 µL aliquot for an activity assay.
4. Add 10 µL of aprotinin, 3.5 µL of β-mercaptoethanol, and 1.5–3.0 mg of UDP-Gal.
5. Incubate the sample on ice for 30–60 min.
6. Add 5.66 mL of prechilled saturated ammonium sulfate in a dropwise manner, and incubate on ice for 30 min.
7. Centrifuge at ≥10,000g for 15 min at 4°C.
8. Resuspend pellet in 5 mL of cold 85% ammonium sulfate and incubate on ice for 30 min.
9. Centrifuge at ≥10,000g for 15 min at 4°C.
10. Resuspend pellet in 1 mL of galactosyltransferase storage buffer.
11. Aliquot enzyme (50 µL).
12. Assay 5 µL of the autogalactosylated and nongalactosylated against a known substrate to determine the activity.
13. Store at −20°C for up to 1 yr.

3.1.5. Control Experiments

3.1.5.1. REMOVING *N*-LINKED SUGARS

Digestion of sample and control proteins with Peptide N: glycosidase F (PNGase F; EC 3.2.2.18.; *see* **Note 22**) is a useful way of showing that reactivity with lectins and galactosyltransferase is not owing to *N*-linked glycans. A positive control for the PNGase F reaction should be included (*see* **Note 23**) as shown in **Fig. 3**.

1. Add 1/10 sample volume of 10X PNGase F denaturing buffer to each sample and heat to 100°C for 10 min (*see* **Note 24**).
2. Add 1/10 sample volume of 10X PNGase F reaction buffer and 10% (v/v) NP-40, mix (*see* **Note 25**).
3. Add 1 µL of PNGase F and incubate samples at 37°C for 1 h to overnight.

3.1.5.2. REMOVING O-GLCNAC

Terminal β-GlcNAc and O-GlcNAc can be removed from proteins using commercial hexosaminidases; these enzymes will also cleave terminal β-GalNAc

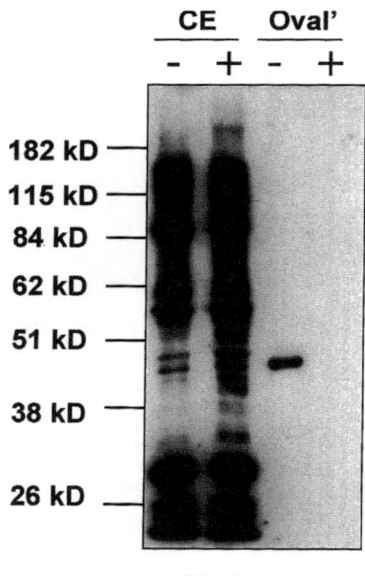

IB:WGA

Fig. 3. sWGA-HRP lectin blot of 20 µg cell extract (CE) and 100 ng of ovalbumin (Oval'), which have (+) and have not (−) been treated with PNGase F before 12% SDS-PAGE and blotting to nitrocellulose.

residues. Unlike O-GlcNAcase, commercial hexosaminidases have low pH optima, typically pH 4.0–5.0.

1. Protein samples (include a positive control such as ovalbumin).
2. Mix sample 1:1 with 2% (w/v) SDS and boil for 5 min.
3. Mix sample 1:1 with reaction mixture and incubate at 37°C for 4–24 h.

3.1.5.3. Increasing the Level of O-GlcNAc Within Cells

The endogenous enzyme that removes O-GlcNAc can be inhibited in cells, resulting in increased levels of total O-GlcNAc within the cell. Increasing stoichiometry can enhance the chance of detecting O-GlcNAc, and provide a specificity control as shown in **Fig. 4**. Dong and colleagues showed that O-(2-acetamido-2-deoxy-D-glucopyranosylidene)-amino-N-phenyl-carbamate (PUGNAc), is a potent inhibitor of O-GlcNAcase in an in vitro assay (K_i = 53 nM) *(19)*. Subsequently, it has been shown that both dividing and stationary cells take up PUGNAc, and cells can be treated by PUGNAc for several days without any apparent cell toxicity *(20)*.

1. Grow cells in monolayer in 100 mm dishes as desired.
2. (Optional) Replace growth media with fresh media containing 40–100 µM PUGNAc (*see* **Note 26**).

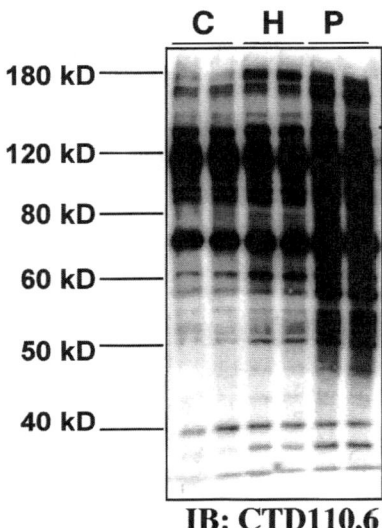

IB: CTD110.6

Fig. 4. Immunoblot with CTD 110.6 of 20 µg of cell extract (C), which has been treated to raise levels of O-GlcNAc by either growth in high glucose (H) or treatment with PUGNAc (P) before 7.5% SDS-PAGE and blotting to nitrocellulose.

3. Incubate cells in an incubator for 6–18 h.
4. At the end of treatment, take out the dishes from the incubator and place on ice.
5. Extract proteins as desired; for example, a nuclear cytoplasmic extraction or a total cell lysate.
6. Immunoblot with antibody as described (**Subheading 3.1.1.**) or sWGA (**Subheading 3.1.2.**). Alternatively, label the protein with galactosyltransferase (**Subheading 3.2.**).

3.2. Site Mapping

Like many posttranslational modifications O-GlcNAc is present at substoichiometric levels, and the detection and subsequent analysis of O-GlcNAc has been challenging. To overcome this, O-GlcNAc has been labeled with [³H]-Gal, using galactosyltransferase, and this label has been used to track proteins and peptides though subsequent purification. In combination with manual Edman degradation (*see* **ref. 5** for a comprehensive method), a number of techniques relying on [³H]-Gal labeling have been used to map glycosylation sites on as little as 10 pmol of starting material (**7**). More recently, it has been shown that automated Edman degradation can be used to map glycosylation sites; but, because approx 10–20 pmol of pure peptide are required, sensitivity is again an issue (**21**).

Mass spectrometry (MS) is very sensitive and several MS-based strategies have been developed for mapping sites of O-GlcNAc attachment. However, until recently, MS provided little advantage over conventional techniques, because the addition of a single GlcNAc to a peptide has been shown to reduce the signal by approximately fivefold; and the presence of the unglycosylated peptide suppresses the signal further *(22)*. Moreover, the β-O-GlcNAc bond is highly labile, and in conventional electrospray ionization-MS, O-GlcNAc is often released at the source and/or at lower collision energies than is required to sequence peptides. To map sites in the mass spectrometer, Greis and co-workers *(7)* analyzed peptides before and after β-elimination. β-elimination of carbohydrates from Ser (89 amu) and Thr (101 amu) residues results in the formation of 2-aminopropenoic acid (69 amu) and α-aminobutyric acid (83 amu), respectively. Because these amino acids have unique masses compared to their parent amino acids, they can be used to map the site of attachment. It should be noted that this method would also release phosphate linked to Ser and Thr residues. In an adaptation of this method, Wells and co-workers add DTT back to the peptide backbone, labeling the glycosylation site *(23)*. Tryptic peptides of the purified proteins are subjected to phosphatase treatment to dephosphorylate the peptides. The peptides then undergo mild β-elimination, followed by Michael addition with DTT (Cleland's reagent, DTT). Derivatizing the β-eliminated peptides with DTT tags the site of modification with a unique molecular weight, facilitating database searching. Also, the DTT tag may be used to enrich for DTT-modified peptides by thiol-affinity chromatography, thereby offering a solution to the issue of low-abundance O-GlcNAc-modified proteins.

3.2.1. BEMAD

3.2.1.1. PERFORMIC ACID OXIDATION (*SEE* **NOTE** 27)

1. Suspend protein sample in 300 μL Performic acid oxidation buffer.
2. Spike with 1–10 pmol of control peptides.
3. Incubate on ice for 1 h.
4. Dry down in Speed-Vac.

3.2.1.2. TRYPSIN DIGESTION

1. Resuspend protein sample in 40 m*M* ammonium bicarbonate.
2. Digest by addition of 1:10–1:100 (w/w) sequencing-grade trypsin overnight at 37°C.
3. Acidify digest by the addition of TFA to 1% (v/v) final concentration.
4. Clean up over a C_{18} reversed-phase column (*see* manufacturer's instructions).
5. Dry peptides using a Speed-Vac concentrator.

3.2.1.3. Phosphatase Treatment

1. Resuspend peptides in 40 mM ammonium bicarbonate, 1 mM MgCl$_2$.
2. Add alkaline phosphatase (1 U/10 µL) and incubate at 37°C for 4 h.
3. Dry peptides using a Speed-Vac concentrator.

3.2.1.4. BEMAD Treatment (*see* **Note 28**)

1. Resuspend peptides in 500 µL BEMAD solution and adjust pH to 12.0–12.5 with triethylamine, if necessary.
2. Incubate reaction at 50°C for 2.5 h.
3. Stop reaction by adding TFA to 1% (v/v) final concentration.
4. Clean up over C$_{18}$ reversed-phase column (*see* manufacturer's instructions).
5. Dry peptides using a Speed-Vac concentrator.

3.2.1.5. Thiol-Affinity Column (*see* **Note 29**)

1. Swell and wash thiopropyl sepharose resin in degassed thiol column buffer.
2. Resuspend peptides in thiol column buffer.
3. Bind peptides to thiol column at room temperature for 1 h.
4. Wash column with 20 mL thiol column buffer.
5. Elute peptides three times sequentially with 150 µL Thiol elution buffer.
6. Acidify peptides by adding TFA to 1% (v/v) final concentration.
7. Clean up over C$_{18}$ reversed-phase column to remove free DTT.
8. Dry down peptides.

3.2.1.6. Liquid Chromatography Tandem Mass Spectrometry (LC/MS/MS), Analysis (*see* **Note 30**)

1. Resuspend peptides in 1% (v/v) acetic acid.
2. Load sample onto nanobore column packed with C$_{18}$, desalted with 1% (v/v) acetic acid.
3. Separate sample over a 75 min linear gradient of increasing acetonitrile at a flow rate of approx 200 nL/min into the MS source (Finnigan LCQ). Data may be collected in automatic mode with a MS scan (2 × 500 ms) followed by two MS/MS scans (3 × 750 ms) of the two highest-intensity peptides with a dynamic exclusion of 2 and a mass gate of 2.0 daltons. Alternatively, MS/MS data may be collected manually by choosing peaks of interest for fragmentation from the MS scans during the run.
4. Turbosequest software may be used to interpret MS/MS data, allowing for a differential mass increase of 136.2 daltons to Ser and Thr residues, 120.2 daltons to Cys residues that may have been derivatized with DTT, 48.0 daltons to performic acid oxidized Cys and Trp residues, and 32.0 daltons to performic acid oxidized Met residues.

4. Notes

1. Ovalbumin (Sigma-Aldrich) is a protein modified by GlcNAc-terminating *N*-linked oligosaccharides that is used as a negative control for immunoblotting with CTD 110.6 (use 100 ng), a positive control for sWGA (use 100 ng) and galactosyltransferase labeling (use 2 µg).
2. BSA (Sigma-Aldrich) is a nonglycosylated protein that can be used as a negative control.
3. BSA-GlcNAc (Sigma-Aldrich) is a protein chemically modified to contain GlcNAc residues, that is used as a positive control for immunoblotting with CTD 110.6 and sWGA (use 1–5 ng).
4. sWGA-HRP can be stored at 1 mg/ml in PBS, pH 7.4, at −20°C for at least 1 y.
5. As a control, sample should be spiked with 1–10 pmol of known phosphorylated and/or O-GlcNAc-modified peptides. A commonly used glycosylated peptide is the BPP peptide (PSVPVS(O-GlcNAc)GSAPGR). Glycosylated peptides can be synthesized as described in *(6)*.
6. The amount of starting material will vary depending on the sensitivity of the LC/MS/MS instrument and the purity of the sample. With the Finnigan LCQ Classic, which is able to reach sensitivities in the final range, pmol amounts of starting protein may be enough, given that its purity is approx 90% and assuming that the stoichiometry of the O-GlcNAc modification is approx 10%. Additional details on protein sample preparation for BEMAD may be found in *(23)*.
7. In order to lessen plastic contamination, we recommend the use of these tubes. All plastic tubes and columns should be rinsed with 50% acetonitrile prior to use and never autoclaved. Also, clear pipet tips should be used whenever possible.
8. Gal-transferase labeling can be performed in conjunction with immunoprecipitation. However, cell extract should be labeled and then the protein of interest should be precipitated, because immunoglobulin contains large amounts of GlcNAc-terminating *N*-linked sugars, which will be preferentially labeled to O-GlcNAc.
9. 20–30 µg of total cytoplasmic, nuclear, or total cell extract is sufficient. For purified proteins, Comer and co-workers found that 25–50 ng of a neoglycoconjugate was sufficient *(12)*.
10. High concentrations of Tween-20 substitute for blocking membranes with milk or (BSA).
11. The antibody often cross-reacts with prestained markers.
12. Before succinylation, WGA will recognize both sialic acid and GlcNAc.

13. To determine changes in levels of O-GlcNAc in total cell extract, approx 20 μg of total protein should be loaded.
14. Note that milk cannot be used as the blocking agent because many of the proteins in milk are modified by glycans that react with sWGA.
15. Acetone is not recommended because free GlcNAc will precipitate.
16. Galactosyltransferase is also active in solutions containing 5 mM DTT, 0.5 M NaCl, up to 2% (v/v) Triton-X 100, up to 2% (v/v) NP-40, and 1 M urea. Digitonin should be used with care because it is a substrate for galactosyltransferase.
17. Ethanol can inhibit the galactosyltransferase reaction, but if less than 4 μL is required, the label can be added directly to the reaction (final reaction volume 500 μL).
18. The 5' AMP is included to inhibit possible phophodiesterase reactions, which might compete for label during the labeling experiment.
19. Free UDP is also an inhibitor, and for studies where complete labeling of the GlcNAc is preferable, such as site mapping, calf intestinal alkaline phosphatase is included in the reaction because it degrades UDP.
20. For studies where complete labeling of the GlcNAc is required, such as site mapping, the reactions are chased with unlabeled UDP-Gal and fresh galactosyltransferase.
21. Size-exclusion chromatography, using Sephadex G50, traditionally is used to desalt samples. However, TCA precipitation, spin filtration/buffer exchange, or other forms of size-exclusion chromatography (e.g., Pharmacia PD10 desalting column) can be used. The addition of carrier proteins such as BSA (approx 67 kDa) and Cytochrome C (approx 12.5 kDa) to samples and buffers will reduce the amount of protein lost owing to nonspecific protein adsorption.
22. PNGase F is distinct from endoglycosidase F (EC 3.2.1.96), which cleaves only a subset of N-linked sugars. In addition, PNGase F will not cleave N-linked sugars with a core α1-3fucose or N-linked sugars at the N- or C-terminus of a protein or peptide.
23. Ovalbumin, which contains one N-linked glycosylation site, will increase in mobility by several kilodaltons on a gel (10–12% SDS-PAGE) after treatment with PNGase F. This mobility shift is difficult to detect on a 7.5% SDS-PAGE gel.
24. Protein samples can contain protease inhibitor cocktails (PIC), such as PIC1 and PIC2 *(11)*.
25. PNGase F is inhibited by SDS. It is essential to add NP-40 to the reaction mixture.

26. The dose-dependence curve for PUGNAc is different in different cell types *(20)*.

27. Performic acid oxidation may be performed before or after trypsin digestion. It is observed that it may help denature samples with many Cys residues. As an alternative, Cys residues may be alkylated using iodoacetamide. If this is done, a mass increase of 57.052 daltons should be allowed for during database searching.

28. This method may be adapted for Ser and Thr phosphorylation sites as follows. Instead of phosphatase treatment, the sample should be acidified to pH 4.5 with TFA and treated with (1 U/20 µL) β-hexosaminidase (New England Biolabs) at 37°C for 16h. Also, the BEMAD solution should be modified to 2% (v/v) Triethylamine, 0.2% (v/v) NaOH, 10 mM DTT, and the reaction allowed to proceed for 5 h at 50°C.

29. Peptides may be bound to the thiol column for longer than 1 h. This is a minimum incubation time.

30. These LC/MS/MS methods should be used as a general guide only. Methods should be optimized according to the specific instrument being used.

Acknowledgments

The authors work is supported by NIH grants HD13563 and CA42486 to GWH, and the National Heart, Lung, and Blood Institute, National Institutes of Health, contract No. N01-HV-28180. Under a licensing agreement between Covance Research Products and The Johns Hopkins University, Dr. Hart receives a share of royalty received by the university on sales of the CTD 110.6 antibody. The terms of this arrangement are being managed by The Johns Hopkins University in accordance with its conflict of interest policies.

References

1. Wells, L., Vosseller, K., and Hart, G. W. (2001) Glycosylation of nucleocytoplasmic proteins: signal transduction and O-GlcNAc. *Science* **291,** 2376–2378.

2. Zachara, N. E. and Hart, G. W. (2002) The emerging significance of O-GlcNAc in cellular regulation. *Chem. Rev.* **102,** 431–438.

3. Griffith L. S. and Schmitz, B. (1999) O-linked N-acetylglucosamine levels in cerebellar neurons respond reciprocally to pertubations of phosphorylation. *Eur. J. Biochem.* **262,** 824–831.

4. Lefebvre, T., Alonso, C., Mahboub, S., et al. (1999) Effect of okadaic acid on O-linked N-acetylglucosamine levels in a neuroblastoma cell line. *Biochim. Biophys. Acta* **1472,** 71–81.

5. Roquemore, E. P., Chou, T. Y., and Hart G. W. (1994) Detection of O-linked N-acetylglucosamine (O-GlcNAc) on cytoplasmic and nuclear proteins. *Methods Enzymol.* **230,** 443–460.

6. Greis, K. D., Hayes B. K., Comer, F. I., et al. (1996) Selective detection and site-analysis of O-GlcNAc modified glycopeptides by beta-elimination and tandem electrospray mass spectrometry. *Anal. Biochem.* **234**, 38–49.
7. Greis, K. D. and Hart, G. W. (1998) Analytical methods for the study of O-GlcNAc glycoproteins and glycopeptides. *Methods Mol. Biol.* **76**, 19–33.
8. Torres, C. R. and Hart, G. W. (1984) Topography and polypeptide distribution of terminal *N*-acetylglucosamine residues on the surfaces of intact lymphocytes. Evidence for *O*-linked GlcNAc. *J. Biol. Chem.* **259**, 3308–3317.
9. Snow, C. M., Senior, A., and Gerace, L. (1987) Monoclonal antibodies identify a group of nuclear pore complex glycoproteins. *J. Cell Biol.* **104**, 1143–1156.
10. Turner, J. R., Tartakoff, A. M., and Greenspan, N. S. (1990) Cytologic assessment of nuclear and cytoplasmic *O*-linked *N*-acetylglucosamine distribution by using anti-streptococcal monoclonal antibodies. *Proc. Natl. Acad. Sci. USA* **87**, 5608–5612.
11. Holt, G. D., Snow, C. M., Senior, A., et al. (1987) Nuclear pore complex glycoproteins contain cytoplasmically disposed *O*-linked *N*-acetylglucosamine. *J. Cell Biol.* **104**, 1157–1164.
12. Comer, F. I., Vosseller, K., Wells, R. L., et al. (2001) Characterization of a mouse monoclonal antibody specific for *O*-Linked GlcNAc. *Anal. Biochem.* **293**, 169–177.
13. Matsuoka, Y., Shibata, S., et al. (2002) Identification of Ewing's sarcoma gene product as a glycoprotein using a monoclonal antibody that recognizes an immunodeterminant containing *O*-linked *N*-acetylglucosamine moiety. *Hybrid. Hybridomics* **21**, 233–236.
14. Haltiwanger, R. S., Blomberg, M. A., and Hart, G. W. (1992) Glycosylation of nuclear and cytoplasmic proteins. Purification and characterization of a uridine diphospho-*N*-acetylglucosamine: polypeptide β-*N*-acetylglucosaminyltransferase *J. Biol. Chem.* **267**, 9005–9013.
15. Fukuda, M. (1990) Characterization of *O*-linked saccharide structures from cell surface glycoproteins. *Methods Enzymol.* **179**, 17–29.
16. Kobata, A. (1994) Size fractionation of oligosaccharides. *Methods Enzymol.* **230**, 200–208.
17. Townsend, R. R., Hardy, M. R., and Lee, Y. C. (1990) Separation of oligosaccharides using high-performance anion-exchange chromatography with pulsed amperometric detection. *Methods Enzymol.* **179**, 65–76.
18. Hardy, M. R. and Townsend, R. R. (1994) High-pH anion exchange chromatography of glycoprotein-derived carbohydrates. *Methods Enzymol.* **230**, 208–225.
19. Dong, D. L.-Y. and Hart, G. W. (1994) Purification and characterization of an *O*-GlcNAc selective *N*-Acety-β-D-glucosaminidase from rat spleen cytosol. *J. Biol. Chem.* **269**, 19,321–19,330.
20. Haltiwanger, R. S., Grove, K., and Philipsberg, G. A. (1998) Modulation of *O*-linked N-acetylglucosamine levels on nuclear and cytoplasmic proteins in vivo using the peptide *O*-GlcNAc-β-*N*-acetylglucosaminidase inhibitor O-(2-acetamido-2-deoxy-D-glucopyranosylidene)amino-*N*-phenylcarbamate. *J. Biol. Chem.* **273**, 3611–3617.
21. Zachara, N. E. and Gooley, A. A. (2000) Identification of glycosylation sites in much peptides by edman degradation. *Methods Mol. Biol.* **125**, 121–128.

22. Hart, G. W., Cole, R. N., Kreppel, L. K., et al. (2000) Glycosylation of proteins-a major challenge in mass spectrometry and proteomics, in *Proceedings of the 4th International Symposium on Mass Spectrometry in the Health and Life Sciences* (Burlingame, A., Carr, S., and Baldwin, M., eds.), Humana Press, Totowa, NJ, pp. 365–382.
23. Wells, L., Vosseller, K., Cole, R. N., et al. (2002) Mapping sites of *O*-GlcNAc modification using affinity tags for serine and threonine post-translational modifications. *Mol. Cell Proteomics* **1,** 791–804.

13

Techniques in Protein Methylation

Jaeho Lee, Donghang Cheng, and Mark T. Bedford

Summary

Proteins can be methylated on the side-chain nitrogens of arginine and lysine residues or on carboxy-termini. Protein methylation is a way of subtly changing the primary sequence of a peptide so that it can encode more information. This common posttranslational modification is implicated in the regulation of a variety of processes including protein trafficking, transcription and protein–protein interactions. In this chapter, we will use the arginine methyltransferases to illustrate different approaches that have been developed to assess protein methylation. Both in vivo and in vitro methylation techniques are described, and the use of small molecule inhibitors of protein methylation will be demonstrated.

Key Words: Protein methylation; lysine; arginine, histones; PABP1; Sam68; AdoMet.

1. Introduction

Signal-transduction pathways commonly use posttranslational modifications to convey information through the cell. Protein methylation is a component of this cellular information network. Proteins can be methylated on carboxy-termini or on the side-chain nitrogens of arginine and lysine residues (1–3). Within signaling pathways, protein methylation occurs both proximal to receptor-mediated responses (4,5) and distal to primary signaling events, where methylation of histones is crucial for laying down the "histone code" and the subsequent activation (or inactivation) of transcriptional loci (6). In addition, methylation is involved in protein trafficking (7), the biogenesis of spliceosomal proteins (8,9) and the regulation of protein–protein interactions (8,10). Arginine residues can

From: *Methods in Molecular Biology, vol. 284:*
Signal Transduction Protocols
Edited by: R. C. Dickson © Humana Press Inc., Totowa, NJ

be dimethylated either asymmetrically (by Type I enzymes) or symmetrically (by Type II enzymes) *(11)*. In mammals, five Type I enzymes (PRMT1–4 and –6) *(12–16)* and a single Type II enzyme (PRMT5) *(17)* have been described. Recently, the SET domain-dependent lysine methyltransferases were discovered *(18)*. The founding member of this family of methyltransferases, Suv39H1, selectively methylates the N-terminus of histone H3. Lysine residues can accept up to three methyl groups forming mono-, di-, and trimethylated derivatives *(19)*. Studies with arginine methyltransferases will be described in this chapter to illustrate different approaches that can be used to investigate protein methylation.

2. Materials

1. pGEX6P-1 Bacterial expression vector (Amersham Biosciences, Piscataway, NJ).
2. Glutathione sepharose 4B (Amersham Biosciences).
3. Glutathione reduced.
4. *Escherichia coli* strain *BL21*.
5. Luria-Bertani (LB) broth: Per liter: 10 g Pancreatic digest of Casein, 5 g of Yeast extract, 10 g of NaCl.
6. Agar: Per liter:10 g of Tryptone, 5 g of yeast extract, 10 g of NaCl, 15 g of Agar.
7. IPTG (isopropyl-β-D-thio-galactopyranoside).
8. Ampicillin.
9. Phosphate-buffered saline (PBS): 137 mM NaCl, 2.7 mM KCl, 4.3 mM Na_2HPO_4, 1.4 mM KH_2PO_4, pH 7.4
10. Acrylamide.
11. Nonfat dry milk.
12. Tween-20.
13. Glacial acetic acid.
14. Ethanol.
15. S-adenosyl-L-[methyl-^{3}H]methionine or [^{3}H]AdoMet; approx 70–85 Ci/mmol from a 12.6 μM stock solution in dilute HCl/ethanol 9:1, pH 2.0–2.5. (Amersham Biosciences).
16. EN[3]HANCE spray (PerkinElmer Life Sciences, Boston, MA).
17. UltraLink immobilized protein A/G (Pierce, Rockford, IL).
18. Elution buffer: 100 mM Tris-HCl, pH 8.0, 120 mM NaCl, 30 mM glutathione reduced.
19. Tris-Glycine running buffer: 25 mM Tris-base, 0.2 M glycine, 0.1% SDS.
20. Transfer buffer: 25 mM Tris-base, 0.2 M glycine, 0.1% sodium dodecyl sulfate (SDS), 20% methanol.
21. Coommassie Blue staining solution: 0.5 g Coomassie Blue, 200 mL methanol 100%.

22. Histones (Sigma, St Louis, MO).
23. 6X SDS protein sample loading buffer: 180 m*M* Tris-HCl, pH 6.8, 30% glycerol, 10% SDS, 0.6 *M* ditriothreitol (DTT), 0.012% bromophenol blue.
24. PVDF (polyvinylidene difluoride) (Milipore, Bedford, MA).
25. Growth Medium A: Dulbecco's Modified Eagle's Medium (DMEM), 10% Fetal Bovine Serum (FBS), cycloheximide (100 µg/mL), chloroamphenicol (40 µg/mL).
26. Growth Medium B: Dulbecco's Modified Eagle's Medium without methionine (DMEM/-Met), 10% FBS (dialysed), cycloheximide (100 µg/mL), chloroamphenicol (40 µg/mL).
27. L-[*methyl*-³H]methionine; approx 70–85 Ci/mmol (Amersham Biosciences).
28. Mild lysis buffer: 10 m*M* Tris-HCl, pH 7.5, 1% (v/v) Triton X-100, 150 m*M* NaCl, 5 m*M* EDTA, protease inhibitor cocktail (Roche, Indianapolis, IN).
29. Adenosine-2, 3-dialdehyde (AdOx) (Sigma).
30. Sonic dismembrater (model 500) with tapered tip (Fisher Scientific, Pittsburgh, PA).
31. Miniprotean three electrophoresis cell (Bio-Rad, Hercules, CA).
32. Semidry electroblotter apparatus (Amersham Biosciences).
33. Kodak Biomax MS X-ray film (Eastman Kodak, Rochester, NY).
34. Antibodies

3. Methods

The analysis of protein methylation can be performed in vitro using recombinant enzymes and substrates, or in vivo in tissue culture. The methods described here detail: (1) the purification of recombinant methyltransferase enzymes, (2) an in vitro methylation assay using enzymes and their substrates, (3) an in vivo methylation assay, (4) the use of methyltransferase inhibitors in combination with in vivo labeling techniques to identify methylated proteins within a cellular context, and (5) a methylation assay using recombinant substrates and cell extracts as the source of enzyme activity.

Certain techniques that are central to the analysis of protein methylation have not been addressed here. Amino-acid analysis of substrates is often used to establish the type of methylation event catalyzed *(12,20)*. In addition, recent studies have demonstrated that methylated peptides are good immunogens and methyl-specific antibody can be produced that track these posttranslational modifications in the cell *(19,21,22)*.

3.1. Purification of Recombinant Methyltransferase Enzymes

The majority of protein arginine methyltransferases (PRMTs) and the SET domain-containing lysine methyltransferases are active as glutathione *S*-transferase (GST) fusion proteins. Thus, facilitating the use of recombinant enzymes to test the substrate specificity of the recombinant enzymes in vitro. Protein methyltransferases were cloned into pGEX vectors using standard molecular-biology techniques. The resulting constructs were transformed into competent bacterial cells (*BL21*).

1. Inoculate single colony of transformed *E. coli* into 5 mL of LB broth containing 50 µg/mL Ampicillin. Grow overnight at 37°C with shaking (250 rpm).
2. Next morning, add the 5 mL overnight bacterial culture to 50 mL of LB broth containing 50 µg/mL Ampicillin (*see* **Note 1**).
3. Culture at 37°C for 1–2 h until optical density (OD_{600}) reaches 0.5.
4. Add 50 µL of 0.1 *M* IPTG to induce recombinant protein expression.
5. Culture at 37°C for an additional 4 h shaking (250 rpm).
6. Centrifuge cells at 3000 *g* for 10 min at 4°C (The pellet can be stored at −70°C).
7. Resuspend pellet in 5 mL of cold 1X PBS.
8. Sonicate for 20 s with pulses of 0.5 s on and 0.5 s off (amplitude 30%).
9. Centrifuge at 10,000 *g* for 15 min at 4°C.
10. Wash Glutathione sepharose beads once with chilled 1X PBS, then put 30 µL of rinsed beads into a 1.5 mL microcentrifuge tube and add the supernatant from **step 9**.
11. Rock the sample tubes for 1 h at 4°C.
12. Wash the beads three times with cold 1X PBS.
13. Add 100 µL of freshly made Elution buffer to the beads and rock for 15 min at 4°C.
14. Centrifuge at 700 *g* for 5 s and remove the supernatant carefully. The supernatant contains the active enzyme.
15. Keep the samples at 4°C for 1–2 d or directly use the enzyme (*see* **Note 2**).
16. An aliquot of the supernatant should be analyzed to confirm that recombinant protein has been purified. Add 2 µL of 6X protein loading buffer to 10 µL of the purified GST-fusion protein sample, heat at 95°C for 5 min and run on a 10% SDS-PAGE gel at 100 volts for 1 h using Tris/Glycine running buffer. Stain with Coommassie Blue staining solution for 30 min, then destain with 30% ethanol and 10% glacial acetic acid for 1 h (*see* **Fig. 1**).

3.2. In Vitro Methylation Assay Using Enzymes and Their Substrates

Both the protein arginine methyltransferases and the SET domain-containing lysine methyltransferases display a large degree of substrate specificity. Different PRMTs can methylate RNA binding proteins, myelin basic protein (MBP), histones, and growth factors (*16,23–26*). Lysine methylated proteins include histones and elongation factor 1A (*27,28*). In vivo methylation assays are used to establish whether a newly discovered methyltransferase is active as a GST

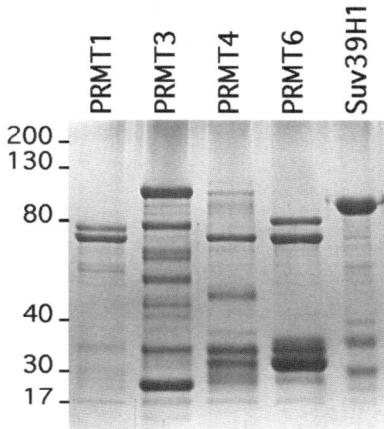

Fig. 1. Purification of recombinant methyltransferase enzymes. Different methyltransferases fused to glutathione *S*-transferase (GST) were expressed in *E. coli* and batch-purified. The GST fusion proteins (1–2 µg) were separated by 10% SDS-PAGE and stained with Coomassie blue. Fusion proteins of the arginine methyltransferase family include PRMT1, 3, 4, and 6. The SET domain-depending lysine methyltransferase Suv39H1 is in the last lane. The molecular-mass markers are shown on the left in kDa.

fusion protein, and to determine the specificity of protein methyltransferases. The substrates used for in vitro methylation reactions can be recombinant proteins (purified as in **Subheading 3.1.**) or histones purified by acid extraction. For best results, always use freshly prepared recombinant enzymes.

1. In a 1.5-mL microcentrifuge tube mix: 1 µg of substrate, 1 µg of recombinant enzyme, and 1 µL of *S*-adenosyl-L-[*methyl*-³H]methionine, 3 µL of 10X PBS, and add H_2O up to 30 µL.
2. Mix the tube by tapping, then centrifuge for 3 s.
3. Incubate sample tubes at 30°C for 1 h.
4. Stop the reaction by adding 6 µL of 6X SDS protein sample loading buffer and heat at 95°C for 5 min.
5. Then run 15 µL of the reaction on a 10% SDS-PAGE gel at 100 volts for 1 h using Tris/Glycine running buffer.
6. The separated samples are then transferred from the gel to a PVDF membrane using a semidry electroblotter.
7. The PVDF membrane harboring the immobilized protein samples is then sprayed with EN³HANCE three times, wait 10 min between each application (*see* **Note 3**).
8. The PVDF membrane is finally left to fully dry for 30 min and then expose to X-ray film overnight. Results of an in vitro methylation assay, using five

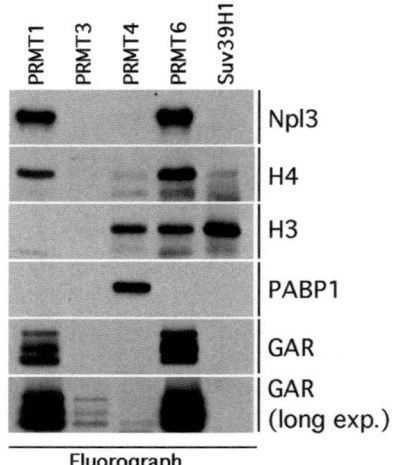

Fig. 2. A demonstration of methyltransferase in vitro substrate specificity. Recombinant PRMT1 (1 µg), PRMT3, PRMT4, PRMT6, and Suv39H1 methyltransferases were incubated with 1 µg of various substrates (Npl3, histone H3, histone H4, PABP1, and GAR) in vitro in the presence of 0.6 µM [³H]AdoMet for 30 min at 30°C in a final volume of 30 µL of PBS. The methylated proteins were separated by SDS-PAGE, transferred to a PVDF membrane, sprayed with EN³HANCE, and the membranes were exposed to X-ray film for 8 h or overnight (long exp.).

different recombinant enzymes and five different substrates, are depicted in **Fig. 2**.

3.3. In Vivo Methylation Assay Using Immunoprecipitation Method

A technique, using tritiated methionine, has been developed that allows the in vivo labeling of cellular proteins that are methylated *(28,29)*. This assay relies on the fact that the methyl group on the universal methyl donor *S*-adenosylmethionine is derived from free methionine in the cell. Cells are labeled with [*methyl*-³H]-ʟ-methionine in the presence of protein-synthesis inhibitor, thus preventing the incorporation of isotope into nascently synthesized proteins, while allowing the labeling of methylated proteins (*see* **Note 4**). This protocol is described here in combination with an immunoprecipitation step to determine the in vivo methylation status of a single endogenous protein.

1. Propagate actively growing HeLa cells on a 10-cm culture plate until they reach 80% confluence.
2. Wash the cells with 1X PBS and add 10 mL of growth medium A.
3. Incubate for 30 min at 37°C in a tissue culture incubator.

4. Wash the cells once with 10 mL of growth medium B.
5. Add 5 mL of medium B containing 50 µCi of L-[*methyl-*³H]methionine.
6. Incubate for 3 h at 37°C in a tissue culture incubator.
7. Wash the cells once with cold 1X PBS.
8. Add 500 µL of cold mild lysis buffer.
9. Dislodge the cells from the culture plate with a plastic scraper.
10. Transfer solution into a 1.5-mL microcentrifuge tube and rock for 30 min at 4°C.
11. Centrifuge at 10,000 *g* for 10 min and keep the supernatant.
12. While the incubation at **step 6** is in progress, wash 20 µL of protein A/G agarose beads with cold mild lysis buffer. Then add 1 µg of each antibody in 300 µL of cold mild lysis buffer to the Protein A/G agarose beads.
13. Incubate for 30 min with rocking at 4°C to allow binding of the antibody to the protein A/G agarose beads.
14. Centrifuge at 700 *g* for 5 s and discard the supernatant.
15. Add cell supernatant from **step 11** to the beads and rock for 2 h at 4°C.
16. Centrifuge at 700 *g* for 5 s and wash the beads three times with cold mild lysis buffer.
17. Add 20 µL of 2X protein sample loading buffer and heat at 95°C for 5 min.
18. The immunoprecipitate is separated by SDS-PAGE at 100 volts for 1 h using Tris/Glycine running buffer.
19. The separated samples are then transferred from the gel to a PVDF membrane using a semidry electroblotter.
20. The PVDF membrane harboring the immobilized protein samples is then sprayed with EN³HANCE three times, with a 10-min delay between each application.
21. The PVDF membrane is finally left to fully dry for 30 min and then expose to X-ray film overnight.

Results of an in vivo methylation assay are depicted in **Fig. 3**. The poly(A)-binding protein (PABP1) is a PRMT4 substrate and is not methylated in *Prmt4*⁻/⁻ cells. The methylation status of the PRMT1 substrate, Sam68, is unaffected by the loss of PRMT4 activity.

3.4. Using Global-Methylation Inhibitors as a Tool to Study Protein Methylation

There are two types of small molecule that inhibit the function of AdoMet-dependent methyltransferases: (1) compounds that are structural analogs of AdoMet and thus compete for the cofactor binding site, and (2) nucleoside inhibitors of AdoHcy hydrolase that cause the accumulation of intracellular AdoHcy levels and ultimately feedback inhibit most methylation reactions.

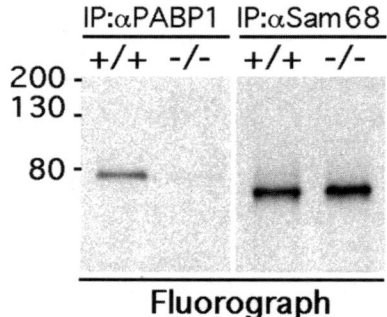

Fig. 3. In vivo methylation assay. Proteins in wild-type ($+/+$) and *Prmt4* mutant ($-/-$) MEF cell lines were labeled in vivo with L-[*methyl*-^{3}H]methionine. Immunoprecipitations (IP) were performed with αPABP1 (a PRMT4 substrate) and αSam68 (a PRMT1 substrate) antibodies. The ^{3}H-labeled proteins were visualized by fluorography. The molecular mass markers are shown on the left in kDa.

Sinefungin is a commonly used AdoMet analogue and adenosine dialdehyde (AdOx) is often used as an AdoHcy hydrolase inhibitor. Because demethylase activity is very low or absent in cells (*see* **Note 5**), endogenous protein substrates are generally fully-methylated. Cells in culture can be treated with AdOx to generate hypomethylated protein extracts that are good all-purpose in vitro substrates for methyltransferases (*12,30*). Here we describe using AdOx in in vivo methylation assays to affirm that tritium labeling is owing to the transfer of a methyl group by an AdoMet-dependent methyltransferase.

3.4.1. AdOx Treatment of Cultured Cells and In Vivo Protein Methylation

A stock solution of AdOx (0.5 *M*) is prepared in dimethyl sulfoxide (DMSO). HeLa cells are grown on a 10 cm culture plate until they are 60% confluent. The cells are then incubated with AdOx at a final concentration of 20 µ*M* (*see* **Note 6**). After 24 h of AdOx treatment the cells are subjected to an in vivo methylation labeling. The procedure is the same as described in **Subheading 3.3.**, except growth medium A and B are now supplemented with 20 µ*M* AdOx.

3.4.2. Immunoprecipitation Using αSam68 Antibodies

The RNA binding protein, Sam68, is a well-described substrate for PRMT1 (*31*), and is frequently used as a positive control in methylation experiments. An immunoprecipitation is performed using αSam68 antibodies and the endogenous Sam68 is separated by SDS-PAGE, transferred to a PVDF membrane and subjected to fluorography as described in **steps 20** and **21** of **Subheading 3.3.** (*see* **Fig 4**). After fluorography the same membrane can be washed (*see*

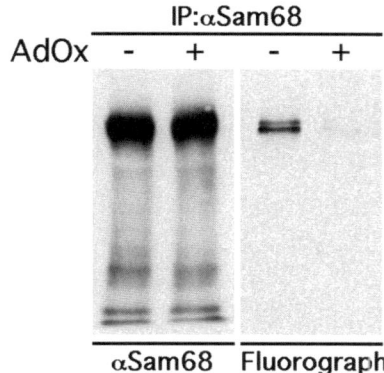

Fig. 4. The methyltransferase inhibitor, AdOx, prevents Sam68 methylation in vivo. HeLa cells were grown for 24 h in the presence of 20 μM AdOx. These cells were then subjected to an in vivo labeling reaction, in the presence of 20 μM AdOx. An IP was performed with a αSam68 antibody. The ^{3}H-labeled proteins were visualized by fluorography (right panel). After fluorography, the same membrane was washed and immunoblotted with an αSam68 antibody (left panel).

Note 3) and reanalyzed by Western with an αSam68 antibody, to ensure equal loading.

3.5. Methylation Assay Using Cell Extracts as an Enzyme Source

Protein methyltransferases are being knocked-out in the mouse by homologous recombination, resulting in the availability of cell lines (ES and MEF) that are null for specific enzyme activities. These cellular reagents provide us with a tool to quickly interrogate if a substrate of interest is indeed methylated as well as query the specificity, or lack of redundancy of a methylation reaction. This technique uses recombinant substrates fused to GST. While these substrates are immobilized on glutathione sepharose, an in vitro methylation reaction is performed using cell extracts as a source of methyltransferase activity to transfer a tritium-labeled methyl group from AdoMet onto GST fusion proteins harboring the methylatable motifs. Here we demonstrate this approach using PABP1 (a PRMT4 substrate [25,32]) and Npl3 (a PRMT1/HMT1 substrate [33,34]). Cell extracts from $Prmt4^{+/+}$ and $Prmt4^{-/-}$ MEF lines were used as the enzyme source.

1. Purify substrates as in **Subheading 3.1.** Stop at **step 12** and do not elute samples with elution buffer.
2. Wash the immobilized substrate once with cold PBS.

3. An aliquot of the immobilized protein is roughly quantitated by SDS-PAGE separation and Coomassie staining, using 1 μg, 2 μg, and 10 μg of BSA as a reference.
4. Propagate actively growing MEFs on a 10-cm culture plate until they reach 80% confluence.
5. Wash once with cold 1X PBS.
6. Add 500 μL of cold 1X PBS.
7. Dislodge the cells from the culture plate with a plastic scraper.
8. Transfer solution into 1.5-mL tube.
9. Sonicate for 10 s with pulses of 0.5 s on and 0.5 s off (amplitude 30%) to disrupt the cells.
10. Centrifuge at 10,000 g for 10 min at 4°C.
11. Transfer the supernatant (300 μL) into the 1.5-mL tube containing 1 μg of GST-fusion protein immobilized on 20 μL of glutathione sepharose from **step 3**.
12. Add 3 μL of *S*-adenosyl-L-[*methyl-*^{3}H]methionine.
13. Incubate for 2 h at 30°C.
14. Wash the beads three times with cold 1X PBS.
15. Add 30 μL of 2X protein sample loading buffer and heat at 95°C for 5 min.
16. The methylated substrates are separated by SDS-PAGE at 100 volts for 1 h using Tris/Glycine running buffer.
17. The separated samples are then transferred from the gel to a PVDF membrane using a semidry electroblotter.
18. The PVDF membrane harboring the immobilized protein samples is then sprayed with EN^{3}HANCE three times, with a 10 min delay between each application.
19. The PVDF membrane is finally left to fully dry for 30 min and then expose to X-ray film overnight.

Results of an in vitro methylation assay using cell extracts as an enzyme source are depicted in **Fig. 5**.

4. Notes

1. Certain GST fusion proteins are produced better than others. The induction volume should be adjust according to the efficiency of recombinant protein production and has to be determined empirically.
2. We have found that the majority of the GST-PRMTs do not retain their activity after freezing at −70°C. The exception to this rule is PRMT4/CARM1, which is stable when stored in 10% glycerol at −70°C. Other recombinant methyltransferases are prepared, stored at 4°C, and used within 2 d.

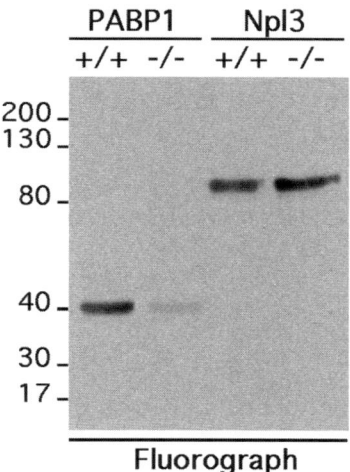

Fig. 5. Cell extracts from *Prmt4−/−* cells are unable to methylate recombinant PABP1. GST-fusion proteins of PABP1 and Npl3 were left bound to glutathione sepharose beads and methylated, in the presence of 0.12 μ*M* [³H]AdoMet for 2 h at 30°C, using the wild-type and mutant cell extracts as an enzyme source. After the methylation reaction the beads harboring the recombinant substrate were washed and eluted by boiling with protein sample loading buffer. Proteins were run on a gel, transferred onto a membrane, sprayed with EN³HANCE, and exposed to X-ray film overnight. The molecular-mass markers are shown on the left in kDa.

3. Traditionally, fluorography for the detection of a tritium signal is performed by soaking a gel in an enhance solution, drying the gel in vacuo, and then exposing the gel to film at −70°C. We have found that our signal intensity is much higher if we transfer the protein onto a PVDF membrane and then spray the PVDF membrane with EN³HANCE aerosol. This approach also has the added advantage of allowing the fluorograph to be analyzed by Western, to ensure equal loading. To do a Western on a PVDF membrane (postfluorography), simply wash it two times with 100% methanol and two times in TBST (Tris-buffered saline with 0.1% Tween-20) to remove the enhance reagent. The washed membrane is then blocked with a skim milk solution followed by a Western blot analysis.

4. It is important to control for the effectiveness of the protein synthesis inhibitors. If protein synthesis is not fully inhibited, then tritium-labeled methionine will be incorporated into newly synthesized proteins and will distort the assay. To confirm the effectiveness of the protein-synthesis inhibitors, one 10-cm culture plate of HeLa cells can be incubated for 30 min in growth medium A and then labeled with 5 mL of medium B

containing 50 µCi of L-[^{35}S]methionine for 3 h. A second plate can be labeled in the absence of protein synthesis inhibitors. Cells from the two plates are lysed in protein loading buffer and an aliquot (25 µL/500 µL) of the protein is separated by SDS-PAGE, and then exposed to X-ray film overnight.

5. The current thinking is that there is no demethylase to counteract the activities of PRMTs and the SET domain-containing lysine methyltransferases *(35)*. The irreversible nature of *N*-arginine and *N*-lysine methylation results in the steady accumulation of these modifications.

6. It has been established that 20 µ*M* of AdOx effectively prevents protein methylation in PC12 *(36)* and RAT1 *(37)* cells. AdOx is cytotoxic at micromolar concentrations and AdOx-treated cells stop growing and lose viability.

Acknowledgments

We wish to thank P. A. Silver for the GST-Npl3 vector, M. Stallcup for the GST-CARM1 vector, S. Clarke for the GST-PRMT1 and GST-GAR vectors, T. Jenuwein for the GST-Suv39H1 vector, and S. Richard for the αSam68 antibody. Mark T. Bedford is supported by NIH grant number DK62248-01.

References

1. Aletta, J. M., Cimato, T. R., and Ettinger, M. J. (1998) Protein methylation: a signal event in post-translational modification. *Trends Biochem. Sci.* **23,** 89–91.

2. Comb, D. G., Sarkar, N., and Pinzino, C. J. (1966) The methylation of lysine residues in protein. *J. Biol. Chem.* **241,** 1857–1862.

3. Paik, W. K. and Kim, S. (1968) Protein methylase I. Purification and properties of the enzyme. *J. Biol. Chem.* **243,** 2108–2114.

4. Mowen, K. A., Tang, J., Zhu, W., et al. (2001) Arginine methylation of STAT1 modulates IFNalpha/beta-induced transcription. *Cell* **104,** 731–741.

5. Abramovich, C., Yakobson, B., Chebath, J., and Revel, M. (1997) A protein-arginine methyltransferase binds to the intracytoplasmic domain of the IFNAR1 chain in the type I interferon receptor. *EMBO J.* **16,** 260–266.

6. Strahl, B. D. and Allis, C. D. (2000) The language of covalent histone modifications. *Nature* **403,** 41–45.

7. McBride, A. E. and Silver, P. A. (2001) State of the arg: protein methylation at arginine comes of age. *Cell* **106,** 5–8.

8. Friesen, W. J., Massenet, S., Paushkin, S., et al. (2001) SMN, the product of the spinal muscular atrophy gene, binds preferentially to dimethylarginine-containing protein targets. *Mol. Cell* **7,** 1111–1117.

9. Brahms, H., Meheus, L., de Brabandere, V., et al. (2001) Symmetrical dimethylation of arginine residues in spliceosomal Sm protein B/B and the Sm-like protein LSm4, and their interaction with the SMN protein. RNA **7,** 1531–1542.

10. Bedford, M. T., Frankel, A., Yaffe, M. B., et al. (2000) Arginine methylation inhibits the binding of proline-rich ligands to Src homology 3, but not WW, domains. *J. Biol. Chem.* **275**, 16,030–16,036.
11. Gary, J. D. and Clarke, S. (1998) RNA and protein interactions modulated by protein arginine methylation. *Prog. Nucleic Acid Res. Mol. Biol.* **61**, 65–131.
12. Frankel, A., Yadav, N., Lee, J., et al. (2002) The novel human protein arginine *N*-methyltransferase PRMT6 is a nuclear enzyme displaying unique substrate specificity. *J. Biol. Chem.* **277**, 3537–3543.
13. Lin, W. J., Gary, J. D., Yang, M. C., et al. (1996) The mammalian immediate-early TIS21 protein and the leukemia-associated BTG1 protein interact with a protein-arginine N-methyltransferase. *J. Biol. Chem.* **271**, 15,034–15,044.
14. Scott, H. S., Antonarakis, S. E., Lalioti, M. D., et al. (1998) Identification and characterization of two putative human arginine methyltransferases (HRMT1L1 and HRMT1L2). *Genomics* **48**, 330–340.
15. Tang, J., Gary, J. D., Clarke, S., and Herschman, H. R. (1998) PRMT 3, a type I protein arginine N-methyltransferase that differs from PRMT1 in its oligomerization, subcellular localization, substrate specificity, and regulation. *J. Biol. Chem.* **273**, 16,935–16,945.
16. Chen, D., Ma, H., Hong, H., et al. (1999) Regulation of transcription by a protein methyltransferase. *Science* **284**, 2174–2177.
17. Branscombe, T. L., Frankel, A., Lee, J. H., et al. (2001) Prmt5 (janus kinase-binding protein 1) catalyzes the formation of symmetric dimethylarginine residues in proteins. *J. Biol. Chem.* **276**, 32,971–32,976.
18. Rea, S., Eisenhaber, F., O'Carroll, D., et al. (2000) Regulation of chromatin structure by site-specific histone H3 methyltransferases. *Nature* **406**, 593–599.
19. Santos-Rosa, H., Schneider R., Bannister A. J., et al. (2002) Active genes are trimethylated at K4 of histone H3. *Nature* **419**, 407–411.
20. Gary, J. D. and Clarke, S. (1995) Purification and characterization of an isoaspartyl dipeptidase from Escherichia coli. *J. Biol. Chem.* **270**, 4076–4087.
21. Chevillard-Briet, M., Trouche, D., and Vandel, L. (2002) Control of CBP co-activating activity by arginine methylation. *EMBO J* **21**, 5457–5466.
22. Li, H., Park S., Kilburn B., et al. (2002) Lipopolysaccharide-induced methylation of HuR, an mRNA-stabilizing protein, by CARM1. Coactivator-associated arginine methyltransferase. *J. Biol. Chem.* **277**, 44,623–44,630
23. Sommer, A., Moscatelli, D., and Rifkin, D. B. (1989) An amino-terminally extended and post-translationally modified form of a 25kD basic fibroblast growth factor. *Biochem. Biophys. Res. Commun.* **160**, 1267–1274.
24. Baldwin, G. S. and Carnegie, P. R. (1971) Isolation and partial characterization of methylated arginines from the encephalitogenic basic protein of myelin. *Biochem. J.* **123**, 69–74.
25. Lee, J. and Bedford, M. T. (2002) PABP1 identified as an arginine methyltransferase substrate using high-density protein arrays. *EMBO Rep.* **3**, 268–273.
26. Davie, J. K. and Dent, S. Y. (2002) Transcriptional control: an activating role for arginine methylation. *Curr. Biol.* **12**, R59–R61.

27. Lachner, M., O'Sullivan, R. J., and Jenuwein, T. (2003) An epigenetic road map for histone lysine methylation. *J. Cell Sci.* **116,** 2117–2124.

28. Coppard, N. J., Clark, B. F., and Cramer, F. (1983) Methylation of elongation factor 1 alpha in mouse 3T3B and 3T3B/SV40 cells. *FEBS Lett.* **164,** 330–334.

29. Liu, Q. and Dreyfuss, G. (1995) In vivo and in vitro arginine methylation of RNA-binding proteins. *Mol. Cell Biol.* **15,** 2800–2808.

30. Frankel, A. and Clarke, S. (1999) RNase treatment of yeast and mammalian cell extracts affects in vitro substrate methylation by type I protein arginine *N*-methyltransferases. *Biochem. Biophys. Res. Commun.* **259,** 391–400.

31. Cote, J., Boisvert, F. M., Boulanger, M. C., et al. (2003) Sam68 RNA binding protein is an in vivo substrate for protein arginine N-methyltransferase 1. *Mol. Biol. Cell* **14,** 274–287.

32. Yadav, N., Lee, J., Kim, J., et al. (2003) Specific protein methylation defects and gene expression perturbations in coactivator-associated arginine methyltransferase 1-deficient mice. *Proc. Natl. Acad. Sci. USA* **100,** 6464–6468.

33. Henry, M. F. and Silver, P. A. (1996) A novel methyltransferase (Hmt1p) modifies poly(A)+-RNA-binding proteins. *Mol. Cell Biol.* **16,** 3668–3678.

34. Siebel, C. W. and Guthrie, C. (1996) The essential yeast RNA binding protein Np13p is methylated. *Proc. Natl. Acad. Sci. USA* **93,** 13,641–13,646.

35. Bannister, A. J., Schneider, R., and Kouzarides, T. (2002) Histone methylation: dynamic or static? *Cell* **109,** 801–806.

36. Najbauer, J. and Aswad, D. W. (1990) Diversity of methyl acceptor proteins in rat pheochromocytoma (PC12) cells revealed after treatment with adenosine dialdehyde. *J. Biol. Chem.* **265,** 12,717–12,721.

37. Frankel, A. and Clarke, S. (2000) PRMT3 is a distinct member of the protein arginine N-methyltransferase family. Conferral of substrate specificity by a zinc-finger domain. *J. Biol. Chem.* **275,** 32,974–32,982.

14

Assaying Lipid Phosphate Phosphatase Activities

Gil-Soo Han and George M. Carman

Summary

Lipid phosphate molecules such as phosphatidate, lysophosphatidate, and diacylglycerol pyrophosphate play roles as signaling molecules in prokaryotic and eukaryotic cells. The cellular processes by which lipid phosphate molecules signal may be attenuated through the action of lipid phosphate phosphatase enzymes. The levels of lipid phosphate phosphatase activities may be used as a marker of signaling events in the cell. In this chapter we describe enzymatic assays that are routinely used to measure the activities of phosphatidate phosphatase, lysophosphatidate phosphatase, and diacylglycerol pyrophosphate phosphatase. These activities are measured by following the release of water-soluble radioactive inorganic phosphate from chloroform-soluble radioactive lipid phosphate substrate following a simple chloroform/methanol/water phase partition.

Key Words: Diacylglycerol pyrophosphate phosphatase; phosphatidate phosphatase; lysophosphatidate phosphatase; lipid signaling.

1. Introduction

Lipid phosphate phosphatases (LPPs) are integral membrane proteins that catalyze the dephosphorylation of a variety of lipid phosphates, including phosphatidate (PA), lysophosphatidate (lysoPA), and diacylglycerol pyrophosphate (DGPP) [1]. These enzymes are Mg^{2+}-independent and N-ethylmaleimide-insensitive, and the genes encoding them have been identified in diverse organisms from bacteria to mammals [2,3]. LPPs contain a three-domain lipid phosphatase motif that is essential for catalytic activity [1,4–6]. LPPs have been

From: *Methods in Molecular Biology, vol. 284:*
Signal Transduction Protocols
Edited by: R. C. Dickson © Humana Press Inc., Totowa, NJ

previously classified as type 2 PA phosphatases (PAP2s) to distinguish them from type 1 PA phosphatases (PAP1s) (*7,8*). PAP1 enzymes are Mg^{2+}-dependent and *N*-ethylmaleimide-sensitive, and they have distinct substrate specificity for PA. These enzymes have been identified in cytosol and membrane fractions, and they are thought to play a major role in phospholipid synthesis. However, the genes encoding them have not yet been identified.

The broad substrate specificity of LPPs on bioactive lipid phosphates suggests that these enzymes are involved in signaling events rather than in phospholipid synthesis (*9,10*). LPPs can play a role in signal transduction by terminating signaling events of lipid phosphates. Because the products of LPPs are also bioactive lipid molecules, they can initiate signal transduction by producing signaling molecules. The expression of LPP activities is likely to modulate the balance of signaling molecules, eliciting differential physiological responses in the organism. In this chapter we describe the methods to measure LPP activity using enzymatically synthesized radioactive PA, lysoPA, and DGPP substrates. These activities are measured by following the release of water-soluble radioactive inorganic phosphate from chloroform-soluble radioactive lipid phosphate substrate following a simple chloroform/methanol/water phase partition.

2. Materials

1. Cardiolipin.
2. DG.
3. DGPP.
4. Monoacylglycerol.
5. PA.
6. 5X DG kinase buffer: 250 m*M* imdazole-HCl, pH 6.6, 250 m*M* octyl-β-D-glucopyranoside, 250 m*M* NaCl, 62.5 m*M* $MgCl_2$, 5 m*M* EGTA, 50 m*M* β-mercaptoethanol, 25 m*M* ATP.
7. 5X PA kinase buffer: 200 m*M* imidazole-HCl, pH 6.1, 50 m*M* $MgCl_2$, 500 m*M* NaCl, 0.5 m*M* EDTA, 2.5 m*M* DTT.
8. *Escherichia coli* (*E. coli*) DG kinase (Sigma, D3065).
9. *Catharanthus roseus* (*C. roseus*) PA kinase (*see* **Note 1**).
10. 3 m*M* PA in 2% Triton X-100.
11. 1 m*M* PA in 10 m*M* Triton X-100.
12. 1 m*M* DGPP in 20 m*M* Triton X-100.
13. 1% Potassium oxalate in methanol/water (2:3, v/v).
14. Anhydrous chloroform (*see* **Note 2**).
15. 0.1 *N* HCl in methanol.
16. 1 *N* NH_4OH in methanol.
17. 1 *M* $MgCl_2$.
18. 100 m*M* NaF.

19. 1 mM ATP.
20. [γ-^{32}P]ATP (3000 Ci/mmol, 5 mCi/mL).
21. Chloroform/methanol (2:1, v/v) containing 1% of concentrated HCl.
22. Chloroform/methanol/1 N HCl (1:2:0.8, v/v).
23. Scintillation fluid for aqueous samples (e.g., Ecoscint H, National Diagnostics LS-275).
24. Thin layer chromatography (TLC) solvent for the purification of PA and lysoPA: chloroform/methanol/water (65:25:4, v/v).
25. TLC solvent for the purification of DGPP: chloroform/acetone/methanol/acetic acid/water (50:15:13:12:4, v/v).
26. TLC plates: silica gel 60, 5 × 20 cm.
27. TLC chambers.
28. Polypropylene tubes (17 × 100 mm).
29. Polypropylene tubes with screw caps (17 × 100 mm).
30. Glass tubes (12 × 75 mm).
31. Scintillation vials.
32. Speed-Vac.

3. Methods

The methods described below outline: (1) the preparation of ^{32}P-labeled phospholipid substrates (PA, lysoPA, and DGPP), (2) assay method, and (3) data analysis.

3.1. Preparation of Substrates

E. coli DG kinase catalyzes the phosphorylation of both DG and monoacylglycerol as substrates. The ^{32}P-labeled PA and lysoPA are therefore enzymatically synthesized from [γ-^{32}P]ATP and DG, and [γ-^{32}P]ATP and monoacylglycerol, respectively (*11*). [β-^{32}P]DGPP is synthesized from PA and [γ-^{32}P]ATP using *C. roseus* PA kinase (*12*).

3.1.1. Synthesis and Purification of [^{32}P]PA and [^{32}P]lysoPA

1. Mix 10 µL of diacylglycerol or monoacylglycerol (25 mg/mL) and 1.8 µL of cardiolipin (5 mg/mL) in a polypropylene tube, and evaporate to dryness in a fume hood (for 10–20 min).
2. Add 40 µL of water and 20 µL of 5X DG kinase buffer. Suspend the lipids thoroughly by vortexing for 5 s.
3. Add 10 µL of DG kinase (4 U/mL) and 30 µL of [γ-^{32}P]ATP (5 mCi/mL). Mix and incubate for 40 min at 30°C.
4. Stop the reaction by adding 0.5 mL of 0.1 N HCl. Add 1.0 mL of chloroform and 1.5 mL of 1 M MgCl$_2$.
5. Mix the solutions by gentle vortexing and centrifuge for 3 min at 100 g.

6. Remove the aqueous phase by aspiration, and add 0.5 mL of 0.1 N HCl in methanol and 1.5 mL of 1 M MgCl$_2$.
7. Mix the solutions and centrifuge as before.
8. Remove the aqueous phase and transfer the chloroform phase to a glass tube.
9. Dry the lipids completely in the Speed-Vac (for about 30 min) and resuspend in 20 µL of anhydrous chloroform.
10. Spot the chloroform solution on the oxalate-treated TLC plate. Rinse the tube with 15 µL of chloroform and spot the samples again (*see* **Note 3**).
11. Develop the plate in a TLC chamber until the solvent front reaches about 2–3 cm from the top of the plate (for about 2 h).
12. Dry the TLC plates in the fume hood, wrap with plastic film, and expose to a photographic film or a phosphorimager screen (for 1–5 min).
13. Develop the film, or scan the image and print it to actual size.
14. Align the unwrapped plate with the film or printed image on a light box and mark the region of [^{32}P]PA or [^{32}P]lysoPA.
15. Moisten the radioactive spot on the TLC plate by using a water sprayer and place a sheet of weighing paper underneath the plate.
16. Scrape the silica off the plate using a new razor blade and transfer the silica into a 15-mL polypropylene centrifuge tube.
17. Add 1 mL of chloroform/methanol/1 N HCl (1:2:0.8, v/v) and mix by vigorous shaking.
18. Centrifuge for 3 min at 100 g and transfer the supernatant to a new 15-mL polypropylene centrifuge tube (*see* **Note 4**).
19. Repeat **steps 16–18**, and combine the extractions.
20. Add 0.5 mL of chloroform and 0.6 mL of 0.1 N HCl, and mix by vortexing.
21. Centrifuge for 3 min at 100 g and remove the aqueous phase by aspiration.
22. Add 1 mL of methanol and 1 mL of 0.1 N HCl, mix and centrifuge as before.
23. Remove the aqueous phase and transfer the chloroform phase to a glass tube.
24. Adjust the pH of the chloroform solution to neutral with 1 N NH$_4$OH in methanol (*see* **Note 5**) and dry completely in a Speed-Vac (for about 30 min).
25. Resuspend the radioactive material in an appropriate volume of 1 mM PA or 1 mM lysoPA in 10 mM Triton X-100 (*see* **Note 6**) to adjust the specific label to 10,000 cpm/nmol. If not for immediate use, store the radioactive substrates at −20°C.

3.1.2. Synthesis of [β-^{32}P]DGPP

1. Add the following reagents in a polypropylene centrifuge tube:

5X PA kinase buffer	20 µL
3 mM PA in 2% Triton X-100	10 µL
1 mM ATP	10 µL
100 mM NaF	5 µL
Water	15 µL

> [γ-^{32}P]ATP (5 mCi/mL)   30 µl
>
> PA kinase (4 U/mL)   10 µL

2. Incubate the reaction mixture overnight at 30°C.
3. Add 1.5 mL of chloroform/methanol (2:1, v/v) containing 1% concentrated HCl, and 0.5 mL of water.
4. Mix by gentle vortexing and centrifuge for 3 min at 100*g*.
5. Remove the aqueous phase by aspiration, and add 0.5 mL of 0.1 *N* HCl in methanol and 0.5 mL of water.
6. Mix by vortexing and centrifuge as before.
7. Remove the aqueous phase by aspiration and transfer the chloroform phase to a glass tube.
8. Follow **steps 9–24** as described for synthesis of PA and lysoPA in **Subheading 3.1.1.**
9. Resuspend the dried radioactive DGPP in an appropriate volume of 1 m*M* DGPP in 20 m*M* Triton X-100 to adjust the specific label of [β-^{32}P] DGPP to 10,000 cpm/nmol. If not for immediate use, store the radioactive substrates at −20°C.

3.2. Assay

PA, lysoPA, and DGPP phosphatase activities are measured for 20 min at 30°C by following the release of water-soluble [^{32}P]P$_i$ from chloroform-soluble [^{32}P]PA, [^{32}P]lysoPA, and [β-^{32}P]DGPP (10,000 cpm/nmol) in a total volume of 0.1 mL. The principle of all three assays is the same. They differ only in the substrates used and the assay conditions (e.g., pH) specific for each enzyme. For example, the yeast *LPP1*-encoded enzyme has different assay conditions depending on substrates (*13*). All assays are conducted in triplicate.

3.2.1. PA and LysoPA Phosphatase Assays

1. Add the following reagents in a polypropylene centrifuge tube:

   500 m*M* Tris-maleate, pH 7.0   10 µL

   100 m*M* β-mercaptoethanol   10 µL

   Water   60 µL

   Enzyme (1.0 mg/mL) (*see* **Note 7**)   10 µL
2. Add 10 µL of 1 m*M* [^{32}P]PA (10,000 cpm/nmol) or lysoPA (10,000 cpm/ nmol) in 10 m*M* Triton X-100 and incubate the reaction mixture for 20 min at 30°C. Add the radioactive substrate to other tubes in 15-s intervals.
3. Stop the reaction by adding 0.5 mL of 0.1 *N* HCl in methanol. Add 1 mL of chloroform and 1 mL of 1 *M* MgCl$_2$ (*see* **Note 8**).
4. Mix by gentle vortexing and centrifuge for 3 min at 100 *g*.
5. Transfer 0.5 mL of aqueous phase to a scintillation vial and add 4 mL of scintillation fluid (*see* **Note 9**).
6. Mix by shaking and measure the radioactivity.

3.2.2. DGPP Phosphatase Assay

1. Add the following reagents to a polypropylene centrifuge tube:

   500 m*M* Citrate buffer, pH 5.0   10 µL
   Water   60 µL
   100 m*M* β-mercaptoethanol   10 µL
   Enzyme (1.0 mg/mL)   10 µL

2. Add 10 µL of 1 m*M* [β-^{32}P]DGPP (10,000 cpm/nmol) in 20 m*M* Triton X-100 and incubate the reaction mixture for 20 min at 30°C. Add the radioactive substrate to other tubes in 15-s intervals.

3. Follow **steps 3–6** described for PA phosphatase assay in **Subheading 3.2.1.**

3.3. Analysis of Data

The specific activity of an enzyme is expressed by U/mg protein. One unit of PA, lyso PA, or DGPP phosphatase is defined as the amount of enzyme that catalyzes the formation of 1 nmol of product per min. Apply the following formula to calculate the specific activity:

Specific activity (nmol/min/mg protein) = {3.2 (correction factor) × cpm (corrected for background)} /{specific label (cpm/nmol) × time (min) × protein volume (mL) × protein concentration (mg/mL)} (*see* **Note 10**).

4. Notes

1. PA kinase is not commercially available. PA kinase can be purified from *C. roseus* cells as described by Wissing and Behrbohm (**12**). If pure PA kinase is not available, membrane preparations from red beet or broccoli can be used as alternative sources of PA kinase.

2. Anhydrous chloroform is prepared by adding solid sodium sulfate to a small bottle of chloroform. Shake well and allow the sodium sulfate to settle. This reagent can be used for several weeks.

3. TLC plates are pretreated with potassium oxalate by soaking them for few seconds in methanol/water (2:3) containing 1% potassium oxalate. The oxalate-treated TLC plates are dried at room temperature and then incubated for at least 30 min at 100–115°C to remove the residual moisture. The hot TLC plates are cooled to room temperature in a dessicator before use.

4. If phase separation occurs and silica is present in the aqueous phase, this is caused by excess water present in the silica. Add a small volume of methanol, mix, and centrifuge again.

5. 1 *N* NH$_4$OH in methanol is stored in the refrigerator and should be used within 2 wk after its preparation.

6. PA in chloroform (100–200 µL) is transferred into a glass tube and dried completely in a Speed-Vac (for about 30 min). The dried PA is measured

and resuspended in an appropriate volume of 10 mM Triton X-100 to the final concentration of 1 mM. If a chloroform solution of PA is out of the freezer for longer than a few minutes, it is kept on ice. Before returning to the freezer, the container is purged with nitrogen gas and sealed. To prevent light-induced oxidation, the container is wrapped with aluminum foil.

7. Crude enzyme preparations used in the assay are generally at a protein concentration of 1.0 mg/mL for a final concentration of 0.1 mg/mL. Be sure that assays are linear with time and protein by varying reaction time and protein concentration.

8. If many samples are routinely assayed, a bottle-top dispenser is efficient to deliver solutions (0.1 N HCl in methanol, chloroform, and 1 M MgCl$_2$). During the incubation of reaction mixtures, prime the dispenser for proper delivery.

9. Make sure that the scintillation fluid is for aqueous samples.

10. In calculating the specific activity, a correction factor of 3.2 is applied because 0.5 mL out of 1.6 mL (total sample volume) is used for the measurement.

Acknowledgment

We wish to thank June Oshiro for helpful comments during the preparation of this chapter.

References

1. Brindley, D. N. and Waggoner, D. W. (1998) Mammalian lipid phosphate phosphohydrolases. *J. Biol. Chem.* **273**, 24,281–24,284.

2. Carman, G. M. (1997) Phosphatidate phosphatases and diacylglycerol pyrophosphate phosphatases in *Saccharomyces cerevisiae* and *Escherichia coli*. *Biochim. Biophys. Acta* **1348**, 45–55.

3. Kanoh, H., Kai, M., and Wada, I. (1997) Phosphatidic acid phosphatase from mammalian tissues: discovery of channel-like proteins with unexpected functions. *Biochim. Biophys. Acta* **1348**, 56–62.

4. Stukey, J. and Carman, G. M. (1997) Identification of a novel phosphatase sequence motif. *Protein Sci.* **6**, 469–472.

5. Neuwald, A. F. (1997) An unexpected structural relationship between integral membrane phosphatases and soluble haloperoxidases. *Protein Sci.* **6**, 1764–1767.

6. Hemrika, W., Renirie, R., Dekker, H. L., et al. (1997) From phosphatases to vanadium peroxidases: a similar architecture of the active site. *Proc. Nat. Acad. Sci. USA* **94**, 2145–2149.

7. Jamal, Z., Martin, A., Gomez-Munoz, A., and Brindley, D. N. (1991) Plasma membrane fractions from rat liver contain a phosphatidate phosphohydrolase distinct from that in the endoplasmic reticulum and cytosol. *J. Biol. Chem.* **266**, 2988–2996.

8. Brindley, D. N. (1984) Intracellular translocation of phosphatidate phosphohydrolase and its possible role in the control of glycerolipid synthesis. *Prog. Lipid Res.* **23**, 115–133.

9. Brindley, D. N., English, D., Pilquil, C., et al. (2002) Lipid phosphate phosphatases regulate signal transduction through glycerolipids and sphingolipids. *Biochim. Biophys. Acta.* **1582,** 33–44.

10. Sciorra, V. A. and Morris, A. J. (2002) Roles for lipid phosphate phosphatases in regulation of cellular signaling. *Biochim. Biophys. Acta* **1582,** 45–51.

11. Walsh, J. P. and Bell, R. M. (1986) sn-1,2-diacylglycerol kinase of *Escherichia coli.* Structural and kinetic analysis of the lipid cofactor dependence. *J. Biol. Chem.* **261,** 15,062–15,069.

12. Wissing, J. B. and Behrbohm, H. (1993) Phosphatidate kinase, a novel enzyme in phospholipid metabolism. Purification, subcellular localization, and occurrence in the plant kingdom. *Plant Physiol.* **102,** 1243–1249.

13. Furneisen, J. M. and Carman, G. M. (2000) Enzymological properties of the *LPP1*-encoded lipid phosphatase from *Saccharomyces cerevisiae. Biochim. Biophys. Acta* **1484,** 71–82.

15

Assaying Phosphoinositide Phosphatases

Gregory S. Taylor and Jack E. Dixon

Summary

The roles of phosphoinositide second messengers as signaling molecules in a vast array of cellular processes including cell growth, metabolism, vesicular transport, programmed cell death, and responses to extracellular signals are only beginning to be understood. The recent identification of novel phosphoinositide signaling molecules underscores the need for methodology with which to characterize the enzymes responsible for regulating cellular phosphoinositide levels. One of the ways in which cells control these lipids is through dephosphorylation by phosphoinositide phosphatases, which oppose and regulate the actions of phosphoinositide kinases. We describe herein two rapid and simple assays for characterizing phosphoinositide phosphatases that can be used to provide a basis for understanding the activity and specificity of these enzymes.

Key Words: Phosphatase; phosphoinositide; phosphatidylinositol; Ymr1p; myotubularin; PTEN; PTP.

1. Introduction

The role of phosphoinositides as second messengers in cellular signaling processes has been intensively studied over the last approx 25 yr. Perhaps the most widely known example of phosphoinositide signaling is the receptor-mediated hydrolysis of phosphatidylinositol 4,5-bisphosphate ($PI(4,5)P_2$) by phospholipase C to release diacylglycerol, an activator of protein kinase C (PKC), and inositol 1,4,5-trisphosphate, which causes the release of intracellular Ca^{2+}. The study of phosphoinositide signaling has recently enjoyed a dramatic resurgence owing to the identification of novel inositol lipids that are critical for a fantastic array of physiological functions that include such diverse

From: *Methods in Molecular Biology, vol. 284:*
Signal Transduction Protocols
Edited by: R. C. Dickson © Humana Press Inc., Totowa, NJ

cellular processes as growth, development, apoptosis, membrane trafficking, and vesicular transport, as well as signaling in the cell nucleus *(1–3)*. As might be expected for such important signaling molecules, abnormal phosphoinositide regulation has also been associated with several human diseases *(4,5)*.

In order to understand the mechanisms by which different phosphoinositides are able to carry out such a wide variety of signaling tasks, it is necessary to identify how these lipids are regulated both spatially and temporally. The levels of cellular phosphoinositides can be controlled by at least three types of enzymes including phospholipases, phosphoinositide kinases, and phosphoinositide phosphatases. Of these three groups of inositol lipid-modifying enzymes, the phospholipases and phosphoinositide kinases have historically enjoyed the more intensive scrutiny *(2,7)*. However, recent attention has been focused on the roles of phosphoinositide phosphatases as regulators of signaling lipids owing to their involvement in human diseases. This interest has been fueled in part by the discovery that PTEN and myotubularin family protein tyrosine phosphatase-like enzymes actually utilize inositol lipids rather than phosphoproteins as their physiological substrates *(8–10)*. In addition, Sac1 domain-containing lipid phosphatase and inositol polyphosphate 5′-phosphatase families have also been shown to play essential roles in regulating phosphoinositide-dependent cellular processes *(11,12)*. These examples provide compelling evidence that a complete understanding of inositol lipid-signaling pathways will require further insight into not only the mechanisms by which phospholipases and phosphoinositide kinases are regulated, but also by which the lipid phosphatases contribute to the overall regulation of cellular phosphoinositide levels. To this end, we describe herein methods for carrying out the in vitro assay of phosphoinositide phosphatase activity. We have found these approaches useful not only for quantitating the specific activities of these enzymes, but also for determining their substrate preferences among the different phosphoinositide species.

2. Materials

2.1. Fluorescent Phosphoinositide Substrate Assay

1. Fluorescent di-C6-NBD6 synthetic phosphoinositide substrates (PI, PI(3)P, PI(4)P, PI(5)P, PI(3,4)P_2, PI(3,5)P_2, PI(4,5)P_2, and PI(3,4,5)P_3 at a concentration of 1 µg/µL in dH_2O) (Echelon Biosciences, Salt Lake City, UT).
2. Fluorescent substrate assay buffer composed of 50 m*M* ammonium acetate and/or 50 m*M* ammonium carbonate buffer containing 0.1% (v:v) 2-mercaptoethanol (Sigma Chemicals, St. Louis, MO).
3. 1.2% Potassium oxalate in dH_2O/MeOH (6:4) (Sigma).
4. Chloroform, methanol, acetone, glacial acetic acid, 2-propanol (Fisher Scientific, Pittsburgh, PA).

5. Speed-Vac concentrator.
6. Dry bath incubator.
7. Glass-thin layer chromatography (TLC) tank.
8. Glass-backed Whatman silica gel TLC plates; 60Å, 250 µm thickness, 20 cm × 20 cm (Fisher Scientific).
9. DNA gel UV illumination/camera system.

2.2. Malachite Green-Based Assay for Inorganic Phosphate

1. Di-C_{16} synthetic phosphoinositide substrates (PI, PI(3)P, PI(4)P, PI(5)P, PI(3,4)P_2, PI(3,5)P_2, PI(4,5)P_2, and PI(3,4,5)P_3 at a concentration of 1 mM in $CHCl_3$/MeOH (9:1) (single phosphate group), or $CHCl_3$/MeOH/dH_2O (5:5:1) (two or more phosphate groups) (Echelon Biosciences, Inc.).
2. Chloroform, methanol (Fisher Scientific).
3. Dioleoylphosphatidylserine at a concentration of 10 mM in $CHCl_3$/MeOH (9:1) (Sigma Chemicals).
4. Malachite green assay buffer: 100 mM sodium acetate, 50 mM bis-Tris, 50 mM Tris (pH range from 5.0–8.0) containing 10 mM dithiothreitol (DTT).
5. 20 mM sodium orthovanadate in dH_2O.
6. Malachite green reagent (prepared as described in **Subheading 3.2.1.**) (Sigma Chemicals).
7. KH_2PO_4 inorganic phosphate standard (40 µM in dH_2O).
8. Speed-Vac concentrator.
9. Dry bath incubator.
10. Refrigerated microfuge.
11. Spectrophotometer with microcuvet (50 µL volume).
12. Probe sonicator: 130 watt ultrasonic cell disruptor with 13 mm probe tip (PGC Scientifics, Frederick, MD).

3. Methods

3.1. Fluorescent Phosphoinositide Substrate Assay

When testing a putative phosphoinositide phosphatase for lipid phosphatase activity, we have found it useful to conduct the preliminary assays with fluorescent phosphoinositide substrates *(13)*. This procedure is advantageous for an initial enzymatic characterization because it allows the detection of even low-level lipid phosphatase activity in a manner that is relatively insensitive to reaction conditions. This occurs primarily because many lipid phosphatases possess a high degree of substrate specificity such that the reaction can be allowed to proceed to a point where, even under nonoptimal conditions, sufficient substrate is hydrolyzed to allow visualization of the reaction product. In addition, because this technique employs TLC to follow the lipid moiety of the substrate rather than inorganic phosphate, it is useful for assessing not only the general substrate specificity of a lipid phosphatase (i.e., PIP vs PIP$_2$ vs PIP$_3$), but also for determining the specific phosphate group(s) on the inositol ring that are hydrolyzed when using multiply-phosphorylated phosphoinositide substrates. This procedure is carried out as follows:

1. Prepare a silica-gel TLC plate by completely immersing it in 1.2% potassium oxalate solution. Place the plate in a fume hood until visibly dry, followed by drying for 1 h in a baking oven (80–100°C).
2. In separate 0.5-mL microfuge tubes, combine 1 μL (1 μg total) of each phosphoinositide substrate [PI(3)P, PI(4)P, PI(5)P, PI(3,4)P$_2$, PI(3,5)P$_2$, PI(4,5)P$_2$, or PI(3,4,5)P$_3$] with 17 μL fluorescent substrate assay buffer (usually from pH 5.0–8.0) and prewarm the samples for 10 min in a dry bath (usually from 30 to 37°C).
3. Initiate the reactions by adding 2 μL of enzyme (usually from 5 ng to 5 μg of enzyme per sample) and incubate for 30–60 min.
4. Terminate the reactions by adding 100 μL acetone and dry the samples under vacuum in a Speed-Vac concentrator (approx 15–20 min using "medium" heat setting).
5. While the samples are drying, prepare the TLC tank by rinsing it twice with approx 100 mL of organic mobile phase (chloroform/methanol/acetone/glacial acetic acid/water, 75:50:20:20:20 v/v) as described by Okada et al. *(14)*. After rinsing, fill the tank to approx 1 cm deep with mobile phase (about 150 mL).
6. Dissolve the reaction products in 20 μL of spotting solution (chloroform/2-propanol/methanol/glacial acetic acid, 5:5:5:2 v/v) and adsorb (spot) each sample onto the prepared TLC plate. By spotting each sample across an area approx 0.5–1 cm wide at the origin, "bands" that are easier to visualize when compared to "spots" will be obtained. Be sure to also spot a sample corresponding to approx 1 μg of PI as a standard. Allow the spotted samples to dry completely (5–10 min in a fume hood).
7. Develop the thin layer plate in mobile phase until the solvent front has traveled approx three-fourths of the distance from the origin to the top of the plate. Remove the plate and allow it to dry completely in a fume hood.
8. Visualize fluorescent phosphoinositides by placing the thin layer plate face up on a DNA gel ultraviolet (UV) light box. Photographs can be taken with either a film or digital camera.

To illustrate the procedure outlined previously, we have tested a bacterial recombinant budding yeast myotubularin (Ymr1p) His-tagged fusion protein against fluorescent lipid substrates. We have previously reported that recombinant Ymr1p can effectively dephosphorylate PI(3)P *(15)*. However, recent re-

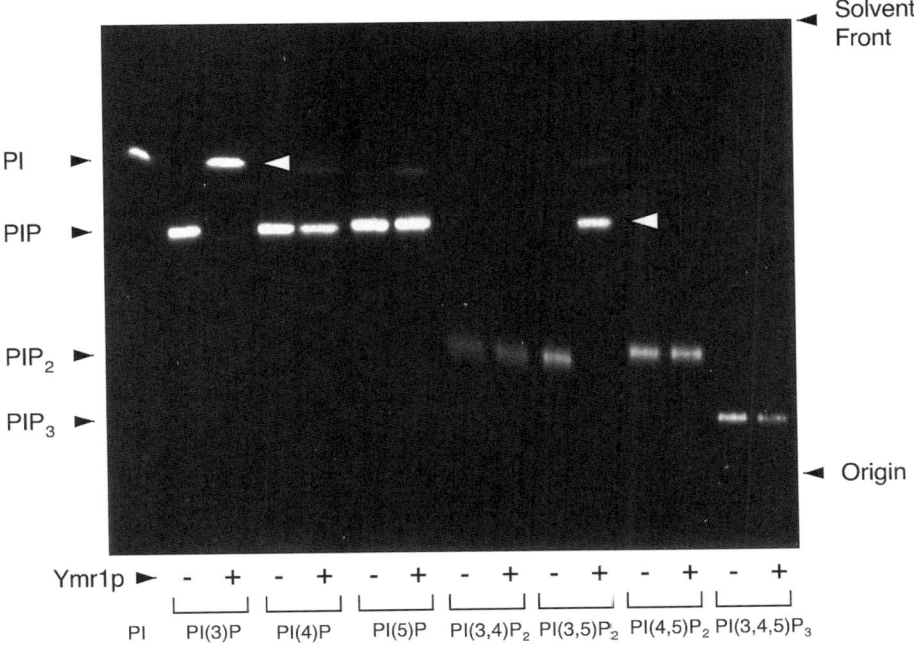

Fig. 1. Determination of the substrate specificity of recombinant Ymr1p using fluorescent phosphinositides. Phosphoinositide phosphatase assays were carried out with 1 μg of each fluorescent (NBD6) phosphoinositide derivative for 30 min at 30°C and either buffer alone (−), or 200 ng Ymr1p fusion protein (+) as described in the Methods section. The fluorescent phosphoinositide reaction products are shown following resolution by TLC and illumination under UV light. The PI and PI(5)P reaction products produced by the action of Ymr1p on PI(3)P and PI(3,5)P$_2$, respectively, are indicated by the white arrows at left and right. The migration positions of the different phosphoinositides are indicated at right and the origin and solvent front positions are shown at left.

ports have revealed that in addition to PI(3)P, myotubularin family enzymes can also utilize PI(3,5)P$_2$ as a substrate in vitro (*16–18*). As expected, Ymr1p completely dephosphorylated PI(3)P to PI (**Fig. 1**, white arrow at left). In addition, Ymr1p was also able to quantitatively convert PI(3,5)P$_2$ to PI(5)P, demonstrating that, like other myotubularin family phosphatases, Ymr1p can efficiently dephosphorylate the D3 position of PI(3,5)P$_2$ (**Fig. 1**, white arrow at right). In addition, a trace of PI can also be seen in the PI(4)P and PI(5)P (+) lanes, however, it is approaching the lower limit of detection in this assay and indicates that these are relatively poor substrates for Ymr1p. The usefulness of this approach for assessing the substrate specificity of phosphoinositide phosphatases is clearly shown in **Fig. 1**. However, additional information, such as enzyme specific activity or, in the case of Ymr1p, relative activity between two different inositol lipid substrates, is often desirable. To this end, we will describe next a malachite green-based assay for inorganic phosphate that we have adapted for use with phosphoinositide phosphatases.

3.2. Malachite Green-Based Assay for Inorganic Phosphate

Although the fluorescent phosphoinositide phosphatase assay described earlier is a useful tool for dissecting the specificity of lipid phosphatases, the primary drawback of this approach is that it is generally a qualitative rather than a quantitative assay *(13)*. To obtain a more quantitative assessment of lipid phosphatase activity, we routinely employ a malachite green-based assay for inorganic phosphate as a complement to the fluorescent phosphoinositide assay. We have adapted this method from the procedure of Harder et al. specifically for use with phosphoinositide phosphatases *(19,20)*. In this method, the enzyme-catalyzed release of inorganic phosphate from synthetic phosphoinositide substrates is quantitated by comparison to a standard curve of inorganic phosphate using a colorimetric assay. This procedure is carried out as follows:

1. Prepare the malachite green reagent by adding one volume of 4.2% (w:v) ammonium molybdate in 4 N HCl to 3 volumes of 0.045% (w:v) aqueous malachite-green solution. Stir the solution for approx 30 min prior to filtration through a 0.22 μm filter. This solution is stable for several months at 4°C. Just before use, add Tween 20 to a final concentration of 0.01% (v:v).

2. In separate 1.5-mL microfuge tubes, combine an aliquot of each 1 mM phosphoinositide substrate (PI(3)P, PI(4)P, PI(5)P, PI(3,4)P$_2$, PI(3,5)P$_2$, PI(4,5)P$_2$, or PI(3,4,5)P$_3$) organic stock solution with a fivefold molar excess of dioleoylphosphatidylserine (PS). Be sure to prepare an excess of substrate for the number of samples to be tested. For example, to carry out 10 × 20 μL reactions at a final concentration (apparent) of 100 μM phosphoinositide and 500 μM PS carrier lipid, combine enough of each lipid organic stock solution for 12–15 samples (i.e., 24–30 μL of 1 mM phosphoinositide, and 12–15 μL of 10 mM PS). This prevents running out of substrate preparation for the last one or two samples to be tested. Dry the lipid mixtures in a Speed-Vac concentrator (approx 15–20 min using "medium" heat setting).

3. Prepare the lipid suspension by adding malachite-green assay buffer (18 mL per each sample to be tested) and sonication in a water bath with a probe-type sonicator. This is carried out by immersing the tip of the sonicator probe just under the surface in a small (approx 400-mL) glass beaker. With the sonicator running at maximum output, hold the microfuge tube by the hinge and immerse the buffer/dried lipid mixture in the water bath di-

rectly under the probe tip for 30 s. Be sure to move the tube around underneath the probe tip to obtain complete dispersal of the lipid mixture. After sonication, the suspension should have a uniform, pearlescent/translucent appearance. If large particles or aggregates are visible, continue the sonication until complete dispersal of the lipid suspension is achieved. After the sonication is complete, transfer 18 µL of substrate preparation to a 0.5-mL microfuge tube for each sample to be tested and prewarm the samples for 10 min in a dry bath (usually at 30 or 37°C).

4. Initiate the reactions by the adding 2 µL of enzyme (usually from 1–500 ng of enzyme) and incubate for 2–30 min.

5. Terminate the reactions by adding an equal volume (20 µL) of ice-cold 20 mM sodium orthovanadate followed by centrifugation at 18,000 × g for 10 min at 4°C. Alternatively, the reactions can be terminated by the addition of 20 µL of 100 mM n-ethylmaleimide. The choice of termination reagent should be based on which compound is the most efficient inhibitor of your enzyme. The purpose of the centrifugation step is to sediment the lipid aggregates, which cause light-scatter and interfere with subsequent spectrophotometric measurements.

6. Remove 25 µL of the supernatant and combine it with 50 µL of malachite-green reagent (containing 0.01% Tween 20) in a new microfuge tube. Measure the absorbance of the samples at 620 nm.

7. Create a standard curve of inorganic phosphate using the 40 µM KH_2PO_4 stock solution, which is prepared with KH_2PO_4 that has been thoroughly dried in a vacuum oven to remove all traces of moisture. The standard curve consists of 25 µL samples of inorganic phosphate (0, 100, 200, 400, 600, 800, and 1000 pmol per 25 µL sample) combined with 50 µL malachite-green reagent as carried out for the lipid-phosphatase reactions. Typically, these samples are prepared in triplicate, and the mean absorbance at 620 nm plotted versus pmol inorganic phosphate per 25 µL sample as shown in **Fig. 2**. A line equation can be calculated from most graphing software applications and is used to determine the total amount of phosphate released in each lipid phosphatase reaction supernatant. The total phosphate liberated in each sample will be 1.6 times that calculated from the standard curve because the total sample volume after termination of the reactions is 40 µL (i.e., 25 µL × 1.6 = 40 µL).

A typical inorganic phosphate standard curve obtained using this procedure is shown in **Fig. 2**. For this standard curve, we have used a fixed total amount of inorganic phosphate in a volume of 25 µL because this is identical to the sample size removed and measured from the actual lipid phosphatase reactions. Alternatively, a concentration-based standard curve can also be used. As

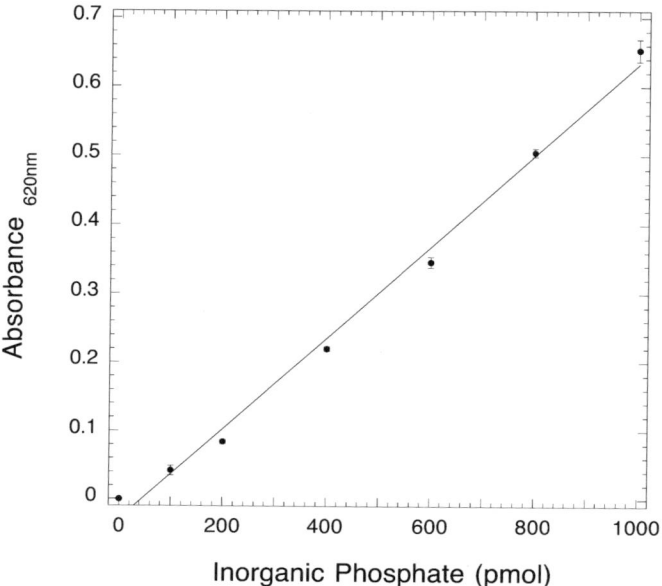

Fig. 2. Inorganic phosphate standard curve. A standard curve of inorganic phosphate was prepared from a 40 μM stock solution of KH_2PO_4 as described in **Subheading 3.** Samples (25 μL) containing 0, 100, 200, 400, 600, 800, or 1000 pmol inorganic phosphate were combined with 50 μL of malachite-green reagent and the absorbance at 620 nm measured. The graph depicts absorbance at 620 nm vs pmol of inorganic phosphate per 25 μL sample. Each data point represents the mean $\pm$ S.E.M of triplicate determinations. The graph and fitted line was generated using Kaleidagraph™ 3.0 software.

a demonstration of this technique, we have employed it to test the lipid-phosphatase activity of recombinant Ymr1p utilizing synthetic di-C_{16}-phospho-inositide substrates. As illustrated in **Fig. 3**, Ymr1p exhibits comparable activity toward both PI(3)P and PI(3,5)P$_2$ substrates. In addition to corroborating its specificity toward fluorescent lipid substrates, the malachite green-based assay for inorganic phosphate also provides quantitative values for the specific activity of Ymr1p against PI(3)P and PI(3,5)P$_2$. In addition to providing insight regarding the relative efficacy of an enzyme toward different phosphoinositide substrates, the ability to quantitatively determine lipid-phosphatase activity is also useful for comparison to other bona fide phosphoinositide phosphatases.

4. Notes

1. As with all enzymatic assays, an important consideration when testing the activity of a putative phosphoinositide phosphatase is the optimiza-

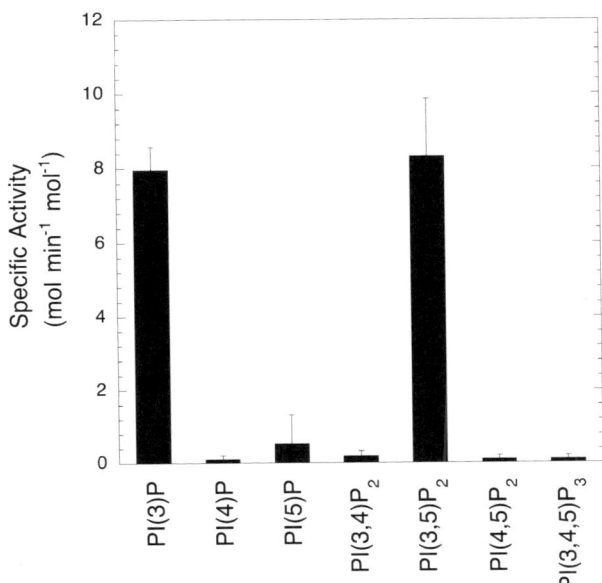

Fig. 3. Determination of Ymr1p specific activity toward synthetic di-C$_{16}$-phosphoinositide substrates utilizing a malachite green-based assay of inorganic phosphate. The activity of recombinant Ymr1p (200 ng) was tested against each phosphoinositide/PS substrate suspension preparation (100 μ*M*/500 μ*M* apparent concentration) in a volume of 20 μL for 15 min at 30°C. After termination of the reactions with 20 μL of 20 m*M* sodium orthovanadate and centrifugation, 25 μL of each reaction supernatant was added to 50 μL malachite-green reagent and the absorbance at 620 nm was measured. Phosphate release was calculated using the inorganic phosphate standard curve shown in **Fig. 2**. The specific activity values are denoted as moles of phosphate released per minute per mole of enzyme. Each bar represents the mean of triplicate determinations ± S.E.M.

tion the reaction parameters. We have pointed out that the fluorescent phosphoinositide substrate assay is relatively insensitive to reaction conditions, because the reaction proceeds essentially to completion. However, it is still a good practice to optimize the conditions for your enzyme of interest. More specifically, time, temperature, pH, and amount of enzyme are the most convenient parameters to alter. Owing to the simplicity of this approach, it is relatively easy to repeat the assays as necessary over a range of time, pH, temperature, or amount of enzyme. The goal of such optimization is to be able to clearly detect substrate hydrolysis, but under conditions where nonspecific activity is minimized.

In the case of the malachite-green assay, it is critical to limit the substrate dephosphorylation to $\leq 10\%$ of the total substrate present in the reaction in order to maintain the reaction kinetics.

2. Another consideration when carrying out lipid-phosphatase assays is the specific form in which the substrate is presented. The fluorescent substrates are completely water-soluble, regardless of acyl-chain length or fluorophore, which may restrict their use in micelle or vesicle preparations. Moreover, the shorter acyl chain nonfluorescent phosphoinositide derivatives (di-C_4 and di-C_8) are also water soluble, which allows their use in determining kinetic parameters. We generally use the di-C_6-NBD6 derivatives in the fluorescent phosphoinositide substrate assay, and both di-C_8 and di-C_{16} nonfluorescent derivatives in the malachite green-based assay. It is most important to test different acyl chain-length derivatives when using the nonfluorescent phosphoinositides, because we have noted that there can be differences in specificity between them. For example, myotubularins do not efficiently dephosphorylate di-C_8-PI(3,5)P_2 under the conditions described here, however, they actually exhibit a slight preference for this substrate over PI(3)P when the di-C_{16} forms are used. This is most likely because of an effect of the acyl-chain length on the conformation of the substrate head group within the lipid-suspension preparations. It is also worth mentioning that there are many different methods for the preparation of lipid suspensions that vary parameters such as lipid content (i.e., PC, PE, PI, cholesterol, and sphingolipids) and specific ways in which the actual suspension is physically carried out. Obviously, such parameters may have a profound effect on the activity of a lipid-modifying enzyme and many of these methods for preparing lipid suspensions are perfectly compatible with the malachite green-based asssay for inorganic phosphate release described here.

Acknowledgments

This work has been supported by grants from the National Institutes of Health and the Walther Cancer Institute. G.T. is supported by a grant from the Muscular Dystrophy Association (MDA).

References

1. Simonsen, A., Wurmser, A. E., Emr, S. D., and Stenmark, H. (2001) The role of phosphoinositides in membrane transport. *Curr. Opin. Cell Biol.* **13,** 485–492.
2. Cantley, L. C. (2002) The phosphoinositide 3-kinase pathway. *Science* **296,** 1655–1657.
3. Toker, A. (2002) Phosphoinositides and signal transduction. *Cell Mol. Life Sci.* **59,** 761–779.

4. Martelli, A. M., Tabellini, G., Borgatti, P., (2003) Nuclear lipids: new functions for old molecules? *J. Cell Biochem.* **88,** 455–461.
5. Irvine, R. F. (2003) Nuclear lipid signalling. *Nature Rev. Mol. Cell Biol.* **4,** 1–12.
6. Pendaries, C., Tronchere, H., Plantavid, M., and Payrastre, B. (2003) Phospho-inositide signaling disorders in human diseases. *FEBS Lett.* **546,** 25–31.
7. Kanaho, Y. and Suzuki, T. (2002) Phosphoinositide kinases as enzymes that pro-duce versatile signaling lipids, phosphoinositides. *J. Biochem.* **131,** 503–509.
8. Maehama, T., Taylor, G. S., and Dixon, J. E. (2001) PTEN and myotubularin: novel phosphoinositide phosphatases. *Annu. Rev. Biochem.* **70,** 247–279.
9. Nandurkar, H. H. and Huysmans, R. (2002) The myotubularin family: novel phos-phoinositide regulators. *IUBMB Life* **53,** 37–43.
10. Wishart, M. J. and Dixon, J. E. (2002) PTEN and myotubularin phosphatases: from 3-phosphoinositide dephosphorylation to disease. *Trends Cell Biol.* **12,** 579–585.
11. Hughes, W. E., Cooke, F. T., and Parker, P. J. (2000) Sac phosphatase domain pro-teins. *Biochem. J.* **350,** 337–352.
12. Whisstock, J. C., Wiradjaja, F., Waters, J. E., and Gurung, R. (2002) The structure and function of catalytic domains within the inositol polyphosphate 5-phos-phatases. *IUBMB Life* **53,** 15–23.
13. Taylor, G. S. and Dixon, J. E. (2001) An assay for phosphoinositide phosphatases utilizing fluorescent substrates. *Anal. Biochem.* **295,** 122–126.
14. Okada, T., Hazeki, O., Ui, M., and Katada, T. (1996) Synergistic activation of Pt-dIns 3-kinase by tyrosine-phosphorylated peptide and beta gamma-subunits of GTP-binding proteins. *Biochem. J.* **317,** 475–480.
15. Taylor, G. S., Maehama, T., and Dixon, J. E. (2000) Myotubularin, a protein tyro-sine phosphatase mutated in myotubular myopathy, dephosphorylates the lipid sec-ond messenger, phosphatidylinositol 3-phosphate. *Proc. Natl. Acad. Sci. USA* **97,** 8910–8915.
16. Walker, D. M., Urbe, S., Dove, S. K., et al. (2001) Characterization of MTMR3, an inositol lipid 3-phosphatase with novel substrate specificity. *Curr. Biol.* **11,** 1600–1605.
17. Berger, P., Bonneick, S., Willi, S., et al. (2002) Loss of phosphatase activity in my-otubularin-related protein 2 is associated with Charcot-Marie-Tooth disease type 4B1. *Hum. Mol. Genet.* **11,** 1569–1579.
18. Schaletzky, J., Dove, S. K., Short, B., et al. (2003) Phosphatidylinositol-5-phos-phate activation and conserved substrate specificity of the myotubularin phos-phatidylinositol 3-phosphatases. *Curr. Biol.* **13,** 504–509.
19. Maehama, T., Taylor, G. S., Slama, J. T., and Dixon, J. E. (2000) A sensitive assay for phosphoinositide phosphatases. *Anal. Biochem.* **279,** 248–250.
20. Harder, K. W., Owen, P., Wong, L. K. H., et al. (1994) Characterization and kinetic analysis of the intracellular domain of human protein tyrosine phosphatase β (HPTPβ) using synthetic phosphopeptides. *Biochem. J.* **298,** 395–401.

16

Assaying Phospholipase A₂ Activity

Christina C. Leslie and Michael H. Gelb

Summary

Mammalian cells contain many structurally and functionally diverse phospholipases A₂ (PLA₂) that catalyze the hydrolysis of *sn-2* fatty acid from membrane phospholipid. Assays are described for measuring the activity of Group IVA cytosolic PLA₂α (cPLAα) and for secreted PLA₂s (sPLA₂) that are suitable for purified enzymes and for measuring activity in crude cell lysates and culture medium. The assay for cPLA₂α involves measuring the calcium-dependent release of radiolabeled *sn-2* arachidonic acid from small unilamellar vesicles of phosphatidylcholine. Methods are described for distinguishing cPLA₂α activity in cell lysates from other PLA₂s. sPLA₂ activity is monitored using a fluorimetric assay that measures the continuous calcium-dependent formation of albumin-bound pyrene fatty acid from the *sn-2* position of phosphatidylglycerol.

Key Words: Phospholipase A₂; phospholipid; fatty acid; arachidonic acid; calcium; small unilamellar vesicles; phosphatidylcholine; phosphatidylglycerol; liposome; pyrene fatty acid.

1. Introduction

Mammalian cells contain multiple structurally diverse forms of phospholipases A₂ (PLA₂) that catalyze the hydrolysis of the *sn-2* fatty acid from membrane phospholipid. PLA₂s function in dietary lipid breakdown, in phospholipid acyl-chain remodeling, in lipid-mediator production and in host defense against microorganisms. There are at least 15 genes in mammals encoding PLA₂ enzymes that comprise three main types: the secreted PLA₂s (sPLA₂), the Group IV cytosolic PLA₂s (cPLA₂), and the Group VI calcium-independent PLA₂s

From: *Methods in Molecular Biology, vol. 284:*
Signal Transduction Protocols
Edited by: R. C. Dickson © Humana Press Inc., Totowa, NJ

(iPLA$_2$) *(1–3)*. All cells contain members of each of these types of enzymes, suggesting that they play distinct functional roles. This chapter describes assays for measuring the activity of the well-characterized, ubiquitous cPLA$_2\alpha$ (group IVA) that mediates agonist-induced arachidonic-acid release *(4)*, and for measuring the activity of sPLA$_2$s that are secreted into cell culture medium. The assays can be used for purified enzymes or for measuring their activity in crude cell fractions. The cPLA$_2\alpha$ assay involves measuring the hydrolysis of radiolabeled arachidonic acid from the *sn-2* position of phosphatidylcholine, in the form of small unilamellar vesicles *(5)*. Procedures are described for processing the cells, preparing the substrate, incubating cell fractions with substrate, terminating the enzymatic reaction, and separating product. Approaches for distinguishing cPLA$_2\alpha$ activity from other PLA$_2$ enzymes in cell lysates are outlined. The assay for measuring sPLA$_2$ activity involves monitoring the calcium-dependent formation of albumin-bound pyrene fatty acid from the *sn-2* position of phosphatidylglycerol. This continuous fluorimetric assay is carried out in a cuvet or a microtiter plate.

2. Materials

2.1. cPLA$_2\alpha$ Assay

1. Homogenization buffer: 10 mM HEPES, pH, 7.4, 0.34 M sucrose, 10% glycerol, 1 mM EGTA, 1 mM phenylmethylsulfonylchloride, 1 μg/mL leupeptin, 10 μg/mL aprotinin.
2. Dithiothreitol (DTT).
3. 1-Palmitoyl-2[1-^{14}C]arachidonyl-phosphatidylcholine (Perkin Elmer Life Sciences).
4. 1-Palmitoyl-2-arachidonyl-phosphatidylcholine (Avanti Polar Lipids).
5. CaCl$_2$.
6. Dioleoylglycerol.
7. Sodium chloride.
8. Bovine serum albumin (BSA; fatty acid free).
9. Dole reagent: 2-propanol/heptane/0.5 M H$_2$SO$_4$, 20/5/1 (v/v/v).
10. Silica gel LC-SI solid-phase extraction (SPE) tubes (Sulpelco, Bellefonte, PA).
11. Visiprep soldi-phase extraction manifold (Supelco).
12. Oleic acid.
13. Probe sonicator with microprobe tip (Braun Instruments).
14. Arachidonyl trifluoromethylketone (Caymen Chemicals).
15. Bromoenol lactone (Caymen Chemicals).
16. Heptane.

2.2. sPLA₂ Assay

1. BSA (fatty acid free, Sigma).
2. CaCl₂.
3. EGTA.
4. 1-Hexadecanoyl-2-(1-pyrenedecanoyl)-*sn*-glycero-3-phosphoglycerol ammonium salt (Molecular Probes, Eugene, OR).
5. Isopropanol.
6. 1-Pyrenedecanoic acid (Molecular Probes).
7. Sodium chloride.
8. Toluene.
9. Tris.

3. Methods

3.1. Assay for Measuring cPLA₂α Activity

cPLA₂α is present in most cells at sufficient amounts to detect its enzymatic activity in cell lysates. cPLA₂α can also be over-expressed in mammalian cells or in insect cells using baculovirus (*6,7*). The vesicle assay described here is suitable for measuring activity of the purified enzyme or for measuring activity of cPLA₂α in crude cell lysates.

3.1.1. Preparation of Cell Lysates

1. Wash cells twice with homogenization buffer (*see* **Note 1**).
2. Resuspend the cells in a small amount of homogenization buffer that will result in a protein concentration of approx 2 µg/µL after homogenization.
3. Sonicate the cell suspension on ice using a microprobe tip for 10 s, two to four times. Monitor the extent of cell disruption, which is variable depending on the cell type, after each 10-s burst of sonication by light microscopy.
4. Centrifuge the lysate at 100,000 *g* for 1 h at 4°C to separate the soluble (cytosolic) and particulate (membrane) fractions using a Beckman TL100 tabletop ultracentrifuge. Transfer the supernatant to another tube and maintain at 0°C. Gently rinse the pellet and sides of the tube with homogenization buffer to remove the small amount of remaining cytosol (discard the rinse), and then resuspend the pellet in a small volume of homogenization buffer (*see* **Note 2**).
5. Determine the protein concentration in the cytosol and membrane fractions, and then add DTT to a final concentration of 1 m*M* to protect cPLA₂ from oxidative inactivation (*see* **Note 3**).

3.1.2. Concentration of Assay Components

The final concentration of components in the assay (50 μL final volume/reaction) are as follows.

1. 30 μM 1-palmitoyl-2-[1-^{14}C]arachidonyl-phosphatidylcholine (100,000 dpm/reaction) (*see* **Notes 4** and **5**).
2. 9 μM dioleoylglycerol (*see* **Note 6**).
3. 150 mM NaCl (*see* **Note 6**).
4. 1 mg/mL BSA (*see* **Note 6**).
5. 50 mM HEPES, pH 7.4.
6. 1 mM $CaCl_2$ (*see* **Note 7**).

3.1.3. Preparation of Substrate

1. Prepare a concentrated stock of substrate by mixing in a glass tube (12 mm × 75 mm) 1-palmitoyl-2[1-^{14}C]arachidonyl-phosphatidylcholine and unlabeled 1-palmitoyl-2-arachidonyl-phosphatidylcholine to achieve a ratio of 100,000 dpm/1.5 nmol. Mix enough for several reactions and then add dioleoylglycerol at 30 mol% (30% of lipid is dioleoylglycerol and 70% is phosphotidylcholine) of the phosphatidylcholine concentration (this results in 0.45 nmol dioleolyglycerol/reaction).
2. Evaporate the solvents from the lipid mixture under a stream of nitrogen. Add 50 mM HEPES buffer, pH 7.4, to make a concentrated substrate solution that is 5–10 times the final assay concentration.
3. Sonicate the substrate mixture for 4 s on ice using a microprobe (Braun Instruments) to form small unilamellar vesicles. Count an aliquot of the solution after sonication to determine the efficiency of liposome formation. Usually >90% of the radioactivity is solubilized. Prepare liposomes fresh (if they are not used immediately, then store on ice for no more than 4 h).

3.1.4. Incubation Conditions

1. Aliquot the substrate mixture into assay tubes (round bottom disposable glass screw cap tubes, 13 mm × 100 mm).
2. Add the NaCl, albumin and calcium.
3. Start the reaction by adding the cytosol or membrane fractions (approx 20 μg protein or more depending on the amount of cPLA$_2$α present in the cells). The final reaction volume after adding all the components is 50 μL. Incubate at 37°C with shaking for five min.
4. Terminate the reaction by adding 2.5 mL of Dole reagent (2-propanol: heptane: 0.5 M H_2SO_4, 20:5:1, v/v/v) to each tube. After each reaction is terminated with Dole reagent, add 1.5 mL heptane and 1 mL water to each tube.

5. Add unlabeled oleic acid (20 µg of a 2 mg/mL stock in chloroform) to each tube to aid in the extraction efficiency.

6. Vortex each tube for approx 20 s and then centrifuge the tubes at 1000 *g* at room temperature for 5 min to cleanly separate the two phases. The radio-labeled fatty acid product is extracted into the heptane phase. The reactions do not need to be immediately processed at this point but can be refrigerated and processed the next day.

7. Remove the upper heptane phase with a pasteur pipet being careful not to remove any of the lower phase. Pass the heptane phase through an SPE tube using a Visiprep SPE manifold. Elute the radiolabeled fatty acid with 2.5 mL of chloroform and collect the eluent in scintillation vials. Contaminating radiolabeled phospholipid substrate in the heptane phase is not eluted by chloroform and is retained on the silica gel.

8. Dry the eluent thoroughly under a stream of nitrogen to remove chloroform, which is a strong quenching agent. Add 0.5 mL heptane, vortex the vials, and let the vials stand at room temperature for about 10 min to solubilize the dried radiolabeled fatty acid. Add scintillation fluid, vortex briefly, and count by liquid-scintillation spectrometry.

9. Control reactions that contain substrate but no enzyme are similarly incubated and processed, and the dpms are subtracted from the experimental results.

3.2. Methods for Distinquishing cPLA$_2$α Activity From Other PLA$_2$ Enzymes

3.2.1. Group VI Calcium-Independent PLA$_2$s (iPLA$_2$)

iPLA$_2$s are widely distributed in tissues and are found in both the soluble and membrane fractions of cells (*8*). They are commonly-assayed using micelles of phospholipid mixed with Triton X-100, however, the vesicle assay described for measuring cPLA$_2$α activity will also detect iPLA$_2$s (*9*). To determine if some of the activity measured using the vesicle assay in crude cell fractions is owing to iPLA$_2$, omit calcium from the assay and include excess EGTA (1 m*M*). cPLA$_2$α and sPLA$_2$s are inactive in the absence of calcium and remaining activity is likely owing to iPLA$_2$. Additionally, bromoenol lactone (Cayman Chemical) selectively inhibits iPLA$_2$s in the 0.5–5.0 µ*M* range without affecting cPLA$_2$α activity (*10,11*) (*see* **Note 8**).

3.2.2. Secreted PLA$_2$ Enzymes

There are 10 distinct mammalian sPLA$_2$ enzymes that are synthesized and secreted by cells or stored in intracellular granules of secretory cells (*2*). An assay for measuring the activity of purified sPLA$_2$, or sPLA$_2$ secreted into cell-culture media, is described in **Subheading 3.3.** The vesicle assay described for measuring cPLA$_2\alpha$ activity in crude cell fractions is not optimal for measuring sPLA$_2$ enzymes, especially if the vesicle assay is carried out in the presence of 1–2 µ*M* free calcium with phosphatidylcholine substrate (*see* **Note 9**), which is optimal for cPLA$_2\alpha$ but suboptimal for sPLA$_2$ enzymes. Many sPLA$_2$s (Groups IB, IIA, IID, IIE, IIF) preferentially hydrolyze anionic phospholipids, however, Group V and X sPLA$_2$s hydrolyze phosphatidylcholine vesicles and may be present in crude cell fractions *(12)*.

Although the sPLA$_2$s are rich in disulfide linkages, the susceptibility of these enzymes to inactivation by treatment with DTT varies widely *(12)*. For example, groups IIE, IIF, and X sPLA$_2$ retain significant activity even when treated for 30 min at 50°C with 10 m*M* DTT, whereas some sPLA$_2$s like Group IIA are fully inactivated after a 10-min treatment at 37°C *(12)*. DTT sensitivity is a commonly reported method for evaluating whether the PLA$_2$ activity in a cell lysate is owing to an sPLA$_2$, but caution is now advised. However, treating cell fractions with the commercially available inhibitor, arachidonyl trifluoromethylketone at a concentration of 10 µ*M* will inactivate cPLA$_2\alpha$ but not sPLA$_2$s *(13)*.

3.2.3. Other Group IV cPLA$_2$ Enzymes

Group IVC cPLA$_2\gamma$ is constitutively associated with membrane and is calcium-independent because it lacks the C2 domain *(14–16)*. Consequently, cPLA$_2\alpha$ can be distinguished from cPLA$_2\gamma$ by omitting calcium from the vesicle assay. Group IVB cPLA$_2\beta$ is calcium-dependent because it contains a C2 domain, but preliminary characterization of its enzymatic properties suggests that it does not exhibit specificity for *sn*-2 arachidonic acid, a characteristic of cPLA$_2\alpha$ *(17)*. Therefore cPLA$_2\alpha$ can be distinguished from cPLA$_2\beta$ by demonstrating that activity in cell fractions is greater (10- to 20-fold) using vesicles composed of 1-palmitoyl-2-arachidonyl-phosphatidylcholine compared to dipalmitoyl-phosphatidylcholine.

3.3. Assay for Measuring sPLA$_2$ Activity

The sPLA$_2$ assay described here is a modification of the original procedure first described by Radvanyi et al. *(18)*. The assay makes use of a phosphatidylglycerol substrate (*see* **Note 9**) with a pyrene fluorophore on the terminal end of the *sn*-2 fatty acyl chain. When these phospholipids pack into the membrane bilayer, the close proximity of the pyrenes, from neighboring phospholipids, causes the spectral properties to change relative to that of monomeric pyrene (owing to exciplex formation). BSA is present in the aqueous phase and captures the pyrene fatty acid when it is liberated from the glycerol backbone owing to the sPLA$_2$-catalyzed reaction. These features allow for a sensitive sPLA$_2$ assay by monitoring the fluorescence of monomeric, albumin-bound pyrene fatty acid. This fluorimetric assay monitors product formation continuously in the presence of calcium (*see* **Note 10**), and the sensitivity of the assay approaches that of conventional fixed time-point assays with radiolabeled phospholipids.

3.3.1. Preparation of Assay Solutions and Substate

1. Dissolve 1 mg of 1-hexadecanoyl-2-(1-pyrenedecanoyl)-*sn*-glycero-3-phosphoglycerol (PPyrPG) in 1 mL of toluene: isopropanol (1:1) and vortex until all solid dissolves (may require slight warming).
2. Measure the concentration of PPyrPG in the stock solution by determining the absorbance at 342 nm in methanol using the extinction coefficient of 40,000 M^{-1} cm^{-1}. Store the stock solution at $-20°C$ in a vial with a teflon-lined screw cap (tightly secure the cap in place with parafilm).
3. Pipet 200 µL of PPyrPM stock solution into an Eppendorf tube, and completely remove the solvent with a stream of nitrogen. Add ethanol (1 mL), and vortex the sample until the lipid dissolves (may require slight warming). Centrifuge the sample (full speed in a microfuge for 2 min) to remove a small amount of particulate that may be present. Transfer the supernatant to a new tube and determine the absorbance of the solution at 342 nm (as above in **step 2**). The concentration of PPyrPM should be approx 100–200 µ*M*. Store the ethanol stock solution at $-20°C$ (as in **step 2** above).
4. Prepare assay buffer (50 m*M* Tris-HCl, pH 7.4, 100 m*M* NaCl, 1 m*M* EGTA) and pass the solution through a 0.2–0.4 micron syringe filter to remove dust particles. Use high quality water to make the buffer (i.e., Milli-Q, Millipore Inc.). Store assay buffer at 4°C.
5. Prepare a solution of BSA (fatty acid-free, Sigma) in purified water at a concentration of 100 mg/mL and filter through a 0.2–0.4 micron syringe filter. Store in small aliquots at $-20°C$.
6. Prepare 1 *M* CaCl₂ in purified water and store at 4°C.
7. To prepare a working solution of assay cocktail for 30 assays (1 mL each), add 300 µL of 100 µ*M* PPyrPG in ethanol (or the appropriate volume of a more concentrated substrate stock) (warm the stock solution in your hands and vortex it before taking the aliquot) to 29.7 mL of assay buffer. Add the PPyrPG drop-wise over approx 1 min to the continuously vortexing solution of assay buffer. This gives a working solution of 1 µ*M* PPyrPG as unilamellar liposomes. This solution is stable at room temperature for up to 24 h.

3.3.2. sPLA₂ Assay Procedure

1. Into a quartz fluorescence cuvette fitted with a small magnetic stir bar, add 0.98 mL of assay cocktail and then add 10 µL of 10% BSA. Finally add the desired amount of sPLA₂ (*see* **Notes 11** and **12**) to the cuvette. Place the

cuvette in the fluorimeter. For most accurate kinetics, the cuvette holder should be thermostatted at 37°C, but reasonably reliable data can be obtained at ambient temperature. Set the fluorimeter excitation wavelength to 342 nm and the emission wavelength to 395 nm. The fluorescence is recorded for approx 1 min in the absence of calcium to obtain the background rate. The reaction is initiated by adding 10 µL of 1 M CaCl$_2$ stock solution. The initial reaction velocity is recorded for approx 2–4 min with magnetic stirring (*see* **Note 13**).

3.3.3. Assay Calibration

In order to determine the specific activity from the observed assay slope, the assay has to be calibrated by adding a known amount of product, 1-pyrenedecanoic acid, to the complete assay cocktail in the absence of sPLA$_2$.

1. Prepare a stock solution of 1-pyrenedecanoic acid (Molecular Probes) in absolute ethanol at a concentration of 50 µM. Store at −20°C in a glass vial with a teflon-lined screw cap.
2. Prepare an assay reaction (*see* **Subheading 3.3.1.**) with all components except enzyme. Record the initial fluorescence and the increase in fluorescence after adding 250 pmole of 1-pyrenedecanoic acid to the cuvette. Use the measured increase in fluorescence per pmole to convert the observed slope measured in the sPLA$_2$ assay to µmol product produced per min per mg of sPLA$_2$.

3.3.4. Microtiter Plate Assay of sPLA$_2$

The aforementioned assay with PPyrPG can also be carried out in a 96-well microtiter-plate using a microtiter plate fluorimeter.

1. Prepare a stock solution of 3% (w/v) BSA in assay buffer (50 mM Tris, pH 8.0, 50 mM KCl, 1 mM CaCl$_2$) and filter (*see* **Subheading 3.3.1.**). Store in aliquots at −20°C.
2. Prepare working solution A (60 µL of 3% BSA plus 940 µL of assay buffer).
3. Prepare solution B (1 mL of solution A plus an aliquot of sPLA$_2$, typically 5–2000 ng depending on the specific activity of the sPLA$_2$). Solution B should be prepared immediately before use to minimize loss owing to absorption of sPLA$_2$ to the tube.
4. Prepare solution C (4.2 µM PPyrPG in assay buffer, prepared (*see* **Subheading 3.3.1.**) for the 1 µM solution).
5. Add 100 µL solution A to each well followed by 100 µL solution B to each well. With a multichannel pipetor, deliver 100 µL of solution C to each well

to start the reaction (*see* **Note 14**). Place the plate in the plate reader to monitor the fluorescence in each well (*see* **Note 15**).

4. Notes

1. Protease inhibitor-cocktail tablets are commercially available (Roche Diagnostics, Mannheim, Germany) and can be used in place of the individually added protease inhibitors (phenylmethysulfonylfluoride [PMSF], leupeptin, aprotinin). However, if the PLA_2 assay is carried out at calcium concentrations in the nM–μM range, which requires the use of calcium/EGTA buffers, use protease inhibitor cocktails without EDTA.

2. Small plastic Pellet Pestles (Kontes Glass Company, Vineland, NJ) are very useful for resuspending the compact membrane pellet into a homogeneous suspension. It is important not to vortex the cell fractions, which will subject $cPLA_2$ to oxidation.

3. $cPLA_2\alpha$ exhibits calcium-dependent translocation from cytosol to membrane in cells treated with agonists that increase intracellular calcium *(19,20)*. However, we have found that this membrane association is not stable and can be reversed if cells are homogenized in the presence of calcium chelators *(21)*. However, some reports have suggested that there is stable association of $cPLA_2\alpha$ with membrane, although the basis for this is not understood *(22,23)*. The relative amount of $cPLA_2\alpha$ in the soluble and particulate fraction can be determined by Western blotting using commercially available antibodies (Santa Cruz Biotechnology). $cPLA_2\alpha$ activity associated with membrane can be measured using the vesicle assay, but it is important to note that the specific activity can not be directly compared to the specific activity in the soluble fraction. Because $cPLA_2\alpha$ will hydrolyze the unlabeled phospholipid in the membrane, this effectively dilutes the labeled substrate in the vesicles, resulting in a lower specific activity.

4. 1-stearoyl-2[1-^{14}C]arachidonyl-phosphatidylcholine (Amersham Biosciences, Piscataway, NJ) or the sn-1 ether-linked substrate 1-O-hexadecyl-2[1-^{14}C] arachidonyl-phosphatidylcholine (American Radiolabeled Chemicals) can also be used as substrates to measure $cPLA_2\alpha$ activity. When using a diacyl-phospholipid substrate, $cPLA_2\alpha$ will sequentially cleave the sn-2 and then the sn-1 fatty acid because it has both PLA_2 and lysophospholipase activity *(24,25)*. However, $cPLA_2\alpha$ cannot hydrolyze the sn-1 aliphatic group from the 1-O-alkyl-linked substrate, 1-O-hexadecyl-2-arachidonyl-phosphatidylcholine.

5. Arachidonic acid is highly susceptible to oxidation and the arachidonyl-containing substrates (labeled and unlabeled) should be stored under nitrogen. Make a concentrated stock solution of unlabeled 1-palmitoyl-2-arachidonyl-phosphatidylcholine in chloroform: methanol (90:10) and store

aliquots at $-80°C$ in glass vials with Teflon-lined caps and with the caps additionally sealed with Teflon tape. Chloroform is very volatile and any evaporation will affect the concentration of the stock solution.

6. $cPLA_2\alpha$ has unusual kinetic properties on sonicated vesicles, exhibiting an initial burst of activity that ceases abruptly in about 10–20 min *(24,26,27)*. This is not owing to depletion of substrate, product inhibition, or irreversible inactivation of $cPLA_2\alpha$. Rather $cPLA_2\alpha$ becomes "trapped" on the vesicles as product accumulates. The addition of NaCl and albumin to the reaction helps to prevent trapping and leads to more linear reaction-progress curves. The presence of dioleolyglycerol in the vesicle also enhances activity and improves the reaction progress *(26–28)*. Dioleolyglycerol acts to perturb the packing density of the phosphatidylcholine bilayer, thereby allowing greater ability of $cPLA_2\alpha$ to penetrate the bilayer and access the substrate. Dioleolyglycerol may also help prevent trapping on the vesicle.

7. Calcium is required in the assay to promote binding of $cPLA_2\alpha$ to phospholipid vesicles *(6,29,30)*. A final concentration of 1 mM free calcium in the assay is saturating. To achieve this final calcium concentration, the amount of EGTA in the homogenization buffer added to the assay (from the cell fractions) must be taken into account. At 37°C, 1 mol of EGTA chelates 1 mol of calcium. A characteristic of $cPLA_2\alpha$ is that concentrations of calcium in the low micromolar range promote its binding to phospholipid vesicles *(29)*. To evaluate the effect of physiological concentrations of free calcium in the vesicle assay, calcium/EGTA buffers are used. A computer program (available at http://www.stanford.edu/~cpatton/webmaxcSR.htm) is used to determine the amount of $CaCl_2$ to add to the reaction mixture, containing a known final concentration of EGTA, which will result in free calcium in the range of 0.1–5 μM. Do not include albumin in the vesicle assay for measuring $cPLA_2$ activity at low micromolar concentrations of free calcium because it can bind calcium.

8. The inhibitors, arachidonyl trifluoromethylketone and methylarachidonyl fluorophosphonate, cannot be used to distinguish $cPLA_2\alpha$ and $iPLA_2$ because both enzymes are inhibited at similar concentrations of these compounds *(10,31,32)*.

9. A detailed study of the interfacial kinetic properties of the full set of mammalian $sPLA_2$s has revealed that all of these enzyme display relatively high activity on anionic phosphatidylglycerol vesicles, whereas only the group V and X $sPLA_2$s also display high activity on phosphatidylcholine vesicles *(12)*. The use of phosphatidylcholine vesicles is not recommended for detecting $sPLA_2$s in general, because many enzymes such as the human and mouse group IIA $sPLA_2$s display very low activity on these vesicles owing to poor interfacial binding to the zwitterionic interface. However, even

using phosphatidylglycerol, which is the most preferred sPLA₂ substrate, the specific activity of the mammalian sPLA₂s varies considerably. For example, the specific activity of human group IB sPLA₂ is approx 2000-fold higher than the specific activity of human group IIE sPLA₂ *(12)*. The physiological significance of this variation in specific activity for the various mammalian sPLA₂s remains to be understood, and it raises the possibility that glycero-phospholipids may not always be the physiological substrates for all of the enzymes. At the present time, phosphatidylglycerol remains the most generally preferred substrate for mammalian sPLA₂s.

10. It is often stated that sPLA₂s require millimolar concentrations of calcium for maximal activity. This is not the case *(12)*. The concentration of calcium required for optimal enzymatic activity of an sPLA₂ depends on the enzyme species and the type of phospholipid substrate used in the assay. This is because calcium is required for the binding of a single phospholipid molecule in the active site of the sPLA₂ at the membrane interface and substrate binding cannot occur if the enzyme is in the water layer. Thus, interfacial binding of sPLA₂ to the membrane surface and calcium binding in the active site are coupled. For example, human group V sPLA₂ binds tighter to phosphatidylglycerol vesicles than to phosphatidylcholine vesicles, and the observed apparent K_{Ca} for this enzyme in the presence of the anionic vesicles is 1 µM, whereas the apparent K_{Ca} increases to 225 µM in the presence of zwitterionic vesicles *(12)*. The statement made in **Subheading 3.2.1.** that cPLA₂α but not sPLA₂ is maximally active in the presence of 1–2 µM calcium holds as long as phosphatidylcholine is the substrate. In the presence of phosphatidylglycerol vesicles, the apparent K_{Ca} varies from about 1 µM (group V sPLA₂s) to about 100 µM (group IIE sPLA₂s) *(12)*.

11. Purified sPLA₂s readily absorb on the walls of containers at concentrations below about 10 µg/mL. Highly dilute sPLA₂ stock solutions, appropriate for adding nanogram amounts to the PPyrPG assay are best prepared by fresh dilution into buffer with approx 1 mg/mL BSA. For the most active sPLA₂s (i.e., group IB and IIA), as little as 0.2–0.5 ng of enzyme gives a readily measured initial velocity. The specific activities of the full set of human and mouse sPLA₂s acting on phosphatidylglycerol vesicles has been measured *(12)*.

12. It is also possible to add crude samples containing sPLA₂ to the PPyrPG assay. For example, we have been able to add up to 50 µL of tissue culture medium containing sPLA₂ (from transfected cells) to the assay.

13. If the amount of sPLA₂ added to the PPyrPG assay is too high, the reaction-progress curve will display significant curvature, and all of the substrate will be consumed in less than a few minutes. A sufficient amount of enzyme should be added to give a progress curve that remains linear for the

first few minutes. It is best to confirm that the initial velocity is proportional to the amount of enzyme.

14. The use of a multichannel pipetor allows the reaction to be initiated in all wells at the same time.

15. Typically eight wells are assayed at a time (more wells can be assayed if the plate reader is capable of measuring the fluorescence from multiple wells several times per minute). At least one well should be reserved for a minus enzyme control. The plate reader needs to be equipped with the appropriate filters to deliver the desired excitation and emission wavelengths (see the cuvette assay for wavelengths). The microtiter plate assay can be calibrated with 1-pyrenedecanoic acid as for the cuvette assay.

Acknowledgments

The authors work is supported by NIH grants HL34303 and HL61378 (to C. Leslie) and HL50040 and HL36236 (to M. Gelb).

References

1. Six, D. A. and Dennis, E. A. (2000) The expanding superfamily of phospholipase A_2 enzymes: classification and characterization. *Biochim. Biophys. Acta* **1488,** 1–19.
2. Valentin, E. and Lambeau, G. (2000) Increasing molecular diversity of secreted phospholipases A_2 and their receptors and binding proteins. *Biochim. Biophys. Acta* **1488,** 59–70.
3. Kudo, I. and Murakami, M. (2002) Phospholipase A_2 enzymes. *Prostaglandins Other Lipid Mediat.* **68–69,** 3–58.
4. Leslie, C. C. (1997) Properties and regulation of cytosolic phospholipase A_2. *J. Biol. Chem.* **272,** 16,709–16,712.
5. Leslie, C. (1990) Macrophage phospholipase A_2 specific for sn-2 arachidonic acid in *Methods in Enzymology* vol. 187 (Murphy, R. and Fitzpatrick, F., eds.), Academic Press, Orlando, FL, pp. 216–225.
6. Clark, J. D., Lin, L.-L., Kriz, R. W., et al. (1991) A novel arachidonic acid-selective cytosolic PLA_2 contains a Ca^{2+}-dependent translocation domain with homology to PKC and GAP. *Cell* **65,** 1043–1051.
7. de Carvalho, M. S., McCormack, F. X., and Leslie, C. C. (1993) The 85-kDa, arachidonic acid-specific phospholipase A_2 is expressed as an activated phosphoprotein in Sf9 cells. *Arch. Biochem. Biophys.* **306,** 534–540.
8. Winstead, M. V., Balsinde, J., and Dennis, E. A. (2000) Calcium-independent phospholipase A_2: structure and function. *Biochim. Biophys. Acta* **1488,** 28–39.
9. Tang, J., Kriz, R. W., Wolfman, N., et al. (1997) A novel cytosolic calcium-independent phospholipase A_2 contains eight ankyrin motifs. *J. Biol. Chem.* **272,** 8567–8575.

10. Ackermann, E. J., Conde-Frieboes, K., and Dennis, E. A. (1995) Inhibition of macrophage Ca^{2+}-independent phospholipase A$_2$ by bromoenol lactone and trifluoromethyl ketones. *J. Biol. Chem.* **270,** 445–450.

11. Jenkins, C. M., Han, X., Mancuso, D. J., and Gross, R. W. (2002) Identification of calcium-independent phospholipase A$_2$ (iPLA$_2$)β, and not iPLA$_2$γ, as the mediator of arginine vasopressin-induced arachidonic acid release in A-10 smooth muscle cells. *J. Biol. Chem.* **277,** 32,807–32,814.

12. Singer, A. G., Ghomashchi, F., Le Calvez, C., et al. (2002) Interfacial kinetic and binding properties of the complete set of human and mouse groups I, II, V, X, and XII secreted phospholipases A$_2$. *J. Biol. Chem.* **277,** 48,535–48,549.

13. Street, I. P., Lin, H. K., Laliberté, F., et al. (1993) Slow- and tight-binding inhibitors of the 85-kDa human phospholipase A$_2$. *Biochemistry* **32,** 5935–5940.

14. Stewart, A., Ghosh, M., Spencer, D. M., and Leslie, C. C. (2002) Enzymatic properties of human cytosolic phospholipase A$_2$γ. *J. Biol. Chem.* **277,** 29,526–29,536.

15. Pickard, R. T., Strifler, B. A., Kramer, R. M., and Sharp, J. D. (1999) Molecular cloning of two new human paralogs of 85-kDa cytosolic phospholipase A$_2$. *J. Biol. Chem.* **274,** 8823–8831.

16. Underwood, K. W., Song, C., Kriz, R. W., et al. (1998) A novel calcium-independent phospholipase A$_2$, cPLA$_2$-γ, that is prenylated and contains homology to cPLA$_2$. *J. Biol. Chem.* **273,** 21,926–21,932.

17. Song, C., Chang, X. J., Bean, K. M., et al. (1999) Molecular characterization of cytosolic phospholipase A$_2$-β. *J. Biol. Chem.* **274,** 17,063–17,067.

18. Radvanyi, F., Jordan, L., Russo-Marie, F., and Bon, C. (1989) A sensitive and continuous fluorometric assay for phospholipase A$_2$ using pyrene-labeled phospholipids in the presence of serum albumin. *Anal. Biochem.* **177,** 103–109.

19. Glover, S., de Carvalho, M. S., Bayburt, T., et al. (1995) Translocation of the 85-kDa phospholipase A$_2$ from cytosol to the nuclear envelope in rat basophilic leukemia cells stimulated with calcium ionophore or IgE/antigen. *J. Biol. Chem.* **270,** 15,359–15,367.

20. Evans, J. H., Spencer, D. M., Zweifach, A., and Leslie, C. C. (2001) Intracellular calcium signals regulating cytosolic phospholipase A$_2$ translocation to internal membranes. *J. Biol. Chem.* **276,** 30,150–30,160.

21. Channon, J. and Leslie, C. C. (1990) A calcium-dependent mechanism for associating a soluble arachidonoyl-hydrolyzing phospholipase A$_2$ with membrane in the macrophage cell line, RAW 264.7. *J. Biol. Chem.* **265,** 5409–5413.

22. Peters-Golden, M. and McNish, R. W. (1993) Redistribution of 5-lipoxygenase and cytosolic phospholipase A$_2$ to the nuclear fraction upon macrophage activation. *Biochem. Biophys. Res. Commun.* **196,** 147–153.

23. Sheridan, A. M., Sapirstein, A., Lemieux, N., et al. (2001) Nuclear translocation of cytosolic phospholipase A$_2$ is induced by ATP depletion. *J. Biol. Chem.* **276,** 29,899–29,905.

24. Leslie, C. C. (1991) Kinetic properties of a high molecular mass arachidonoyl-hydrolyzing phospholipase A$_2$ that exhibits lysophospholipase activity. *J. Biol. Chem.* **266,** 11,366–11,371.

25. de Carvalho, M. G. S., Garritano, J., and Leslie, C. C. (1995) Regulation of lysophospholipase activity of the 85-kDa phospholipase A$_2$ and activation in mouse peritoneal macrophages. *J. Biol. Chem.* **270,** 20,439–20,446.

26. Leslie, C. C. and Channon, J. Y. (1990) Anionic phospholipids stimulate an arachidonoyl-hydrolyzing phospholipase A$_2$ from macrophages and reduce the calcium requirement for activity. *Biochim. Biophys. Acta.* **1045,** 261–270.

27. Ghomashchi, F., Schuttel, S., Jain, M. K., and Gelb, M. H. (1992) Kinetic analysis of a high molecular weight phospholipase A$_2$ from rat kidney: divalent metal-dependent trapping of enzyme on product-containing vesicles. *Biochemistry* **31,** 3814–3824.

28. Kramer, R. M., Checani, G. C., and Deykin, D. (1987) Stimulation of Ca^{2+}-activated human platelet phospholipase A$_2$ by diacylglycerol. *Biochem. J.* **248,** 779–783.

29. Nalefski, E. A., Sultzman, L. A., Martin, D. M., et al. (1994) Delineation of two functionally distinct domains of cytosolic phospholipase A$_2$, a regulatory Ca^{2+}-dependent lipid-binding domain and a Ca^{2+}-independent catalytic domain. *J. Biol. Chem.* **269,** 18,239–18,249.

30. Hixon, M. S., Ball, A., and Gelb, M. H. (1998) Calcium-dependent and -independent interfacial binding and catalysis of cytosolic group IV phospholipase A$_2$. *Biochem.* **37,** 8516–8526.

31. Lio, Y. C., Reynolds, L. J., Balsinde, J., and Dennis, E. A. (1996) Irreversible inhibition of Ca$^{(2+)}$-independent phospholipase A$_2$ by methyl arachidonyl fluorophosphonate. *Biochim. Biophys. Acta* **1302,** 55–60.

32. Ghomashchi, F., Loo, R. W., Balsinde, J., et al. (1999) Trifluoromethyl ketones and methyl fluorophosphonates as inhibitors of group IV and VI phospholipases A$_2$: Structure-function studies with vesicle, micelle, and membrane assays. *Biochim. Biophys. Acta* **1420,** 45–56.

17

Measurement and Immunofluorescence of Cellular Phosphoinositides

Hiroko Hama, Javad Torabinejad, Glenn D. Prestwich, and Daryll B. DeWald

Summary

Phosphoinositides are a vitally important class of intracellular-signaling molecules that regulate cellular processes, including signaling through cell-surface receptors, remodeling of the cytoskeleton, vesicle-mediated protein trafficking, and various nuclear functions. Methods for the analysis of in vivo phosphoinositide concentration, such as the one described in this chapter enable quantification of all phosphoinositides from a population of cells. This method involves metabolic labeling of cells with myo-[2-^{3}H] inositol, followed by lipid extraction, and quantification by high-performance liquid chromatography (HPLC). It provides improved efficiency and reproducibility when analyzing yeast, plant cells, and is applicable to animal cells as well. In addition, a technique for determining the intracellular location of phosphoinositides is described. When quantification and localization techniques are used in parallel, an investigator can identify cell, and even subcellular concentration changes. The technique described in this chapter uses immunodetection with antiphosphoinositide antibodies to determine the localization and relative concentrations of phosphinositides in fixed cells. The availability of antibodies allows an investigator to perform immunofluorescence and potentially immunoelectron microscopy of phosphoinositide localization on particular cellular, organellar, or vesicular membranes.

Key Words: Signal transduction; HPLC; animal; plant; yeast cells.

1. Introduction

The potential role of phosphoinositides as intracellular signaling molecules was first reported half a century ago by Hokin and Hokin (*1*). Since then, a great

From: *Methods in Molecular Biology, vol. 284:*
Signal Transduction Protocols
Edited by: R. C. Dickson © Humana Press Inc., Totowa, NJ

deal has been learned about the complexity of cellular processes regulated by phosphoinositides. The growing list of phosphoinositide-regulated cellular processes includes signaling through cell-surface receptors *(2–4)*, remodeling of the cytoskeleton *(5,6)*, vesicle-mediated protein trafficking *(7,8)*, and various nuclear functions *(9–11)*. Each process can be simultaneously regulated by one or more specific phosphoinositides, owing to temporally and spatially restricted formation of phosphoinositides. In addition to phosphatidylinositol (PtdIns) itself, there are seven known phosphoinositides in eukaryotic cells with phosphate monoesters attached to the 3, 4, or 5 positions of the inositol ring of phosphatidylinositol: PtdIns(3)P, PtdIns(4)P, PtdIns(5)P, PtdIns(3,4)P$_2$, PtdIns(3,5)P$_2$, PtdIns(4,5)P$_2$, and PtdIns(3,4,5)P$_3$. The remodeling of phosphoinositide phosphorylation in space and time is precisely coordinated via regulated actions of phosphatidylinositol and phosphoinositide kinases, phosphoinositide phosphatases, and phospholipases.

Following synthesis at a specific site of action, phosphoinositides recruit and bind effector (phosphoinositide-binding) proteins and activate a variety of downstream signaling cascades. Unique among the phosphoinositides, PtdIns(4,5)P$_2$ either binds effector proteins and/or serves as a precursor for the second messengers inositol trisphosphate and diacylglycerol. However, binding to proteins and altering their localization and/or activity appears to be the primary function of phosphoinositides. Effector protein relocalization involves specific interactions between phosphoinositides and phosphoinositide-binding domains effector-protein. Among the 10 or more recognition motifs characterized are the PH (pleckstrin homology) domains, PX (Phox homology) domains, and FYVE (Fab1p, YOTB, Vac1, EEA1) domains. These phosphoinositide-binding motifs facilitate relocalization, conformational changes, and activation or inactivation of numerous downstream-effector proteins. This is significant because hundreds of eukaryotic proteins contain motifs that are likely to confer phosphoinositide binding. For example, in the human genome, there are approx 150–200 proteins containing predicted PH domains.

A number of studies have been conducted to measure the activities of phosphatidylinositol kinases in cell extracts to demonstrate the roles of phosphoinositides in signaling pathways (reviewed in **ref. *12***). Although these studies provided evidence for involvement of phosphoinositides and phosphoinositide kinases in regulatory systems, further analyses were necessary to demonstrate in vivo formation of specific phosphoinositides. Methods for the analysis of in vivo phosphoinositide concentrations such as the one described in this chapter enable quantification of all seven phosphoinositides from a population of cells. This method involves metabolic labeling of cells with *myo*-[2-^{3}H] inositol, followed by lipid extraction, and quantification by high-performance liquid chromatography (HPLC). The procedure described here is a modified version of the one developed by Cantley et al. *(13)*. It provides improved efficiency and reproducibility when analyzing yeast and plant cells *(14,15)* and is applicable to animal cells as well *(16)*.

In addition to whole-cell and tissue quantification of phosphoinositides, techniques have been developed to determine the intracellular localization of phosphoinositides. When quantification and localization techniques are used in parallel, an investigator can identify whole-tissue, cell, and even subcellular concentration changes. This is valuable, because phosphoinositide synthesis or degradation usually does not occur uniformly throughout the cell. Instead, it occurs at distinct membrane sites like the Golgi or plasma membrane.

Localization of specific phosphoinositides in living cells can be determined using fluorescence microscopy of cells transfected with gene constructs encoding phosphoinositide-binding proteins fused to the green fluorescent protein (GFP) *(17)*. Colocalization of the GFP fusion and phosphoinositides occurs via association of the phosphoinositide-binding domain with phosphoinositides. This powerful technique has been described several times, and will not be covered here. A less commonly-used approach is the use of fluorescently labeled rhodamine to visualize PtdIns(4,5)P$_2$ in cells, which localizes intracellular and intranuclear PtdIns(4,5)P$_2$ that cannot be detected by the GFP-PH domain constructs *(18)*. An alternative method described in this chapter involves phosphoinositide immunodetection with antiphosphoinositide antibodies *(19–23)* to determine the localization and relative concentrations in fixed cells. The availability of antibodies allows an investigator to perform immunofluorescence and potentially immunoelectron microscopy of phosphoinositide localization on particular cellular, organellar, or vesicular membranes.

1.1. Measurement of Phosphoinositides

Analysis of relative phosphoinositide concentrations using HPLC is sometimes called "headgroup analysis," because chromatographic separation of the water-soluble moiety of phosphoinositides (glycerophosphoinositol phosphates; gPIPs) is done after removal of the hydrophobic acyl chains (deacylation). Cells are first metabolically labeled with *myo*-[2-^{3}H] inositol for 12–24 h. Before lipid extraction, cells are treated with trichloroacetic acid (TCA) to inactivate enzymes that might degrade phosphoinositides during the extraction process. [^{3}H]-Labeled phosphoinositides are extracted with organic solvents, and deacylated by alkaline treatment. The resulting gPIPs are separated using strong anion-exchange HPLC. Small fractions are collected and radioactivity is measured by scintillation counting. Here we present protocols for labeling tissue-culture cells (NIH3T3 fibroblasts and 3T3-L1 preadipocytes), yeast (*Saccharomyces cerevisiae*), and plant cells (*Arabidopsis thaliana* submerged and hydroponic cultures). Lipid extracts from these cells are deacylated and analyzed by HPLC using the same procedure.

1.2. Immunofluorescence of Phosphoinositides

Monoclonal antibodies (MAbs) directed against PtdIns(4,5)P$_2$ and PtdIns (3,4,5)P$_3$ have been developed, and immunofluorescence procedures using these reagents enable the detection of these phosphoinositide species in fixed mammalian cells. When combined with the cellular phosphoinositide analyses described earlier, this approach allows an investigator to visualize in a cell the precise location where modulation of phosphoinositide concentrations is occurring. For the cellular immunolocalization of phosphoinositides, the cells must first be fixed with a chemical fixative (e.g., formaldehyde) and then permeabilized with a detergent (e.g., Triton X-100). Nonspecific binding sites in the cells are then blocked by incubating the cells in a solution containing a blocking reagent such as goat serum. The cells are incubated with the primary antibody directed against PtdIns(4,5)P$_2$ *(20,21)* or PtdIns(3,4,5)P$_3$ *(22)*. The cells are washed with a buffered salt solution and then incubated with a fluorophore-tagged secondary antibody. After a final washing to remove nonspecifically bound secondary antibodies, cells are visualized by epifluorescence or laser-scanning confocal microscopy.

2. Materials

2.1. Growth Media

1. NIH3T3 or 3T3-L1 cells are grown in Dulbecco's Modified Eagle's Medium (DMEM) containing 10% calf serum.
2. A yeast nitrogen base medium containing appropriate supplements and carbon source is used for the growth of yeast cells. The "drop-out mix" *(24)* is prepared without inositol.
3. *Arabidopsis thaliana* is grown submerged in a medium composed of 0.5X Murashige and Skoog (MS) and B5 vitamins. Alternatively, plants can be grown hydroponically on the same medium excluding the vitamins.
4. Falcon 2059 tubes (Becton Dickinson Labware, Lincoln Park, NJ).
5. Disposable cell scrapers (Fisher Scientific, Pittsburgh, PA).

2.2. Radiochemicals

1. *myo*-[2-^{3}H] Inositol (10–25 Ci/mmol) (American Radiolabeled Chemicals, (ARC), St. Louis, MO; Amersham Biosciences, Piscataway, NJ, ICN Biomedical, Irvine, CA; or PerkinElmer Life Sciences, PerkinElmer, Boston, MA).
2. For HPLC standards, PtdIns(4)P [inositol-2-^{3}H] (ARC). PtdIns(4,5)P$_2$ [inositol-2-^{3}H] (ARC or PerkinElmer).

3. [^{32}P] PtdIns(3)P is prepared by in vitro phosphorylation of PtdIns by yeast PtdIns 3-kinase with [γ-^{32}P] ATP, followed by separation by thin-layer chromatography (TLC) and extraction from the TLC media *(15)*.
4. [^{32}P] PtdIns(3,4)P$_2$ and [^{32}P]PtdIns(3,4,5)P$_3$ are prepared by in vitro phosphorylation of PtdIns(4)P and PtdIns(4,5)P$_2$, respectively, by mammalian PI 3-kinase with [γ-^{32}P] ATP, followed by TLC purification.
5. PtdIns(3,5)P$_2$ [inositol-2-^{3}H] is produced in vivo by salt-stressed yeast cells *(15,25)*.

2.3. Chromatography Media

1. Silica gel 60 TLC plates (0.25 mm thick, Merck no. 5724 or equivalent) are used to purify [^{32}P]-labeled standards prepared by in vitro phosphorylation.
2. Strong anion-exchange columns Partisil 5 SAX (4.6 mm × 250 mm) or Partisil 10 SAX (4.6 mm × 250 mm) (Whatman, Clifton, NJ) are used for HPLC headgroup analysis.
3. Ion-exchange columns must be fitted with guard columns (e.g., Phenomenex, SecurityGuard™, Torrance, CA, part no. KJO-4282) containing SAX inserts (Phenomenex, part no. AJO-431).

2.4. Immunofluorescence Reagents

1. Coverslips and slides (Fisher Scientific, or Ted Pella, Redding, CA).
2. Purified RC6F8 anti-PtdIns(3,4,5)P$_3$ IgM and 2C11 anti-PtdIns(4,5)P$_2$ IgM (Echelon Biosciences, Salt Lake City, UT).
3. High quality formaldehyde (Ted Pella, Redding, CA).
4. Fluorophore-tagged (e.g., FITC, Texas Red) antimouse IgM secondary antibodies are available (Jackson ImmunoResearch Laboratories, West Grove, PA or Rockland, Gilbertsville, PA).

2.5. Other Chemicals

1. All eight forms of nonradioactive phosphoinositides, with different acyl-chain lengths and a variety of reporter groups (Echelon Biosciences).
2. Additional natural and synthetic phosphoinositides (Avanti Polar Lipids, Alabaster, AL; Matreya State College, PA; and Calbiochem, San Diego, CA).
3. Organic solvents and all other chemicals can be purchased from several commercial sources (*see* **Subheading 4.**) for methylamine.
4. Scintillation cocktail must be miscible with aqueous solutions in order to provide consistent results when counting fractions.

2.6. Lipid Handling

Lipid extracts and deacylated lipids should be dried under a stream of nitrogen gas or in a Speed-Vac concentrator (Thermo Savant, Holbrook, NY). Dried lipids should be re-suspended using an bath sonicator such as Bransonic table-top cleaners (Branson Ultrasonics, Danbury, CT). A TLC tank is used for standard preparation by in vitro phosphorylation.

2.7. HPLC

Headgroup analysis can be performed with a Beckman System Gold HPLC with 32 Karat software (Beckman Coulter, Fullerton, CA) or equivalent. The system should be equipped with pumps for dual solvent delivery, an ultraviolet (UV) detector, and a fraction collector. The fraction collector is fitted with a rack for liquid-scintillation vials. Alternatively, an on-line radioactivity detector (e.g., β-RAM, INUS, Tampa, FL) can be used to detect separated gPIPs, in lieu of fraction collection and scintillation counting.

2.8. Microscopy

Epifluorescence or laser scanning confocal microscopes are used to examine cells labeled with fluorescently tagged antibodies. For the work presented in this chapter, cells were examined using a Nikon TE-300 inverted microscope interfaced with a Bio-Rad MRC 1024 confocal system (Bio-Rad Laboratories, Hercules, CA). Images are collected using 60X oil-immersion objective. Depending on the fluorophore, appropriate excitation wavelengths and emission filters are used.

3. Methods

3.1. Labeling and Lipid Extraction of Yeast Cells

1. A synthetic medium containing 5 μCi/mL of *myo*-[2-^{3}H] inositol is inoculated with a small amount of fresh culture to give an A_{600nm} of 0.01. A 5 mL culture is sufficient for each sample.
2. Cells are grown with shaking (200 rpm) until the A_{600nm} is 0.6–1.0, typically 14–20 h at 30°C.
3. Growth is terminated by addition of TCA (final concentration 5%), followed by incubation on ice for 1 h in polypropylene tubes (Falcon 2059). This treatment prevents degradation of lipids by lipases during manipulation of the cells *(26)*. It is important to avoid excessive exposure to TCA (higher concentrations, higher temperatures, or prolonged periods), because lipids can be degraded by the acid *(26)*.
4. Cells are harvested by centrifugation, and washed twice with 1 mL of H_2O.
5. The washed pellet can be stored at −80°C or used directly in **step 6**.
6. This procedure relies on a solvent for extraction that is very effective for yeast cells *(26)* and can be used for other cells as well *(15,16)*. The extraction solvent contains 95% ethanol/H_2O/diethyl ether/pyridine/conc. NH_4OH

(15/15/5/1/0.018 v/v). This solution without H_2O (called EEP solvent) can be prepared for use in a series of experiments for a few days, but long-term storage of EEP solvent is not recommended.

7. Washed yeast cells (5 mL of culture in late-log phase growth) are suspended in 0.5 mL of H_2O and 0.75 mL of EEP solvent is added. Extraction is conducted at 57°C for 30 min with occasional mixing. While the mixture is still warm, cell debris is removed by centrifugation for 5 min at 5000 g at room temperature. The supernatant fluid is dried under a stream of N_2 or in a Speed-Vac concentrator with appropriate traps.
8. Extracted lipids can be stored at $-80°C$ or immediately deacylated.

3.2. Labeling and Lipid Extraction of Mammalian Cells

1. For labeling NIH 3T3 fibroblasts or 3T3-L1 preadipocytes, cells should be grown to at least 60% confluency in 75 cm^2 flasks.
2. Cells are washed and then labeled for 36 h with *myo*-[2-^{3}H] inositol (20 μCi/mL) in inositol-free DMEM + 10% calf serum.
3. After 24 h, the medium should be removed and replaced with a fresh *myo*-[2-^{3}H] inositol-containing medium. At this point, if growth-factor stimulation (e.g., with PDGF to activate PI 3-kinase) is part of the experiment, cells are serum deprived for 2 h in inositol-free and serum-free DMEM containing 0.2% BSA and 10 μCi/mL *myo*-(2-^{3}H)-inositol, followed by platelet-derived growth factor (PDGF) (50 ng/mL) stimulation and harvest.
4. For cell harvest, ice-cold TCA is added to the medium in the flasks to a final concentration of 10%. The flasks containing the 10% TCA are incubated on ice for 1 h with the cells immersed in the solution.
5. Cells are released from the flasks by gently scraping with a disposable cell scraper followed by pipeting them into 15- or 50-mL conical screw-cap centrifuge tubes.
6. Centrifuge the samples for 5 min at 5000 g.
7. Remove supernatant fluid and add 5 mL of a 5% TCA, 1 mM EDTA solution to the pellets.
8. After the cells are resuspended, they should be centrifuged as above and the supernatant removed. Pellets can be stored at $-80°C$, or lipids can be extracted immediately.
9. Lipids are extracted from the cell pellet by resuspending cells in 0.75 mL chloroform/methanol/HCl (40/80/1 v/v) and vortexing the cells vigorously every 60 s for 15 s. Cells must be kept on ice between vortexing.
10. Add 0.25 mL of chloroform and 0.45 mL of 0.1 M HCl to the cells and vortex the tube.
11. Samples are centrifuged at 5000 g for 2 min, and the bottom, organic layer is transferred to another tube for continued processing.

12. Ammonia (50 µL of a 1 *M* solution) is added to the cells and the solutions in the tubes are dried in a Speed-Vac concentrator or under nitrogen. The samples can then be deacylated (*see* **Subheading 3.4.**) or stored at −80°C prior to deacylation.

3.3. Labeling and Lipid Extraction of Plant Cells

1. Two-week-old *A. thaliana* plants grown in a liquid medium (0.5X MS basal salt mixture, pH 5.8) containing B5 vitamins are submerged in 1 mL of the same medium with reduced myo-inositol (10 µ*M*) and 50 µCi/mL of *myo*-[2-³H] inositol. Labeling is accomplished in 1.6-mL microcentrifuge tubes for 20 h on a gyratory shaker (80 rpm) at 22–26°C. Alternatively, seeds can be germinated and plants grown hydroponically in 0.5X MS basal salt mixture, pH 5.8. Labeling of the hydroponically grown plants is achieved by placing the roots in a 2.0 mL cup containing the 0.5X MS medium and 100 µCi/mL of *myo*-[2-³H] inositol.
2. Growth of radiolabeled *A. thaliana* plants is terminated by addition of TCA (final conc. = 5%) followed by incubation on ice for 1 h.
3. Plantlets are washed 5 times with 10 mL of H_2O (room temperature) and transferred into a 5-mL Dounce tissue grinder after resuspension in 0.5 mL of H_2O, and 0.75 mL of EEP solvent. Tissues are homogenized and transferred into microcentrifuge tubes and incubated at 57°C for 30 min.
4. Cell debris is removed by centrifugation and the supernatant is dried under a stream of N_2 or in a Speed-Vac concentrator.

3.4. Deacylation of Glycerolipids

Lipids are deacylated by the method of Serunian et al. (*13*) with minor modifications. All the procedures are carried out in 1.6 mL or 2-mL plastic microcentrifuge tubes.

1. Dried lipids are resuspended in 0.5 mL of methylamine reagent (42.8% of 25% methylamine, 45.7% of methanol, 11.4% of *n*-butanol) by bath sonication, incubated at 53°C for 50 min, and dried in a Speed-Vac concentrator (*see* **Note 1**).
2. Deacylated lipids are suspended in 0.75 mL H_2O by sonication and then extracted 3 times with 0.5 mL *n*-butanol/petroleum ether/ethyl formate (20/4/1 v/v) or until the aqueous phase is no longer cloudy. The aqueous phase is dried in a Speed-Vac concentrator and suspended in 200 µL of H_2O.
3. A small portion (10–20 µL) of each sample is used to determine the radioactivity by liquid-scintillation counting. For preparation of loading samples for HPLC, standardization can be done using the [³H] counts, which is a close approximation of the content of PtdIns (phosphoinositides are minor

components). Alternatively, cell number, protein contents, or lipid phosphate radioactivity could be used for the same purpose.

3.5. Headgroup Analysis

1. It is useful to include AMP, ADP, and ATP in each sample in order to monitor the column performance by UV absorption. Typically, a portion of each sample ($1.5–2.5 \times 10^6$ cpm) is mixed with 40 nmoles each of AMP, ADP, and ATP. Glycerophosphoinositol monophosphate species [gPI(3)P, gPI(4)P, and gPI(5)P] elute between AMP and ADP, glycerophosphoinositol bis-phosphate species [gPI(3,5)P$_2$, gPI(3,4)P$_2$, and gPI(4,5)P$_2$] elute between ADP and ATP, and gPI(3,4,5)P$_3$ elutes after ATP.

2. Phosphoinositides are resolved with the following mobile phase of ammonium phosphate (pH 3.8; flow rate of 1 mL/min; *see* **Notes 2–4**).

   a. Gradient I for separation of gPI(3)P, gPI(4)P, gPI(3,4)P$_2$, gPI(3,5)P$_2$, and gPI(4,5)P$_2$.
   5 mL of 10 mM
   40 mL of a linear gradient, 10 mM to 0.7 M
   2 mL of a linear gradient, 0.7 to 1 M
   3 mL of 1 M

   b. Gradient II for separation of gPI(3)P, gPI(4)P, gPI(3,4)P$_2$, gPI(3,5)P$_2$, gPI(4,5)P$_2$, and gPI(3,4,5)P$_3$.
   5 mL of 10 mM
   60 mL of a linear gradient, 10 mM to 0.8 M
   2 mL of a linear gradient, 0.8 to 1 M
   3 mL of 1 M

Fractions are collected every 0.3 min (0.3 mL/fraction), mixed with 2 mL of water-miscible scintillation cocktail, and counted in a liquid scintillation counter (*see* **Notes 5** and **6**).

3.6. HPLC Standards

1. [^{3}H]-Labeled standards obtained from commercial sources are mixed with small amounts of non-radioactive carrier lipids (any phospholipids) and deacylated (*see* **Subheading 3.4.**). In vitro phosphorylation of PtdIns, PtdIns(4)P, and PtdIns(4,5)P$_2$ is performed as previously described (*27*).

2. Desired substrates are sonicated in 20 mM HEPES, pH 7.5, and mixed with 60 μM ATP, 0.2 mCi/mL [γ-^{32}P] ATP, 10 mM MgCl$_2$, and appropriate enzyme sources in a total volume 50 μL. The mixture is incubated for 5 min or longer, depending on the enzyme sources, at room temperature.

3. The reaction is terminated by the addition of 80 μL 1 *M* HCl and the lipids are extracted with 160 μL of chloroform/methanol (1/1). The lower organic phase is dried and re-dissolved in chloroform for TLC.
4. Entire samples are spotted onto silica gel 60 TLC plates that have been pre-treated with *trans*-1,2-diaminocyclohexane-*N,N,N',N'*-tetraacetic acid, and developed with a solvent containing 75 mL of methanol, 60 mL of chloroform, 45 mL of pyridine, 12g of boric acid, 7.5 mL of H_2O, 88 % (v/v) 3 mL of formic acid, 0.375g of 2,6-di-*tert*-butyl-4-methylphenol, and 75 μL of ethoxyquin *(28)*. (**Note:** It is necessary to dissolve the boric acid in water prior to the addition of organic solvents.)
5. After development, TLC plates are completely dried and exposed to X-ray films for 1–2 h at room temperature to identify the phosphoinositide spots (carried-over [γ-^{32}P] ATP stays at the origin).
6. The spots of desired products can be identified by comparison to non-radioactive phosphoinositides run on the same TLC system and stained by iodine vapor. The TLC plate is overlaid on the X-ray film, and the area on the TLC plate corresponding to the radioactive products is marked.
7. The silica matrices are carefully scraped off from the TLC plates and collected in 1.5-mL tubes. It is advisable to remove the silica surrounding the marked spots first, and then recover the spots.
8. Lipids are extracted from the matrices with a solvent containing chloroform/methanol/H_2O (16/16/5). The extracted phosphoinositides are mixed with carrier lipids and deacylated (*see* **Subheading 3.4.**).
9. The enzyme used for preparation of PtdIns(3)P is obtained by ammonium sulfate precipitation (25–30%) of yeast cytosol from a strain overexpressing the PtdIns 3-kinase Vps15p/Vps34p complex (strain TVY614 *[29]* carrying pJSY324.15 *[30]* and pPHY52 *[31]*).
10. Similarly, mammalian cell extracts may also be used for preparation of PtdIns(3,4)P_2 and PtdIns(3,4,5)P_3. The most effective extracts are prepared from tissue-culture cells transfected with an expression vector for the p110 subunit of PI 3-kinase *(32)*.

3.7. Immunolocalization of Phosphoinositides

1. NIH3T3 cells at logarithmic stage on coverslips are serum-starved overnight and stimulated with insulin (100 ng/mL) or PDGF (50 ng/mL). Growth factor stimulation is typically done for 1–15 min.
2. Reactions are stopped by washing the cells with cold tris-buffered saline (TBS) and cells are processed for immunofluorescence.
3. Cells on glass coverslips are fixed with 2% formaldehyde (in cell media) for 20 min, and then permeabilized with 0.5% Triton X-100 in TBS for 15 min at room temperature.

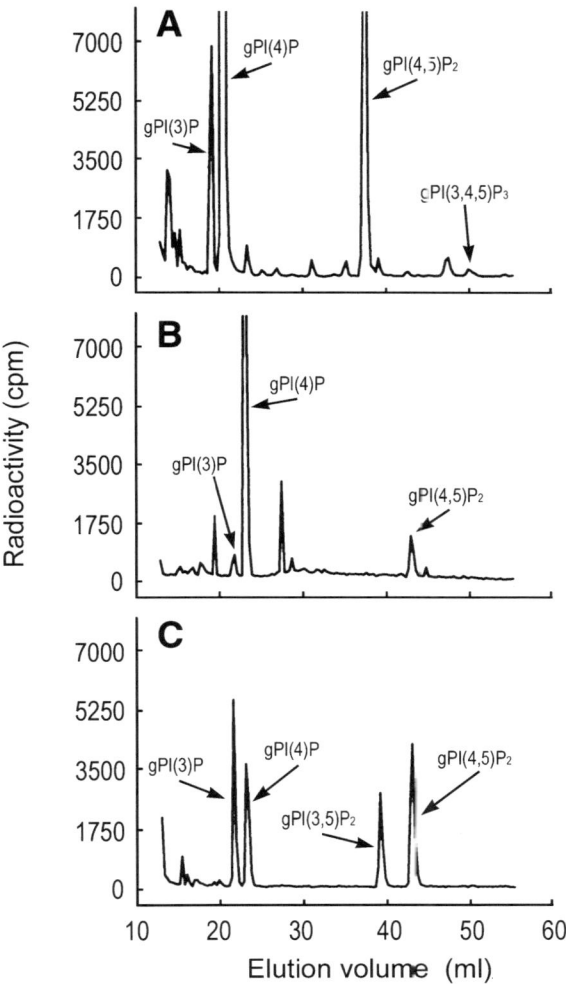

Fig. 1. HPLC chromatograms of glycerophosphoinositols from mammalian, plant, and yeast cells. The NIH3T3 fibroblast cells (**A**) were stimulated with platelet derived growth factor (PDGF) for 5 min prior to harvest and analysis of glycerophosphoinositols (Gradient II). Plant cells (**B**) and yeast cells (**C**) were subjected to 0.25 *M* NaCl for 30 min and 1.0 *M* NaCl shock for 20 min, respectively, prior to lipid extraction and headgroup analysis (Gradient I). Then lipids were extracted, deacylated, and glycerophosphoinositols analyzed by HPLC using a Partisil 10 SAX column and gradients described in the text. Fractions were collected, counted in a scintillation counter, and counts in each fraction were plotted. gPI(3)P is glycerophosphoinositol 3-phosphate, gPI(4)P is glycerophosphoinositol 4-phosphate, and all other glycerophosphoinositols are likewise designated.

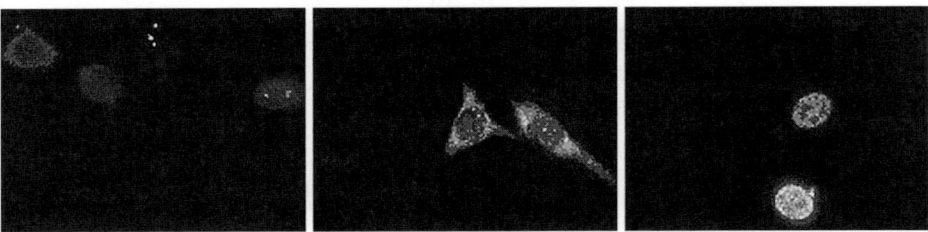

Fig. 2. Immunofluorescence detection of PtdIns(4,5)P$_2$ and PtdIns(3,4,5)P$_3$ in PDGF-stimulated 3T3 L1 preadipocytes. Cells were stimulated with PDGF (50 ng/mL) for 5 min and then prepared for immunofluorescence detection of phosphoinositides as described in the text. The control cells **(left panel)** were not incubated with antiphosphoinositide antibody. Cells that were probed with anti-PtdIns(3,4,5)P$_3$ antibody **(middle panel)** displayed primarily cytosolic and membrane staining, and those probed with anti-PtdIns(4,5)P$_2$ antibody **(right panel)** displayed primarily nuclear staining.

4. After blocking with 10% goat serum in TBS, either RC6F8 MAb (anti-PtdIns(3,4,5)P$_3$ antibody) ascites (1:50 dilution) or 2C11 (anti-PtdIns(4,5)P$_2$ antibody) ascites (1:5000 dilution) is added and incubated at room temperature for 1 h.
5. After washing three times with the blocking solution, fluorophore-labeled antimouse IgM (1:2000 dilution) is added and incubated at room temperature for 1 h.
6. Cells are washed three times with deionized water and observed using a laser scanning confocal microscope or fluorescent microscope.

4. Notes

1. For deacylation of lipids, we have experienced variations in efficiencies depending on the source of methylamine. It is advised to test deacylation with nonradioactive phospholipids, followed by TLC.
2. Ammonium phosphate (used for HPLC mobile phase) contains surprisingly large quantities of impurities. It is necessary to remove the impurities by filtration through nitrocellulose membranes (0.45 μm). Once prepared, the buffers are easily contaminated by molds. It is not recommended to store ammonium phosphate solutions for an extended period.
3. The samples applied onto the HPLC columns contain some H$_2$O-insoluble materials as well as small amount of lipids that are carried over through extractions. After repeated uses, columns start to lose resolution. When the number of counts in the fractions in-between the gPI(3)P and gPI(4)P peaks

is above background (loss of baseline resolution), the guard column should be replaced with a new one.

4. A new SAX column should be conditioned first by running left-over samples or nonradioactive samples. For unknown reasons, a significant portion of sample is so tightly bound to a new column that it never elutes. We have also observed that a very small portion of bound species remains bound throughout the gradient even at 1 *M* ammonium phosphate. When the same gradient is run without injecting a sample, sometimes small glycerophosphoinositol peaks elute. To avoid any interference from a previous column run, we typically apply two "cleaning cycles" of 10 mL linear gradient of 10 m*M* to 1 *M* ammonium phosphate between each HPLC run.

5. Recently, a nonradioactive method was reported *(33)*. In this method, gPIPs are separated by anion-exchange HPLC and detected by conductivity measurements. Although the apparent resolution of this procedure is not as high, it is less expensive and avoids the use of radioactive materials. In addition, a sensitive mass assay for PtdIns(3,4,5)P$_3$ has been reported, but this only works with this phosphoinositide and requires an isotope-dilution method with deacylation and analysis of the Ins(1,3,4,5)P$_4$ head group *(34)*. A third approach is the use of matrix-assisted laser desorption/ionization time-of-flight mass spectrometry (MALDI-TOF MS), which has been used for the detection and quantification of phosphoinositides *(35)*.

6. Phosphoinositides from cell extracts and in vitro enzyme-catalyzed reactions may also be analyzed by competitive displacement reactions using tagged and/or labeled PH domains as probes. Stable complexes are formed between a biotinylated target lipid and an appropriate PH domain, and phosphoinositides present in samples are detected by their ability to compete for binding to the PH domain. The complexes are detected using time-resolved FRET *(36)*. This concept has been independently used to develop a competitive fluorescence polarization (FP)-based assay amenable to high throughput screening. The FP assay has been used to determine activity of phosphoinositide 3-kinase (PI 3-K) and the type-II SH2-domain-containing inositol 5-phosphatase (SHIP2) *(37)*. This assay is based on the interaction of specific phosphoinositide binding proteins with fluorophore-labeled phosphoinositide and inositol phosphate tracers. The enzyme reaction products are detected by their ability to compete with the fluorescent tracers for protein binding, leading to an increase in the amount of free tracer and a decrease in polarization values. The antilipid and competitive assay methodologies offer new opportunities in detection of phosphoinositide abnormalities in cancer cells, discovery of new anticancer agents targeted at inhibition of PI 3-kinase, and targeted monitoring of the effects of these agents in vivo *(38)*.

Acknowledgments

The authors wish to thank the American Cancer Society and NIH (Grant NS29632 to G.D.P.) for support of work at Utah State University and the University of Utah, and Echelon Biosciences, for providing reagents in our laboratories.

References

1. Hokin, M. R. and Hokin, L. E. (1953) Enzyme secretion and the incorporation of P-32 into phospholipids of pancreas slices. *J. Biol. Chem.* **203,** 967–977.
2. Carpenter, C. L. and Cantley, L. C. (1996) Phosphoinositide kinases. *Curr. Opin. Cell Biol.* **8,** 153–158.
3. Stoyanov, B., Volinia, S., Hanck, T., et al. (1995) Cloning and characterization of a G protein-activated human phosphoinositide-3 kinase. *Science* **269,** 690–693.
4. Stephens, L. R., Eguinoa, A., Erdjument-Bromage, H., et al. (1997) The G beta gamma sensitivity of a PI3K is dependent upon a tightly associated adaptor, p101. *Cell* **89,** 105–114.
5. Hall, A. (1998) Rho GTPases and the actin cytoskeleton. *Science* **279,** 509–514.
6. Yin, H. L. and Janmey, P. A. (2003) Phosphoinositide regulation of the actin cytoskeleton. *Annu. Rev. Physiol.* **65,** 761–789.
7. De Camilli, P., Emr, S. D., McPherson, P. S., and Novick, P. (1996) Phosphoinositides as regulators in membrane traffic. *Science* **271,** 1533–1539.
8. Czech, M. P. (2003) Dynamics of phosphoinositides in membrane retrieval and insertion. *Annu. Rev. Physiol.* **65,** 791–815
9. Irvine, R. F. (2002) Nuclear lipid signaling. *Sci. STKE* 2002, RE13.
10. Irvine, R. F. (2003) Nuclear lipid signalling. *Nat. Rev. Mol. Cell Biol.* **4,** 349–361.
11. Martelli, A. M., Tabellini, G., Borgatti, P., et al. (2003) Nuclear lipids: new functions for old molecules? *J. Cell Biochem.* **88,** 455–461.
12. Toker, A. (2002) Phosphoinositides and signal transduction. *Cell Mol. Life Sci* **59,** 761–779.
13. Serunian, L. A., Auger, K. R., and Cantley, L. C. (1991) Identification and quantification of polyphosphoinositides produced in response to platelet-derived growth factor stimulation. *Methods Enzymol.* **198,** 78–87.
14. Stock, S. D., Hama, H., DeWald, D. B., and Takemoto, J. Y. (1999) SEC14-dependent secretion in Saccharomyces cerevisiae. Nondependence on sphingolipid synthesis-coupled diacylglycerol production. *J. Biol. Chem.* **274,** 12979–12983.
15. Hama, H., Takemoto, J. Y., and DeWald, D. B. (2000) Analysis of phosphoinositides in protein trafficking. *Methods* **20,** 465–473.
16. DeWald, D. B. (2003) Measurements of cellular phosphoinositide levels in 3T3-L1 adipocytes. *Methods Mol. Med.* **83,** 145–154.
17. Balla, T., Bondeva, T., and Varnai, P. (2000) How accurately can we image inositol lipids in living cells? *Trends Pharmacol Sci.* **21,** 238–241.
18. Holz, R. W., Hlubek, M. D., Sorensen, S. D., et al. (2000) A pleckstrin homology domain specific for phosphatidylinositol 4, 5- bisphosphate (PtdIns-4,5-P2) and

fused to green fluorescent protein identifies plasma membrane PtdIns-4,5-P2 as being important in exocytosis. *J. Biol. Chem.* **275,** 17,878–17,885.

19. Fukami, K., Matsuoka, K., Nakanishi, O., et al. (1988) Antibody to phosphatidylinositol 4,5-bisphosphate inhibits oncogene-induced mitogenesis. *Proc. Natl. Acad. Sci. USA* **85,** 9057–9061.

20. Thomas, C. L., Steel, J., Prestwich, G. D., and Schiavo, G. (1999) Generation of phosphatidylinositol-specific antibodies and their characterization. *Biochem. Soc. Trans.* **27,** 648–652.

21. Osborne, S. L., Thomas, C. L., Gschmeissner, S., and Schiavo, G. (2001) Nuclear PtdIns(4,5)P$_2$ assembles in a mitotically regulated particle involved in pre-mRNA splicing. *J. Cell Sci.* **114,** 2501–2511.

22. Chen, R., Kang, V. H., Chen, J., et al. (2002) A monoclonal antibody to visualize PtdIns(3,4,5)P$_3$ in cells. *J. Histochem. Cytochem.* **50,** 697–708.

23. Niswender, K. D., Gallis, B., Blevins, J. E., et al. (2003) Immunocytochemical detection of phosphatidylinositol 3-kinase activation by insulin and leptin. *J. Histochem. Cytochem.* **51,** 275–283.

24. Burke, D., Dawson, D., and Stearns, T. (2000) *Methods in Yeast Genetics 2000: A Cold Spring Harbor Laboratory Course Manual.* Cold Spring Harbor Laboratory Press, Cold Spring Harbor, New York.

25. Dove, S. K., Cooke, F. T., Douglas, M. R., et al. (1997) Osmotic stress activates phosphatidylinositol-3,5-bisphosphate synthesis. *Nature* **390,** 187–192.

26. Hanson, B. A. and Lester, R. L. (1980) The extraction of inositol-containing phospholipids and phosphatidylcholine from Saccharomyces cerevisiae and Neurospora crassa. *J. Lipid Res.* **21,** 309–315.

27. Stack, J. H., DeWald, D. B., Takegawa, K., and Emr, S. D. (1995) Vesicle-mediated protein transport: regulatory interactions between the Vps15 protein kinase and the Vps34 PtdIns 3-kinase essential for protein sorting to the vacuole in yeast. *J. Cell Biol.* **129,** 321–334.

28. Walsh, J. P., Caldwell, K. K., and Majerus, P. W. (1991) Formation of phosphatidylinositol 3-phosphate by isomerization from phosphatidylinositol 4-phosphate. *Proc. Natl. Acad. Sci. USA* **88,** 9184–9187.

29. Vida, T. A. and Emr, S. D. (1995) A new vital stain for visualizing vacuolar membrane dynamics and endocytosis in yeast. *J. Cell Biol.* **128,** 779–792.

30. Stack, J. H., Herman, P. K., Schu, P. V., and Emr, S. D. (1993) A membrane-associated complex containing the Vps15 protein kinase and the Vps34 PI 3-kinase is essential for protein sorting to the yeast lysosome-like vacuole. *EMBO J.* **12,** 2195–2204.

31. Herman, P. K. and Emr, S. D. (1990) Characterization of VPS34, a gene required for vacuolar protein sorting and vacuole segregation in *Saccharomyces cerevisiae.* *Mol. Cell Biol.* **10,** 6742–6754.

32. Katagiri, H., Asano, T., Ishihara, H., et al. (1996) Overexpression of catalytic subunit *p110alpha* of phosphatidylinositol 3- kinase increases glucose transport activity with translocation of glucose transporters in 3T3-L1 adipocytes. *J. Biol. Chem.* **271,** 16,987–16,990.

33. Nasuhoglu, C., Feng, S., Mao, J., et al. (2002) Nonradioactive analysis of phosphatidylinositides and other anionic phospholipids by anion-exchange high-performance liquid chromatography with suppressed conductivity detection. *Anal. Biochem.* **301,** 243–254.

34. van der Kaay, J., Batty, I. H., Cross, D. A., et al. (1997) A novel, rapid, and highly sensitive mass assay for phosphatidylinositol 3,4,5-trisphosphate (PtdIns(3,4,5)P$_3$) and its application to measure insulin-stimulated PtdIns(3,4,5)P$_3$ production in rat skeletal muscle in vivo. *J. Biol. Chem.* **272,** 5477–5481.

35. Muller, M., Schiller, J., Petkovic, M., et al. (2001) Limits for the detection of (poly-)phosphoinositides by matrix-assisted laser desorption and ionization time-of-flight mass spectrometry (MALDI-TOF MS). *Chem. Phys. Lipids* **110,** 151–164.

36. Gray, A., Olsson, H., Batty, I. H., et al. (2003) Nonradioactive methods for the assay of phosphoinositide 3-kinases and phosphoinositide phosphatases and selective detection of signaling lipids in cell and tissue extracts. *Anal. Biochem.* **313,** 234–245.

37. Drees, B. E., Weipert, A., Hudson, H., et al. (2003) Competitive fluorescence polarization assays for the detection of phosphoinositide kinase and phosphatase activity. *Comb. Chem. High Throughput Screen* **6,** 321–330.

38. Prestwich, G. D., Chen, R., Feng, L., et al. (2002) *In situ* detection of phospholipid and phosphoinositide metabolism. *Adv. Enzyme Regul.* **42,** 19–38.

18

Measuring Dynamic Changes in cAMP Using Fluorescence Resonance Energy Transfer

Sandrine Evellin, Marco Mongillo, Anna Terrin, Valentina Lissandron, and Manuela Zaccolo

Summary

cAMP is a ubiquitous second messenger that controls numerous cellular events including movement, growth, metabolism, contraction, and synaptic plasticity. With the emerging concept of compartmentalization of cAMP-dependent signaling, a detailed study of the spatio-temporal intracellular dynamics of cAMP is required. Here we describe a new methodology for monitoring fluctuations of cAMP in living cells, based on the use of a genetically encoded biosensor. The regulatory and catalytic subunits of the main cAMP effector, the protein kinase A (PKA), fused with two suitable green fluorescent protein (GFP) mutants is used for measuring changes in fluorescence resonance energy transfer (FRET) that correlate with changes in intracellular cAMP levels. This method allows the study of cAMP fluctuations in living cells with high resolution both in time and in space.

Key Words: cAMP; biosensor; protein kinase A (PKA); signal transduction; green fluorescent proteins; cell imaging; fluorescence resonance energy transfer (FRET).

1. Introduction

cAMP dependent signals represent one of the major intracellular-transduction pathways and are involved in numerous cellular events including movement, growth, metabolism, contraction, and synaptic plasticity *(1)*. Most of cAMP cellular effects are mediated by protein kinase A (PKA), an holotetramer composed of two regulatory (R) and two catalytic (C) subunits. Activation of PKA by

From: *Methods in Molecular Biology, vol. 284:*
Signal Transduction Protocols
Edited by: R. C. Dickson © Humana Press Inc., Totowa, NJ

cAMP induces a conformational change in the R subunits and, consequently, the dissociation of the C subunits from the R subunits *(2)*. The liberated C subunits both phosphorylate a large variety of cytoplasmic substrates and diffuse into the nucleus, where they modify transcription processes by phosphorylating several transcription factors *(3,4)*. cAMP, therefore, is capable of transmitting a large variety of different signals, but how such diversity is encoded by a single second messenger is still largely to be explained. In recent years, evidence has been accumulating that the specificity of response to a given stimulus is conferred not only by a tight regulation of the temporal dynamics of the signaling molecules, but also by a precise spatial organization of the signaling pathways *(5)*. In particular, the view is emerging that cAMP/PKA mediated signaling is highly compartmentalized: PKA is anchored via A kinase anchoring proteins (AKAPs) to specific subcellular structures and cAMP itself can be generated in discrete pools that, in turn, can selectively activate defined PKA subsets *(6–10)*.

1.1. Imaging of cAMP in Single Living Cell

Commonly, total cellular cAMP levels are measured in cell lysates with radioimmunoassays. Such an approach offers very poor temporal resolution and no spatial resolution and is therefore inadequate to study the fine details of the cAMP dependent signaling pathway *(11,12)*. An approach to image the dynamics of free cAMP in single living cells is based on microinjection of fluorescein-labeled PKA C-α subunits and rhodamine-labeled PKA RI-β subunits *(13)*. Such approach relies on fluorescence resonance energy transfer (FRET), a physicochemical phenomenon whereby the excited state energy of a donor fluorophore can be transferred to an acceptor fluorophore, which can then emit its own characteristic fluorescence *(14–17)*. With FRET, the intensity of the donor's fluorescence emission decreases and concomitantly there is an increase in the acceptor's emission intensity. For FRET to occur, the acceptor must absorb at roughly the same wavelengths as the donor emits. Moreover, FRET depends on the antiparallel alignment of the donor and acceptor fluorophores electric dipoles and is highly sensitive to the donor-acceptor distance (between 10 Å and 100 Å). In fact, FRET efficiency decreases with the sixth power of the distance between donor and acceptor, and therefore a minimal perturbation of the spatial relationship between the two fluorophores can drastically alter the efficiency of energy transfer. As a consequence, FRET is one of the few tools available for measuring nanometer scale distances and changes in distances, both in vitro and in vivo. The fluorescein- and rhodamine-labeled probe for cAMP has not found a wide application, owing to some technical drawbacks: it requires microinjection of a very large protein complex (approx 170-kD) and such a procedure is not applicable to all cell types. Moreover, a high amount of the cAMP-biosensor needs to be injected, resulting in unequal distribution of the probe and in toxicity. A further problem is

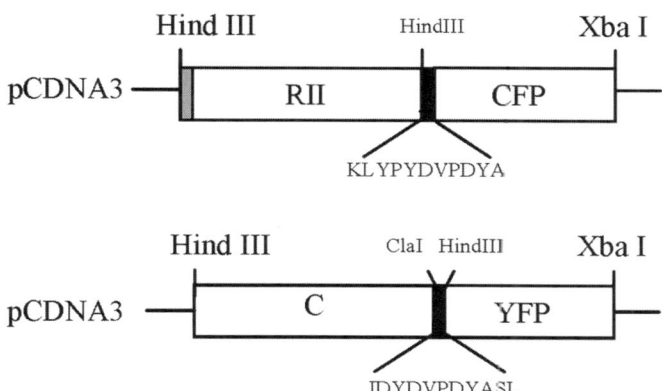

Fig. 1. RII-CFP and C-YFP constructs. Schematic representation of the cDNA encoding the GFP-tagged PKA subunits. Gray box, six-histidine tag, black box, peptide linker, RII, regulatory β-subunit, type II, C, catalytic α-subunit.

the tendency of the probe to aggregate and precipitate. More recently, a new generation of cAMP-biosensors was generated that is totally genetically encoded *(18)*. To generate such a sensor, the RII-β subunit and the C-α subunits of PKA were fused respectively to the cyan (CFP) and yellow (YFP) mutants of the green fluorescent protein (GFP). CFP and YFP are two fluorophores with excitation and emission spectra suitable for FRET. In addition, a peptide linker was introduced between the GFP moieties and the PKA subunits in order to increase the chance that CFP and YFP come close enough for FRET to occur (*see* **Fig. 1**). Upon transfection, a molecularly homogeneous probe is generated and cAMP-induced dissociation of the two PKA-subunits translates into a drop of FRET efficiency. At low cAMP levels, when PKA is in its inactive state, excitation of the donor RII-CFP at 430 nm leads to emission of CFP at 480 nm (*see* **Fig. 2**). As the two fluorophores are in near proximity, part of the excited state energy of CFP is non-radiatively transferred to YFP, which then emits at 545 nm. On the contrary, at high cAMP concentration, R and C subunits dissociate from each other and FRET between CFP and YFP is no more possible, therefore excitation of CFP at 430 nm leads to emission at 480 nm only. FRET changes can be measured as the ratio of donor to acceptor emission intensity and correlate with cAMP changes.

2. Materials
2.1. Cell Culture and Transfection
1. Cells: CHO.
2. DNA constructs: pCDNA3 RII-CFP and pCDNA3 CAT-YFP.

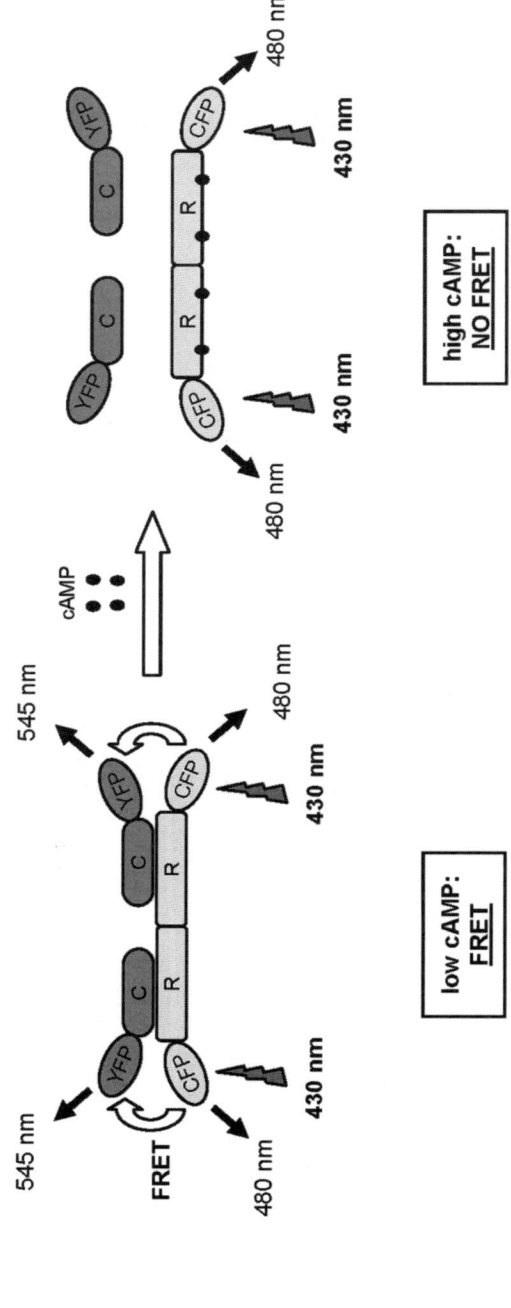

Fig. 2. Schematic representation of the genetically encoded FRET-based biosensors. Gray arrows indicate fluorescence excitation and black arrows indicate fluorescence emission. Excitation and emission peak wavelengths are stated. In conditions of low cAMP, the GFP-tagged PKA is in its inactive holotetrameric conformation and FRET is maximal. When cAMP increases, it binds to the regulatory subunits and the catalytic subunits are released abolishing FRET.

3. Transfection reagents: FuGENE6 (Roche).
4. Serum-free culture medium.
5. TE buffer: 10 mM Tris-HCl, pH 7.4, and 1 mM EDTA, pH 8.0.
6. Tissue-culture plate, 6-well.
7. Glass coverslips (good quality required, 24-mm diameter CoverGlass, cat. no. 406/0189/50, VWR International).
8. Coverslip holder.

2.2. Imaging System for FRET

1. Microscope: Olympus IX50.
2. Monochromator: Polychrome IV (T.I.L.L. Photonics, GmbH, Germany).
3. Oil-immersion objective: 100 × PlanApo; 1.30 NA (Olympus; *see* **Note 1**).
4. 455 nm dichroic mirror: 455DCLP (Chroma).
5. CFP emission filter: 480DF30 (Chroma).
6. YFP emission filter: 545DF35 (Chroma).
7. Beam splitter device: MultiSpec Micro-Imager™ (Optical Insight).
8. Digital camera: PCO SensiCam QE.
9. Computer: PC Pentium 4 processor, 800 MHz. 512 Mb SDRAM 40 Gb hard disk.
10. Image acquisition and processing software: TILLvisION v3.3 (T.I.L.L. Photonics, GmbH, Germany), MetaFluor (Universal Imaging, West Chester, PA), or "ImageJ" (*see* **Subheading 4.1.**).

3. Methods

Here we described a typical experiment in which the change in the intracellular levels of cAMP, generated in CHO cells by forskolin, are monitored by measuring FRET changes.

3.1. Cell Culture and Transfection

The aim of the first part of the experimental procedure is to express the probe in the cells of interest.

Transfection methods vary depending on the cell type. We routinely use Fu-Gene6 transfection reagent.

1. First day: Plate the cells onto glass coverslips and grow them to 50–60% confluence (*see* **Note 2**).
2. Third day: Transfect the cells with the cDNA encoding for the two sub-units of the probe. For each 24-mm glass coverslips, mix 1.5 micrograms of each cDNA, 6 µL FuGene6 and 100 µL of serum-free DMEM. Incubate the DNA-FuGene6 mix for 15 min and subsequently add the mix to the cells.

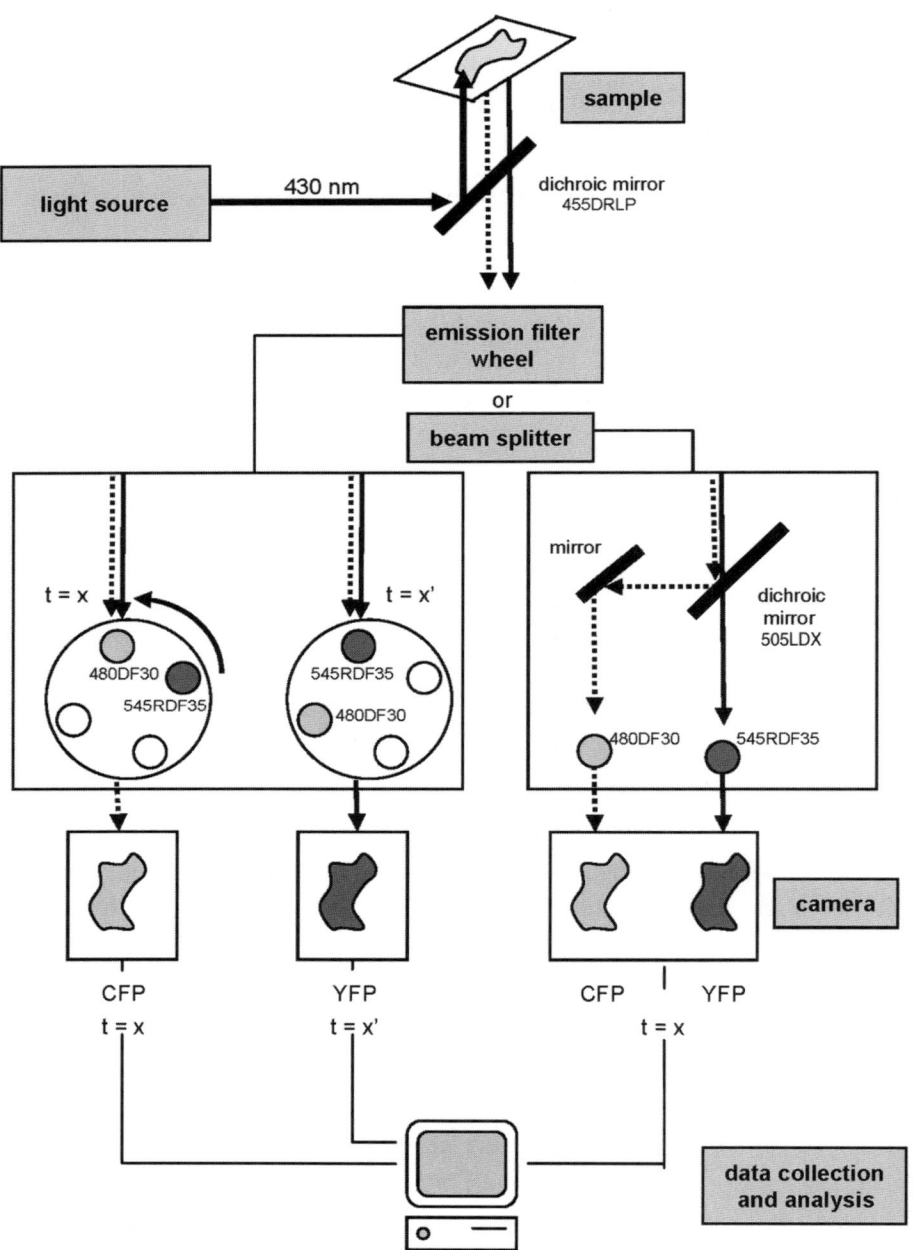

3. Fifth day: Inspect the cells at the fluorescence microscope (*see* **Note 3**). Mount the coverslip on a sealed holder (or chamber) and bath the cells in HEPES-buffered medium or saline (*see* **Note 4**).

3.2. Cell Imaging

A convenient way to estimate FRET changes is to calculate the donor fluorescence to the acceptor fluorescence ratio changes. In this way a rise in cAMP (leading to dissociation of PKA subunits and therefore to reduced FRET between CFP and YFP) translates into a rise in the ratio value. CFP to YFP ratio changes are calculated by collecting the emitted fluorescence within two spectral windows centered respectively on the donor's (CFP, 480 nm) and acceptor's (YFP, 545 nm) emission peaks, upon selective excitation of the donor fluorophore only (430 nm). A setup for FRET imaging is shown schematically in **Fig. 3**. The system is composed of a light source for excitation at 430 nm, a dichroic mirror that reflects the excitation beam to the sample expressing RII-CFP and C-YFP and that transmits the light emitted by the sample to a digital camera. The two fluorophores individual emissions are further defined using bandpass emission filters. CFP is typically excited between 430 and 440 nm. Such excitation can be obtained either with a software-controlled monochromator or with a mercury or xenon bulb and an appropriate band pass filter (e.g., 430DF30; *see* **Note 5**). In both cases, a software-controlled shutter is necessary in order to limit the illumination of the sample only to the time required for image acquisition, thus minimizing bleaching of the fluorophores and photodamage of the sample.

Collection of individual CFP and YFP emission can be achieved in two ways. The first possibility is to use a software-controlled filter wheel that mounts the two bandpass filters appropriate for collecting CFP and YFP emission signals (e.g., 480DF30 for CFP and 545RDF35 for YFP). This configuration presents some flexibility because, as other filters can be added to the wheel, the same setup can be used for imaging several fluorescent indicators. On the other hand, a short delay between the acquisition of the cyan and yellow emission signals will occur, owing to the time required for the wheel to shift from one filter position to the other (*see* **Fig. 3** and **Note 6**). Such delay introduces an artefact, as a consequence of the fact that the CFP/YFP ratio image is calculated using numerator and

---

Fig. 3. FRET imaging setup. Schematic representation of a typical setup used to perform FRET imaging. Black thick lines indicate fluorescence excitation (430 nm). Thin full and dotted lines indicate fluorescence emission for CFP (480 nm) and YFP (545 nm), respectively. Light gray and dark gray circles represents the specific bandpass emission filters for CFP and YFP, respectively.

denominator intensity values that are not acquired at the same time. Such artefact is particularly relevant when fast kinetics for cAMP changes are expected. Also, a very obvious misalignment artefact can be generated when imaging cells that move during the experiment (such as contracting myocytes), resulting in the acquisition of CFP and YFP images that are not perfectly superimposable for ratio calculation. A possible alternative to the filter wheel is the use of a beam splitter. With such a device, it is possible to acquire simultaneously CFP and YFP emissions on the two halves of the camera sensor, by means of a 505 nm dichroic mirror (505LDX) that splits emitted light into two spectrally separate, independent beams. The reflected beam (wavelengths <505 nm) and the transmitted beam (wavelength >505 nm) are then further defined by the bandpass-emission filters 480DF30 and 545DF35 to give, respectively, CFP and YFP emission images on the detector. For image acquisition, a high sensitivity, low intrinsic noise, digital camera must be used, in order to decrease integration time. A scientific-grade, cooled charge-coupled-device (CCD) digital camera, 12 bit, is normally the choice.

3.2.1. Set Protocol

The aim of the experiment is to record a series of images of CFP and YFP emission intensities in a time-course and the following parameters have to be defined:

1. Set the exposure time. This corresponds to the time in which the sample is illuminated and the camera acquires CFP and YFP emission signals. One should aim at generating images with signal to background ratio >3. Indicatively, exposure time is normally between 50 ms and 300 ms (*see* **Note 7**).
2. Set the delay time. This corresponds to the time between two consecutive illuminations of the sample. During this time, the shutter controlling the incident light has to be closed in order to minimize bleaching of the fluorophores. The duration of the delay time depends on the kinetics expected. We generally use delay times >500 ms (*see* **Note 8**).
3. Set the number of acquisitions.
4. Set the camera binning. Binning 2 or 4 is normally the choice. A higher binning increases the intensity of the signal at the expense of resolution.

3.2.2. Run Protocol

During this phase of the experiment, the chosen cell is excited at the appropriate wavelength and the emitted CFP to YFP fluorescence is collected by the camera. Most types of software allow a live display of the mean fluorescence intensities for each fluorophore. A typical time course of CFP and YFP emission intensities, acquired upon stimulation of the transfected cell with forskolin, is illustrated in **Fig. 4D**.

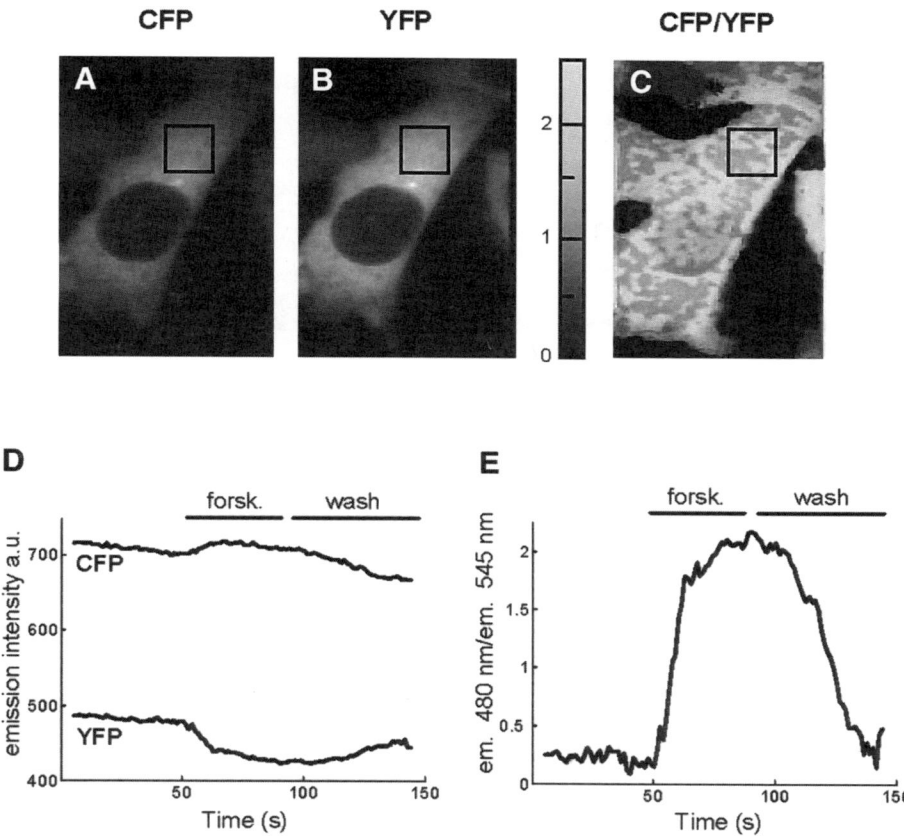

Fig. 4. Imaging of cAMP kinetics in living cells. CHO cells co-transfected with RII-CFP and C-YFP. (**A**) RII-CFP image obtained by exciting at 480 nm and collecting emission through the cyan channel. (**B**) C-YFP image obtained by exciting at 480 nm and collecting emission through the yellow channel. Images shown in (**A**) and (**B**) were acquired 100 s after the addition of 20 μ*M* forskolin to the sample. (**C**) Pseudocolor ratio image (480 nm/545 nm) calculated by dividing the intensity values of the cell in (**A**) and (**B**). (**D**) Kinetics of mean fluorescence intensities of CFP and YFP, calculated in the square region of interest (ROI) shown in (**A**) and (**B**) and recorded before and after addition of forskolin. These values were used to calculate the mean fluorescence emission ratio (480 nm/ 545 nm) shown in (**E**).

3.2.3. Image Processing and Data Analysis

The collected images must be software-processed. Aim of the processing step is to generate a new image representing the pixel by pixel ratio of CFP and YFP emission images.

1. If the system is equipped with a beam splitter: Split and align CFP and YFP images, in order to obtain two distinct and superimposable images. If the system is equipped with a filter wheel, CFP and YFP image acquisition are already in two separate stacks.
2. Subtract the background noise to each of the collected images. Such operation removes the out-of-the-cell fluorescence and the intrinsic electronic noise of the camera, reducing artifacts and increasing the dynamic range of the measure (*see* **Note 9**).
3. Calculate the ratio between CFP and YFP images. The value of each pixel in the ratio image corresponds to the CFP/YFP ratio in the corresponding pixels of the raw images (*see* **Fig. 4C** and **Note 10**). Such ratiomeric images are normally displayed in pseudocolor, according to a user-defined lookup table that assigns to each ratio value a different color. Draw regions of interest (ROIs) in the cell compartment of interest, in order to measure the average ratio kinetics within that region, as shown in **Fig. 4C**, and **E**.

4. Notes

1. Care must be taken in the choice of the objective: Regardless of the magnification, one should choose a high-quality-high-numerical aperture (>1,1 NA) objective, in order to achieve optimal collection of fluorescence emission.
2. Cell concentration should be such that cells are not confluent when the experiment is performed. Confluent cells may downregulate their surface receptors and, as a consequence, may not respond to hormonal stimuli. For adherent cells, enough time should be allowed for the cells to settle on the coverslip, for optimal efficiency; therefore, we normally transfect the cells 36–48 h after plating.
3. Virtually any epifluorescence microscope can be used for FRET imaging. An inverted microscope is usually preferable, because a top-open experimental chamber is more accessible for addition of compounds to the sample. An upright microscope can also be used, but it requires a closed experimental chamber making the setup less flexible.
4. Culture medium containing phenol-red is not recommended because it is autofluorescent and measuring FRET in such medium would therefore increase the background.
5. A new xenon or mercury bulb generates a more intense light beam, as compared to bulbs that have been used for several hours. This can cause very rapid photobleaching and photodamage of the sample. To minimize damage, it is appropriate to use a neutral density filter in the excitation beam path.
6. In the case of a set up equipped with a filter wheel, it is possible to minimize the delay between the acquisition of the cyan and yellow emission signals by mounting the emission filters next to each other.

7. The exposure time depends on the fluorescence intensity of the sample. One should tend to select for observation cells that are bright enough to allow <300 ms exposure time. Higher exposure time and the consequent over illumination of the cell induces a quicker irreversible photodestruction (bleaching) of the fluorophores.

8. Faster acquisition rate can be difficult to achieve. Indeed fast repetitive illumination of the sample accelerates bleaching and may induce a change in the fluorescent properties of the fluorophores (photoisomerisation) that can lead to artefacts.

9. Background fluorescence is owing to intrinsic noise of the camera and to cells' autofluorescence. Background subtraction is particularly important if the analyzed cell presents a low fluorescence level. In this case a large proportion of the total signal is owing to background signal that will not change during the experiment. Because background intensity may be different in the cyan and yellow channels, this results in artefactual ratio values.

10. The CFP/YFP ratio is a pure number, the absolute value of which depends strictly on the experimental conditions and setup. The ratiometric measurement corrects, within a certain limit, for uneven distribution of the probe or changes in the focal plane.

4.1. On-Line Links

MultiSpec Micro-Imager™ E-mail: techsupport@optical-insights.com
Olympus web-site: http://www.olympus.com
TILLvisION web-site: http://www.till-photonics.de/software
   E-mail: service@till-photonics.com
T.I.L.L. Photonics web-site: http://www.till-photonics.de
   E-mail service: service@till-phoyonics.de
Chroma: http://www.chroma.com
Omega Opticals: http://www.omegafilters.com
ImageJ: http://rsb.info.nih.gov/ii/

Acknowledgments

Data in this manuscript is drawn from research funded by Telethon Italy and the European Commission (project QLK3-CT-2002-02149).

References

1. Beavo, J. A. and Brunton, L. L. (2002) Cyclic nucleotide research: still expanding after half a century. *Nat. Rev. Mol. Cell Biol.* **3,** 710–718.
2. Taylor, S. S., Buechler, J. A. and Yonemoto, W. (1990) cAMP-dependent protein kinase: framework for a diverse family of regulatory enzymes. *Annu. Rev. Biochem.* **59,** 971–1005.

3. Walsh, D. A. and Van Patten, S. M. (1994) Multiple pathway signal transduction by the cAMP-dependent protein kinase. *FASEB J.* **8,** 1227–1236.
4. Lee, K. A. (1991) Transcriptional regulation by cAMP. *Curr. Opin. Cell Biol.* **3,** 953–959.
5. Zaccolo, M., Magalhaes, P. and Pozzan, T. (2002) Compartmentalisation of cAMP and Ca^{2+} signals. *Curr. Opin. Cell Biol.* **14,** 160–166.
6. Hempel, C. M., Vincent, P., Adams, S. R, et al.. (1996) Spatio-temporal dynamics of cyclic AMP signals in an intact neural circuitm. *Nature* **384,** 166–169.
7. Rich, T. C., Fagan, K. A., Tse, T. E., et al. (2001) A uniform extracellular stimulus triggers distinct cAMP signals in different compartments of a simple cell. *Proc. Natl. Acad. Sci. USA.* **98,** 13,049–13,054.
8. Griffioen, G. and Thevelein, J. M. (2002) Molecular mechanisms controlling the localisation of protein kinase A. *Curr. Genet.* **41,** 199–207.
9. Kapiloff, M. S. (2002) Contributions of protein kinase A anchoring proteins to compartmentation of cAMP signaling in the heart. *Mol. Pharmacol.* **62,** 193–199.
10. Zaccolo, M. and Pozzan, T. (2002) Discrete microdomains with high concentration of cAMP in stimulated rat neonatal cardiac myocytes. *Science* **295,** 1711–1715.
11. Zaccolo, M. and Pozzan, T. (2003) CAMP and Ca^{2+} interplay: a matter of oscillation patterns. *Trends Neurosci.* **26,** 53–55.
12. Schwartz, J. H. (2001) The many dimensions of cAMP signaling. *Proc. Natl. Acad. Sci. USA.* **98,** 13,482–13,484.
13. Adams, S. R., Harootunian, A. T., Buechler, Y. J., et al. (1991) Fluorescence ratio imaging of cyclic AMP in single cells. *Nature* **349,** 694–697.
14. Miyawaki, A. and Tsien, R. Y. (2000) Monitoring protein conformations and interactions by fluorescence resonance energy transfer between mutants of green fluorescent protein. *Methods Enzymol.* **327,** 472–500.
15. Selvin, P. R. (2000) The renaissance of fluorescence resonance energy transfer. *Nat. Struct. Biol.* **7,** 730–734.
16. Zaccolo, M. and Pozzan, T. (2000) Imaging signal transduction in living cells with GFP-based probes. *IUBMB Life* **49,** 375–379.
17. Pozzan, T., Mongillo, M., Rudolf, R. (2003) Investigating signal transduction with genetically encoded fluorescent probes. *Eur. J. Biochem.* **270,** 2343–2352.
18. Zaccolo, M., De Giorgi, F., Cho, C. Y., et al. (2000) A genetically encoded, fluorescent indicator for cyclic AMP in living cells. *Nat. Cell Biol.* **2,** 25–29.

19

In Vivo Detection of Protein–Protein Interaction in Plant Cells Using BRET

Chitra Subramanian, Yao Xu, Carl Hirschie Johnson, and Albrecht G. von Arnim

Summary

The emerging technique of bioluminescence resonance energy transfer (BRET) allows us to detect protein interactions in live cells and in real time, thus providing a new window into cellular signal transduction processes. We present experimental protocols for expressing fusion proteins between luciferase and fluorescent proteins that are the basis for BRET measurement, as well as for measuring and imaging BRET in a variety of cell types. Despite our focus on plant cells, the techniques described here are easily adaptable to other cell systems that have yet to benefit from the BRET technique.

Key Words: Bioluminescence; protein–protein interaction; live imaging; Arabidopsis; yellow fluorescent protein; Renilla luciferase; coelenterazine.

1. Introduction

Protein interactions are important for coordinating cellular signaling events as well as metabolic functions in any cell. Numerous techniques have been developed to detect and study protein–protein interactions either in vitro or in vivo. Among the more widely used in vitro methods are co-immunoprecipitation, filter binding, surface plasmon resonance spectroscopy, and pull-down assays (*1*). Unfortunately, it is often difficult to verify one underlying premise for such in vitro assays, namely, that the interacting proteins are extracted or present in their

From: *Methods in Molecular Biology, vol. 284:*
Signal Transduction Protocols
Edited by: R. C. Dickson © Humana Press Inc., Totowa, NJ

native state. Among the in vivo methods, the yeast two-hybrid assay has become immensely popular because it lends itself to the selection of interacting proteins from a library of partner proteins. However, even this elegant and powerful technique has its drawbacks. For example, the interaction is tested only in the nucleus of a yeast cell. Moreover, interactions that depend on cellular compartmentalization or on more than two interaction partners are not always detectable by this method. Finally, the yeast two hybrid assay is based on reporter-gene expression as an indirect readout. The exquisite sensitivity afforded by this feature also renders the assay less than quantitative and prone to false positive results.

A number of in vivo assays are variations on the theme of fragment complementation *(2–4)*. These assays may be less sensitive and thus less prone to false-positive results than assays including a reporter gene. However, fragment complementation may not provide a real-time readout of the interaction.

Interaction assays based on direct and physical resonance energy transfer between two compatible interaction partners can eliminate some of the disadvantages of indirect two-hybrid based methods. Bioluminescence resonance energy transfer (BRET) is a nondestructive in vivo assay for the real time detection of protein interactions *(5)*. In the jellyfish *Aequorea victoria*, the blue luminescence of the "donor" protein aequorin is quenched by resonance energy transfer to an associated "acceptor" protein, green fluorescent protein (GFP), followed by emission of a photon according to the emission spectrum of GFP *(6)*.

In general, BRET is defined as radiationless energy transfer between a luciferase donor and a fluorescent protein acceptor. BRET was first developed into a tool for protein interaction studies by fusing one partner protein to the blue light-emitting luciferase from the sea pansy *Renilla* (RLUC) and a second partner protein to yellow fluorescent protein (YFP); *(5,7)*. BRET is detected only when RLUC and YFP are brought into close proximity by a protein interaction between their covalently attached fusion partners (**Fig. 1**).

Since its establishment in *Escherichia coli*, BRET has been applied elegantly in mammalian cells to monitor protein–protein interactions in vivo *(8)*, in particular to follow temporal changes in the ligand-dependent conformation, dimerization, or early signal-transduction events of G–protein coupled receptors *(9–18)*. However, applications of BRET in mammalian cells beyond the G–protein coupled receptors are rare *(19)*.

In principle, BRET is closely related to fluorescence resonance energy transfer (FRET) between fluorescent proteins such as cyan fluorescent protein (CFP) and YFP, with the exception that the photon donor for FRET is a fluorescent protein (e.g., CFP) rather than a luciferase *(20)*. For both BRET and FRET alike, the acceptor needs to be located within about 50Å of the photon donor and the emission spectrum of the donor must overlap with the excitation spectrum of the acceptor. Both FRET and BRET are suitable for real time in vivo measurements

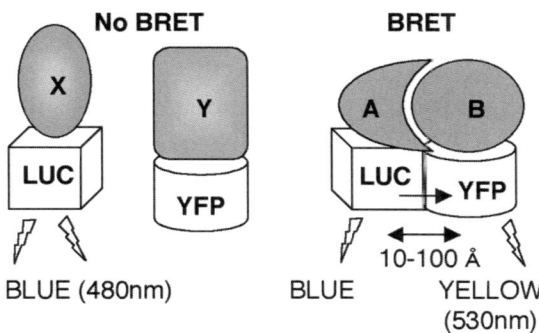

Fig. 1. Schematic of BRET. Proteins X and Y do not interact; therefore, YFP cannot be excited by the luminescence of RLUC. Proteins A and B interact, bringing the RLUC and YFP within the critical distance and allowing YFP to be excited by the luminescence of RLUC.

of protein interaction. However, there is one key difference between BRET and FRET. FRET requires an excitation light source, whereas BRET does not. Instead, BRET requires a substrate. In the case of RLUC, this substrate is coelenterazine, which is nontoxic and membrane permeable. This difference has important consequences *(21)*. First, the excitation light source needed for FRET may cause photobleaching or phototoxicity, as well as autofluorescence and unintended biological effects owing to cellular photoreceptors. Second, not only the photon donor (often CFP), but even the photon acceptor (often YFP) may be partially activated by the excitation light source, contributing to background signal. Therefore, solid quantitative imaging skills and expensive instrumentation are needed to collect reliable FRET data under in vivo conditions. For comparison, no excitation light is needed for BRET. In fact, BRET is measured in a background of complete darkness. Thus, all photons emitted by the YFP acceptor originate from the luciferase and are indicative of BRET.

We have adapted the BRET technique to investigate cellular-signaling events in plants in response to light and the circadian clock (*see* **Table 1**). Most plant cells possess phytochrome and cryptochrome photoreceptors, which could complicate FRET-based interaction assays. Moreover, we were interested to explore the utility of BRET among soluble cytosolic and nuclear proteins, i.e., outside the arena of plasma membrane-associated receptors. Finally, FRET has been used for imaging, and we are intent on expanding similar BRET applications by developing imaging protocols for BRET.

Here we present proof of concept for BRET interaction in plant cells on the basis of data from two proteins, the cyanobacterial clock protein, KaiB, for which BRET data had been collected in *E. coli (5)*, and the light-regulatory basic leucine zipper (bZip) transcription factor, HY5 *(22)*. BRET data suggest that both pro-

Table 1. BRET Expression Vectors and Plasmids

Construct Name	Remarks	Genbank Accession #
35S:RLUC	*Renilla* luciferase (RLUC), bluescript (Stratagene) vector with 35S promoter and terminator	AY189980
35S:EYFP	Enhanced yellow fluorescent protein (EYFP)	AY189981
35S:Ala-RLUC	Alanine linker (AAAPVAAAAAA)-RLUC (*see* **Note 5**)	AY189982
35S:RLUC-Ala	RLUC-Alanine linker (AAAARS)	AY189983
35S:Ala-EYFP	Alanine linker (AAAPVAAAAAA)-EYFP	AY189984
35S:EYFP-Ala	EYFP-Alanine linker (AAAARS)	AY189985
35S:RLUC-EYFP	The RLUC coding region is fused to EYFP	
35S:RLUC-KaiB	KaiB is a cyanobacterial clock protein	
35S:EYFP-KaiB		
35S:RLUC-Ala-HY5	HY5 is an Arabidopsis bZip protein	
35S:EYFP-Ala-HY5		
pBin19-RLUC	*Renilla* luciferase (RLUC) expression cassette in a T-DNA transformation vector	
pBin19-hRLUC	Humanized RLUC cDNA	

teins homodimerize in plant cells (**Fig. 2**). HY5 is nuclear-localized, whereas KaiB is both nuclear and cytoplasmic, confirming that the utility of BRET in eukaryotic cells is not restricted to membrane-associated cell-surface receptors.

We also found that human codon-optimization improves the expression of RLUC in Arabidopsis. Furthermore, using a covalent fusion of RLUC and YFP as a positive control, we established BRET assay conditions in three different plant cell settings, namely, transiently transformed onion epidermis, stably transformed tobacco bright-yellow (BY-)2 suspension culture cells and stably transgenic *Arabidopsis thaliana* plants (**Fig. 3** and **Table 2**). Finally, we demonstrate that BRET data can be collected by imaging entire live seedlings with a cooled CCD camera (**Fig. 4**). Hence, BRET is a promising technique to monitor in vivo protein interaction in a variety of plant systems.

2. Materials

2.1. Plasmids, Vectors, and Strains

1. pRL-null (Promega, Madison,WI) contains the Renilla luciferase (RLUC) coding region.
2. pEYFP (CLONTECH Laboratories Inc., Palo Alto, CA) contains the coding region of enhanced YFP.

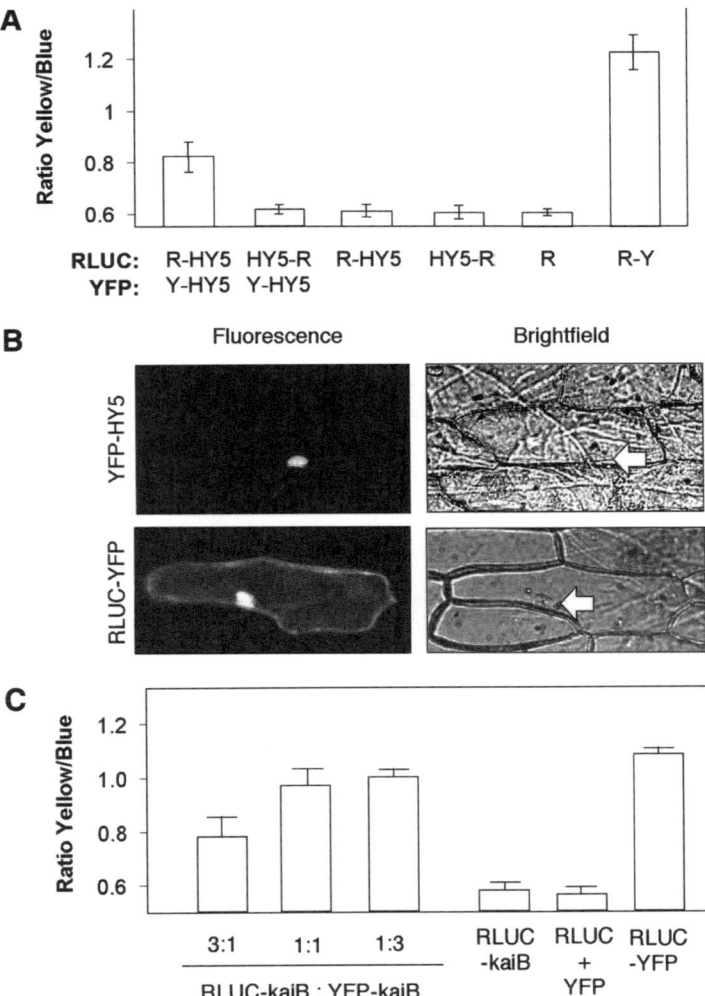

Fig. 2. BRET in onion epidermal cells. (**A**) BRET graph showing interaction of RLUC (R)-HY5 with YFP (Y)-HY5. HY5-RLUC with YFP-HY5 does not show BRET. RLUC-HY5, HY5-RLUC, and RLUC are negative controls, whereas RLUC-YFP is a positive control showing BRET between RLUC and YFP. (**B**) The top panel shows nuclear enrichment of YFP-HY5. The bottom panel shows the localization of RLUC-YFP in both cytoplasm and nucleus. RLUC-YFP (64.5 kDa) presumably equilibrates between nucleus and cytoplasm by diffusion. The nuclei are highlighted with arrows in the brightfield images on the right. (**C**) BRET graph showing homodimerization of cyanobacterial KaiB protein. Note the dependence of BRET on the ratios of expression plasmids used. Also note the absence of BRET upon co-expression of unfused RLUC and YFP proteins (*see* **Notes 12** and **15**).

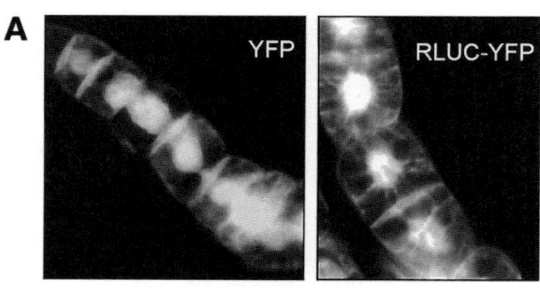

		Blue	Yellow	Ratio
RLUC	#1	192.5	101.9	0.53
	#2	93.9	53.3	0.57
	#3	83.1	46.4	0.56
RLUC-YFP	#1	271.6	246.2	0.90
	#2	49.0	38.8	0.80
	#3	20.8	19.9	0.96
Control		0.00	0.00	-

Fig. 3. BRET in tobacco BY-2 cells. (**A**) Fluorescence of BY-2 cells expressing YFP or RLUC-YFP. (**B**) Luminescence and yellow to blue ratio of RLUC and RLUC-YFP from about 0.1 mL of cells taken from three independently transformed cell lines. Control shows untransformed cells. RLUC-YFP is the positive control and RLUC is the negative control where no BRET is observed.

3. hRLUC (Packard/Perkin-Elmer) is a human codon-optimized version of the RLUC cDNA.
4. pT7/RLUC-KaiB and pT7/EYFP-KaiB contain translational fusions of kaiB to RLUC and YFP, respectively *(5)*.
5. pUC-HY5 contains the HY5 cDNA *(22)*.
6. pBin19 *(23)* binary T-DNA transformation vector (Genbank accession, cat. no. U12540;).
7. pPZP222 *(24)* binary T-DNA vector (Genbank accession Cat. no. U10463).

2.2. Reagents, Buffers, and Media

1. Murashige-Skoog (MS) medium: 4.4 g/L of MS salts (Sigma, St. Louis, MO cat. no. M5524), 0.5 g/l (3 mM) MES buffer, pH 5.7, 1% sucrose, 0.8% bacto agar.
2. Coelenterazine (BioSynth, Naperville, IL or Biotium, Hayward, CA): 100 μM stock solution (*see* **Note 1**).

Table 2. Comparison of Native (RLUC) and Humanized RLUC (hRLUC) in Arabidopsis Seedlings[a]

Transgene	Line	Blue	Yellow	Ratio
hRLUC	#1	127.5	83.5	0.66
	#2	22.9	13.8	0.60
	#3	9.5	6.3	0.66
RLUC	#1	4.84	3.19	0.66
	#2	4.68	2.93	0.63
	#3	0.25	0.16	0.62
RLUC-YFP	#1	1.81	3.70	2.04
	#2	1.72	3.73	2.16
	#3	1.56	2.73	1.75
None (control)		0.02	0.01	—

[a]Absolute luminescence values (averaged from four 10 s readings) and yellow to blue ratios are shown for three independent transgenic lines with a range of expression levels. **Note:** BRET signal from the RLUC-YFP fusion protein (*see* **Note 14**).

3. LB Medium: 10 g/L bacto-tryptone, 5 g/L Bacto-yeast extract, 10 g/L NaCl, pH 7.0.
4. Infiltration medium: 4.3 g/L MS salts, 5% sucrose, 44 nM benzylaminop-urine, 0.01% Silwet L-77 detergent (pH 5.6 with KOH).
5. BY2 medium: 4.3 g/L MS salts, 0.1 g/L thiamine, 10 mg/L myo-inositol, 0.21 g/L KH_2PO_4, 0.2 µg/mL 2,4-D, 3% sucrose (pH 5.6 with KOH).
6. Eppendorf plasmid DNA prep kit (Eppendorf, Hamburg, Germany).
7. Tungsten particles (Sigma, Aldrich, St. Louis, MO).
8. Silwet L-77 (Lehle Seeds, Round Rock, TX).

2.3. Apparatus

1. Turnerdesigns TD20/20 luminometer with the Dual-Color Accessory™ (*see* **Note 2**).
2. Cooled ($-50°C$) charge-coupled device (CCD) camera (TE/CCD512BKS, Princeton Instruments, Trenton, NJ) (*see* **Note 3**).
3. Epifluorescence microscope.
4. PDS-He biolistic particle gun (Bio-Rad, Hercules, CA).

3. Methods
3.1. Construction of BRET Gene Fusions

Recombinant expression plasmids were generated to express protein fusions to RLUC or YFP under the control of the cauliflower mosaic virus 35S promoter and terminator (**Table 1**). The expression vectors are derived from a previous

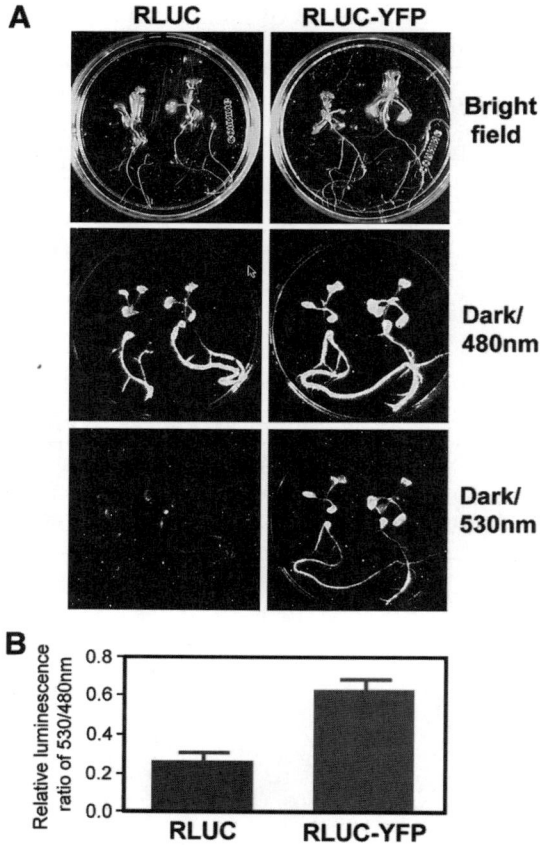

Fig. 4. BRET imaging in transgenic Arabidopsis expressing RLUC or RLUC-YFP.
(A) Upper panels, 10-d-old seedlings photographed with a regular camera in brightfield. Middle and bottom panels, same seedlings as above imaged for 180 s with a
CCD camera through interference filters transmitting light of 480 nm (middle panels)
or 530 nm (bottom panels). **(B)** Relative average 530 nm/480 nm ratios of the quantified luminescence intensity in the seedling images from (A).

series designed to express GFP fusion proteins *(25)*. The 35S:RLUC-YFP construct is a fusion of RLUC and YFP with a short peptide sequence (RDPRVP-
VAT) in between. In this case, the RLUC and YFP are in close proximity to
serve as a positive control for BRET *(5)*. 35S:RLUC was used as a negative control. In order to test for homo/hetero-dimerization between protein partners, fusion proteins were created either as N-or C-terminal fusions to RLUC or to
EYFP *(see* **Notes 4** and **5**). Fusion proteins were co-expressed transiently in

onion epidermal cells or stably in transgenic *Arabidopsis thaliana* or tobacco BY-2 suspension culture cells.

3.2. Co-Transformation and Co-Expression of Fusion Proteins

Both green and unpigmented plant tissues are suitable for BRET experiments. However, it is critical that the pigmentation level is the same between tissue co-expressing the RLUC-and YFP-tagged proteins and tissue expressing the RLUC-only control. It is important to include the RLUC-tagged protein alone as a negative control in every experiment.

3.2.1. Onion Epidermis

1. High-quality DNA of the expression plasmids is obtained using the Eppendorf Mini Prep Kit.
2. 500 ng of each DNA (RLUC fusion and YFP fusion) are coated onto tungsten particles. For each bombardment 20 μL of the coated tungsten particles are bombarded into a single layer of onion epidermal cells using the PDS-He biolistic particle gun (*see* **Note 6**).
3. For biolistic transformation of expression plasmids into onion epidermal cells, the inner epidermis of an onion scale is peeled and placed on a petri plate containing MS medium (*see* **Note 7**).
4. The bombardment is performed according to the procedure provided with the particle gun from Bio-Rad.
5. The transformed onion epidermis is incubated at room temperature in darkness overnight.
6. YFP expression is confirmed by examining the yellow fluorescence on an epifluorescence microscope as shown in **Fig. 2B**.
7. The RLUC expression is measured using a TD-20/20 luminometer (**Fig. 2A, 2C**; *see* **Subheading 3.3.** for details; also *see* **Note 8**).

3.2.2. Arabidopsis

1. In order to perform luminescence assays in *Arabidopsis* the expression cassette is first transferred to a binary vector, for example pBin19 or pPZP222 (**Table 1**).
2. Agrobacterium strain GV3101 is transformed with the binary vector containing the transgene, which is then transformed into *Arabidopsis* (*see* **Note 9**).
3. Transformed seedlings are selected by their ability to grow on antibiotic medium (kanamycin or gentamycin). YFP expression is checked by fluorescence microscopy.
4. Seedlings are tested for expression of RLUC using the TD20/20 luminometer as shown in **Table 2**.

3.2.3. Tobacco BY-2 Cells

1. Agrobacterium harboring the appropriate plasmid is grown to an OD_{600} of 1.0 in LB in the presence of appropriate antibiotics, namely kanamycin for pBin19 and spectinomycin for pPZP222 T-DNA vectors.
2. For each transformation 5 mL of 3-d-old BY-2 cells are transferred to an empty petri dish.
3. 5 µL of 20 m*M* acetosyringone is added and swirled gently to mix.
4. Two transformations are set up for each plasmid containing the transgene. 25 µL or 100 µL of Agrobacteria are added to each petri dish. The petri dish is wrapped with parafilm and incubated at 27°C for 2 d.
5. The cells are transferred to a fresh tube and washed three times with 10 mL of BY2 medium. A final wash is done with 10 mL of BY2 medium containing 500 µg/mL of carbenicillin or 125 µg/mL of cefotaxime to control Agrobacterium growth.
6. 1 mL of the cell suspension is plated on to petri plates containing BY2 medium and carbenicillin/cefotaxime and other appropriate antibiotics (kanamycin for pBin19 and gentamycin for pPZP222 T-DNAs).
7. The petri plates are wrapped with parafilm and incubated at 27°C for 10–14 d.
8. Transformed calli are transferred to fresh antibiotic plates and checked for YFP (**Subheading 3.2.1., step 6**) and RLUC expression (**Fig. 3**).
9. BRET is measured using the TD-20/20 luminometer as shown in **Fig. 3B**.

3.3. BRET Assays

3.3.1. Luminescence Assays in Plant Cells

1. Once YFP expression has been confirmed, the tissues are examined for BRET. The tissue is placed in an Eppendorf tube or a 12 × 50 mm round bottom polypropylene tube.
2. In case of onion epidermal cells, the tissue is immersed in 900 µL of distilled water. Half-strength MS medium (2.15 g/L MS salts) containing 0.02% Silwet L-77 is used for *Arabidopsis* seedlings. For tobacco BY-2 cells, a small amount of cells from the calli is aseptically removed and suspended in 500 µL of BY2 medium.
3. Freshly diluted coelenterazine is added to a final concentration of 1 µ*M* (*see* **Note 10**).
4. The luminescence is measured in the TD-20/20 luminometer with the dual-color accessory (*see* **Note 2**).
5. Four individual pairs of luminescence readings are collected through blue and yellow filters (*see* **Notes 8, 10,** and **11**).

6. Background readings are obtained using untransformed tissue in the presence of coelenterazine substrate. The average background value is typically between 0.02–0.05 for the blue filter and 0.00–0.01 for the yellow filter.
7. BRET is determined by calculating the yellow to blue ratio from the background-subtracted readings (for data *see* **Figs. 2A, 2C,** and **3B;** *see* **Subheading 3.3.2.**).

3.3.2. Calculations for BRET

1. Four pairs of readings are taken for each tissue sample and average background is subtracted from each individual reading (*see* **Note 13**).
2. The BRET ratio is calculated by taking the sum of background corrected yellow (Y) readings and dividing them by the sum of background corrected blue (B) readings. (*see* **Note 14**)

$$\text{BRET} = [\Sigma \, (Y_{sample} - Y_{background})] \, / \, [\Sigma \, (B_{sample} - B_{background})]$$

3. Results in onion cells indicate homodimerization of HY5 and KaiB (**Fig. 2A, 2C**). In the case of a BRET interaction the yellow to blue ratio is usually between 0.8 and 1.1. For comparison, the yellow to blue ratio for the negative control, i.e. the RLUC fusion alone, lies between 0.55–0.69. This variation may be owing to subtle differences in the pigmentation of the host tissue samples. It is important to note that RLUC and YFP do not interact with each other; only if the fusion protein interacts with its partner the RLUC and YFP are in close proximity to perform BRET. The positive control is the RLUC-YFP fusion, which gives a yellow to blue ratio of 1 to 1.3 (*see* **Notes 12** and **15**).
4. Results in Arabidopsis seedlings suggest that about fivefold higher luciferase levels can be achieved with a human codon optimized version of RLUC (Perkin-Elmer, **Table 2**).
5. Results in tobacco BY-2 cells confirm that BRET is detectable in this cell type (**Fig. 3**). The yellow to blue ratios in BY-2 cells are usually lower than in onion epidermis, perhaps owing to the slight pigmentation of the cells.

3.3.3. BRET Imaging In Vivo

1. Transgenic Arabidopsis seedlings expressing RLUC or RLUC-YFP are grown in the light for 10 d and transferred to a 35-mm Petri dish containing 1 mL of MS salts with 3 μM of coelenterazine.
2. Seedlings are imaged in darkness with a cooled-CCD camera through either a 475–485 nm or a 525–535 nm interference filter. For comparison, seedlings are photographed under regular brightfield conditions (**Fig. 4A**).
3. The 530/480 nm luminescence ratio in different parts of the seedlings is calculated. We used custom software created by Dr. Takao Kondo (CCD

focus *[26]*). Many image-analysis programs should be adequate for this purpose (**Fig. 4B**).

4. Notes

1. The substrate coelenterazine is sensitive to light and oxygen. Coelenterazine is dissolved in ethanol to a final concentration of 250 μM. Small aliquots are made and dried in a Speed-Vac. Each tube is gently flushed with N_2 gas to remove the O_2 and stored at −80°C until further use. Prior to the assay the substrate is redissolved in a small volume of ethanol and distilled water is added for a final concentration of 100 μM.

2. The TD20/20 Luminometer is sensitive to 0.1 femtogram of luciferase and has a wide dynamic range of greater than 10^5. The Dual-Color Accessory™ contains alternative filters, a blue band-pass filter (333–463 nm) with a >90% transmittance over a bandwidth of 370–410 nm, and a yellow long-pass filter (520 nm and greater) with a >80% transmittance above 550 nm. These filters together capture about 10% of the total luminescence. Alternatively, plate-reading luminometers that have been designed for BRET include the Fusion (Perkin-Elmer/Packard), the Mithras (Berthold Technologies, Oak Ridge, TN), and the PolarStar (BMG Lab Technologies, Durham, NC).

3. We used 475–485 nm and 525–535 nm interference filters (Ealing Electro-Optics, Holliston, MA) to detect blue and yellow luminescence, respectively.

4. In order to test BRET between two protein partners, one has to construct the partner proteins as either RLUC or YFP fusions. When constructing N-terminal fusions to RLUC/YFP, the stop codon in the cDNA to be tagged must be removed. Care must be taken that the fusion partners are in frame. Errors during synthesis of oligonucleotides intended for subcloning are a common source of unintended frameshifts. It is advisable to create both N-and C-terminal fusions of RLUC or YFP and both partner proteins in order to test all possible combinations of heterodimers (eight combinations) or homodimers (four combinations). This increases the probability of a tight juxtaposition between RLUC and YFP, which is crucial for efficient BRET. However, sometimes BRET cannot be observed in any of the possible combinations even though the two proteins may be well-established as interaction partners using other methods; perhaps the optimal distance between RLUC and EYFP still cannot be achieved owing to steric hindrance or protein-folding problems. As shown in **Fig. 2A**, only one of the two combinations that we tested for HY5 homodimerization showed BRET.

5. 35S:RLUC-Ala and 35S:EYFP-Ala add four to nine alanine residues in between the RLUC or YFP and the fusion protein partner in an attempt to minimize the risk of steric hindrance within and among hybrid fusion proteins. Genbank accession numbers for the BRET vectors are given in **Table 1**.

6. Following standard procedure, tungsten particles are prepared by suspending them at 50 mg/mL in 100% ethanol. The slurry is sonicated for 30 minutes and washed with ethanol four times. The particles are stored in $-20°C$ as 1-mL aliquots until further use. Before the bombardment the aliquot of tungsten particle is vigorously vortexed for 15 min. Per bombardment, 20 µL is placed into a separate Eppendorf tube and washed twice with water. 500 ng of DNA in up to 5 µL TE is added along with 20 µL 2.5 *M* $CaCl_2$ and 8 µL 0.1 *M* spermidine, free base. The sample is incubated on ice for 15 min and then washed three to five times with 100% ethanol. The DNA-coated tungsten particles are dried onto a plastic disk (macrocarrier, Bio-Rad). Until bombardment, the disk is kept in a petri dish containing a filter paper and some drying agent.

7. The onion is cut into four quarters and the inner and outer most leaves (scales) of the onion bulb are discarded. Using a fine pair of forceps the inner epidermis is peeled and placed, inside up, on a petri plate containing MS medium. The petri plate is kept covered until bombardment to prevent drying.

8. The luminometer is set to read after a 3-s delay in case of onion and BY-2 cell samples and a 20-s delay in case of *Arabidopsis* to allow the decay of delayed chlorophyll autofluorescence in green tissues. For each data point, four pairs of luminescence readings are taken through blue and yellow filters. The integration time is set to 10-s in each case.

9. For Arabidopsis transformation, we routinely transfer RLUC fusions into pPZP222 (which confers plant gentamycin-resistance) and YFP fusions into pBIN19 (plant kanamycin-resistance) to facilitate selection of double-transformed *Arabidopsis*. Strain GV3101 is transformed by electroporation (capacitance 50 microFarad; load resistance 150 Ohm; gap width 1 mm) and transformants are selected on LB plates containing kanamycin (pBin19) or spectinomycin (pPZP222), as well as gentamycin and rifampicin.

10. It is important to add the coelenterazine substrate immediately before taking the reading in case of onion epidermal cells as the time when the luminescence peaks varies from sample to sample (5–30 min). Moreover, some interactions could be transient and so results could be flawed if there is a delay between the treatment and analysis of the sample.

11. Before measurement, it is advisable to keep the samples in darkness or very dim light, because bright light can excite delayed chlorophyll fluorescence that might obscure the true luminescence signal.

12. Transient transformation with equal amounts of RLUC- and-YFP fusion plasmid may still result in unequal amounts of expression of each protein, in part owing to differences in the expression or stability of the proteins in the cell. The optimal ratio should be determined empirically. In the case of

RLUC-kaiB and YFP-kaiB fusions we found that RLUC:YFP ratios of 3:1, 1:1 and 1:3 yielded BRET with 40, 90, and 100% relative efficiency (**Fig. 2C**). When working with transgenic plants one could optimize the dosage of the BRET-tagged genes. For example, given F1 hybrid plants heterozygous (+/−; +/−) for the two genes of interest, their selfed F2 progeny will contain individual plants with three different dosages of the two genes (1:2, 1:1, and 2:1).

13. Luminescence typically drifts over time in part because the luciferase is unstable in the presence of substrate. If the drift is substantial, data points may be extrapolated to the same time point for blue and yellow readings.

14. There are several ways to calculate the magnitude of BRET. According to the "BRET ratio" recommended by Angers and coworkers *(8)*, the BRET signal of the "RLUC-only" control is set to near zero. Instead, we simply express BRET as the ratio of background corrected yellow (Y) and blue (B) luminescence readings.

15. BRET is primarily an in vivo technique for learning more about protein interactions that have previously been identified by other methods. However, this does not preclude utilizing BRET to screen for new interactions *(5,21)*. In our experience only about 10% of protein pairs tested have shown BRET, even though in vitro or yeast two-hybrid interaction data had been collected for the majority of these pairs.

16. Supporting website at http://fp.bio.utk.edu/vonarnim/BRET.html

Acknowledgments

This work was supported by NSF grant MCB-0114653 (to A. G. von Arnim and C. H. Johnson) and DOE grant DE-FG02-96ER20223 (to A. G. von Arnim). We thank Kristin Kolberg and Turnerdesigns for their cooperation in building a filter accessory for luminometry.

References

1. Golemis, E. (2002) *Protein-Protein Interactions: A Molecular Cloning Laboratory Manual.* Cold Spring Harbor Laboratory Press, Cold Spring Harbor, NY.

2. Hu, C. D., Chinenov, Y., and Kerppola, T. K. (2002) Visualization of interactions among bZIP and Rel family proteins in living cells using bimolecular fluorescence complementation. *Mol. Cell* **9,** 789–798.

3. Subramaniam, R., Desveaux, D., Spickler, C., et al. (2001) Direct visualization of protein interactions in plant cells. *Nat. Biotechnol.* **19,** 769–772.

4. Spotts, J. M., Dolmetsch, R. E., and Greenberg, M. E. (2002) Time-lapse imaging of a dynamic phosphorylation-dependent protein-protein interaction in mammalian cells. *Proc. Natl. Acad. Sci. USA* **99,** 15,142–15,147.

5. Xu, Y., Piston, D. W., and Johnson, C. H. (1999) A bioluminescence resonance energy transfer (BRET) system: application to interacting circadian clock proteins. *Proc. Natl. Acad. Sci. USA* **96,** 151–156.

6. Prendergast, F. G. (1999) Biophysics of the green fluorescent protein. *Methods Cell Biol.* **58,** 1–18.

7. Xu, Y., Johnson, C. H., and Piston, D. (2002) Bioluminescence resonance energy transfer assays for protein-protein interactions in living cells. *Methods Mol. Biol.* **183,** 121–133.

8. Angers, S., Salahpour, A., Joly, E., et al. (2000) Detection of beta 2-adrenergic receptor dimerization in living cells using bioluminescence resonance energy transfer (BRET). *Proc. Natl. Acad. Sci. USA* **97,** 3684–3689.

9. Cheng, Z. J. and Miller, L. J. (2001) Agonist-dependent dissociation of oligomeric complexes of G protein-coupled cholecystokinin receptors demonstrated in living cells using bioluminescence resonance energy transfer. *J. Biol. Chem.* **276,** 48,040–48,047.

10. Kroeger, K. M., Hanyaloglu, A. C., Seeber, R. M., et al. (2001) Constitutive and agonist-dependent homo-oligomerization of the thyrotropin-releasing hormone receptor. Detection in living cells using bioluminescence resonance energy transfer. *J. Biol. Chem.* **276,** 12,736–12,743.

11. Hanyaloglu, A. C., Seeber, R. M., Kohout, T. A., et al. (2002) Homo-and hetero-oligomerization of thyrotropin-releasing hormone (TRH) receptor subtypes. Differential regulation of beta-arrestins 1 and 2. *J. Biol. Chem.* **277,** 50,422–50,430.

12. Issafras, H., Angers, S., Bulenger, S., et al. (2002) Constitutive agonist-independent CCR5 oligomerization and antibody-mediated clustering occurring at physiological levels of receptors. *J. Biol. Chem.* **277,** 34,666–34,673.

13. Jensen, A. A., Hansen, J. L., Sheikh, S. P., and Brauner-Osborne, H. (2002) Probing intermolecular protein-protein interactions in the calcium-sensing receptor homodimer using bioluminescence resonance energy transfer (BRET). *Eur. J. Biochem.* **269,** 5076–5087.

14. Lavine, N., Ethier, N., Oak, J. N., et al. (2002) G protein-coupled receptors form stable complexes with inwardly rectifying potassium channels and adenylyl cyclase. *J. Biol. Chem.* **277,** 46,010–46,019.

15. Mercier, J. F., Salahpour, A., Angers, S., et al. (2002) Quantitative assessment of beta 1 and beta 2-adrenergic receptor homo and hetero-dimerization by bioluminescence resonance energy transfer. *J. Biol. Chem.* **277,** 44,925–44,931.

16. Ramsay, D., Kellett, E., McVey, M., et al. (2002) Homo- and hetero-oligomeric interactions between G-protein-coupled receptors in living cells monitored by two variants of bioluminescence resonance energy transfer (BRET): hetero-oligomers between receptor subtypes form more efficiently than between less closely related sequences. *Biochem. J.* **365,** 429–440.

17. Wang, Y., Wang, G., O'Kane, D. J., and Szalay, A. A. (2001) A study of protein-protein interactions in living cells using luminescence resonance energy transfer (LRET) from Renilla luciferase to Aequorea GFP. *Mol. Gen. Genet.* **264,** 578–587.

18. Yoshioka, K., Saitoh, O., and Nakata, H. (2002) Agonist-promoted heteromeric oligomerization between adenosine A(1) and P2Y(1) receptors in living cells. *FEBS Lett.* **523,** 147–151.

19. Germain-Desprez, D., Bazinet, M., Bouvier, M., and Aubry, M. (2003) Oligomerization of TIF1 transcriptional regulators and interaction with ZNF74 nuclear matrix protein revealed by BRET in living cells. *J. Biol. Chem.* **278,** 22,367–22,373.

20. Boute, N., Jockers, R., and Issaad, T.. (2002) The use of resonance energy transfer in high-throughput screening: BRET versus FRET. *Trends Pharmacol Sci.* **23,** 351–354.

21. Xu, Y., Kanauchi, A., von Arnim, A. G., et al. (2003) Bioluminescence resonance energy transfer: monitoring protein-protein interactions in living cells. *Methods Enzymol.* **360,** 289–301.

22. Oyama, T., Shimura, Y., and Okada, K. (1997) The Arabidopsis HY5 gene encodes a bZIP protein that regulates stimulus-induced development of root and hypocotyl. *Genes Dev.* **11,** 2983–2995.

23. Bevan, M. (1984) Binary Agrobacterium vectors for plant transformation. *Nucleic Acids Res.* **12,** 8711–8721.

24. Hajdukiewicz, P., Svab, Z., and Maliga, P. (1994) The small, versatile pPZP family of Agrobacterium binary vectors for plant transformation. *Plant Mol. Biol.* **25,** 989–994.

25. von Arnim, A. G., Deng, X. W., and Stacey, M. G. (1998) Cloning vectors for the expression of green fluorescent protein fusion proteins in transgenic plants. *Gene* **221,** 35–43.

26. Kondo, T., Tsinoremas, N. F., Golden, S. S., et al. (1994) Circadian clock mutants of cyanobacteria. *Science* **266,** 1233–1236.

20

Revealing Protein Dynamics by Photobleaching Techniques

Frank van Drogen and Matthias Peter

Summary

Green fluorescent proteins (GFPs) are widely used tools to visualize proteins and study their intracellular distribution. One feature of working with GFP variants, photobleaching, has recently been combined with an older technique known as fluorescence recovery after photobleaching (FRAP) to study protein kinetics in vivo. During photobleaching, fluorochromes get destroyed irreversibly by repeated excitation with an intensive light source. When the photobleaching is applied to a restricted area or structure, recovery of fluorescence will be the result of active or passive diffusion from fluorescent molecules from unbleached surrounding areas. Fluorescence loss in photobleaching (FLIP) is a variant of FRAP where an area is bleached, and loss of fluorescence in surrounding areas is observed. FLIP can be used to study the dynamics of different pools of a protein or can show how a protein diffuses, or is transported through a cell or cellular structure. Here, we discuss these photobleaching fluorescent imaging techniques, illustrated with examples of these techniques applied to proteins of the *Saccharomyces cerevisiae* pheromone response MAPK pathway.

Key Words: Fluorescence recovery after photobleaching (FRAP); fluorescence loss in photobleaching (FLIP); GFP; fluorescence; kinetics; dynamics; microscopy.

1. Introduction

Green fluorescent proteins (GFPs), identified in the jellyfish *Aequora victoria*, are widely used tools to tag proteins and study their intracellular distribution *(1,2)*. A wide variety of techniques have been developed to study the

From: *Methods in Molecular Biology, vol. 284:*
Signal Transduction Protocols
Edited by: R. C. Dickson © Humana Press Inc., Totowa, NJ

three-dimensional distribution of GFP-tagged proteins in cells and organisms and color variants of GFP have been developed to observe several proteins at the same time *(3,4)*. Microscopes have also been enhanced to facilitate time-lapse experiments *(5)*. During the last few years one feature of GFP variants, photobleaching, has been combined with an older technique known as fluorescence recovery after photobleaching (FRAP) to study protein kinetics in vivo *(6)*.

Photobleaching is a phenomenon in which fluorochromes can be destroyed irreversibly by repeated excitations with an intensive light source. When the photobleaching is applied to a restricted area, generally achieved by using a laser, recovery of fluorescence in this area will be the result of active or passive diffusion from fluorescent molecules from unbleached surrounding areas. When a distinct subcellular structure is bleached, recovery will represent passive or active transport of fluorescent molecules to the bleached structure *(8)*. Although photobleaching destroys the fluorochrome of the bleached protein, the rigid structure of GFP itself is not compromised. Hence, when a protein fused to GFP is shown to be functional and behaving like the wild-type protein, photobleaching will not change its properties.

Fluorescence loss in photobleaching (FLIP) is a variant of FRAP where an area is bleached, and loss of fluorescence in surrounding areas is observed. FLIP can be used to study the dynamics of different pools of a protein or can show how a protein diffuses, or is transported through a cell or cellular structure.

FRAP historically has been used to study the dynamics and mobility of membrane components labeled with fluorophores, or fluorescent antibodies in various cells and artificial membranes. FRAP with GFP fusion proteins was initially also used to study membrane proteins *(8)*. However, photobleaching techniques using GFP-tagged proteins are now used to show kinetic properties of proteins, the exchange between different cellular departments and changes in mobility. For example, we have used FRAP and FLIP to study the dynamics of MAPK signaling cascades, including nuclear turnover and dynamics between membrane associated pools and the nucleus *(9–11)*. Others have used FRAP to study recruitment of proteins to centrosomes during the cell cycle *(12,13)*, or to study cytoskeletal architecture and arrangements *(14)*. In a very elegant study, the Ellenberg lab used different GFP variants, one of which was bleached, to show chromosomal positioning in nuclei *(15)*.

2. Materials

1. Expression vectors with GFP.
2. α-GFP-antibody or antibody specific for the protein of interest.
3. cDNA of protein of interest.
4. Oligonucleotide primers.
5. Molecular biology reagents (restriction enzymes, T4 DNA ligase, etc.).

6. Material for minipreps (*see* **ref. *16*** or one of the many commercially available kits).
7. Appropriate growth medium for system used.
8. Microscopy slides and coverslips.
9. Microscope suitable for photobleaching experiments. (We used a Zeiss LSM510 confocal laser scanning microscope.)
10. Cycloheximide.
11. Formaldehyde.

3. Methods

The methods described here outline the construction of the GFP fusion vector (**Subheading 3.1.**), the control of expression and functionality of the GFP fusion construct (**Subheading 3.2.**), optimization of detection with a confocal laser-scanning microscope (**Subheading 3.3.**), fluorescence recovery after photobleaching (**Subheading 3.4.**), and fluorescence loss in photobleaching (**Subheading 3.5.**). These methods will be illustrated with the examples of the yeast MAPK Fus3p and Kss1p.

3.1. GFP-Fusion Proteins

Vectors with the different GFP variants are available from different sources; care should be taken to choose a vector with optimal codon usage for the particular organism used. Although all GFP variants (GreenFP, CyanFP, YellowFP, RedFP, etc.) can be used in photobleaching experiments, the procedures are described here are essentially for Enhanced-GFP (EGFP) (*see* **Note 1**). EGFP has two mutations that improve its characteristics. First, a phenylalanine to leucine mutation at position 64 reduces temperature sensitivity *(17)*, whereas mutating the serine at position 65 to a threonine increases fluorescence intensity and protein stability *(18)*. The choice of the promoter to drive expression of the fusion product is dependent on a number of variables; in general it is desirable to choose a promoter, which results in levels close to the levels of the endogenous protein of interest under the circumstances studied if expression from such a promoter is sufficient for visualization (*see* **Note 2**).

The cDNA sequence of interest should be cloned in frame with GFP. The GFP sequence can either be 5 or 3 of the cDNA, and will lead to an amino-terminal or a carboxyl-terminal localization of GFP, respectively (*see* **Notes 3** and **4**).

3.1.1. Vector Construction

Throughout this chapter examples of previously described experiments will be presented, mainly using two MAPK from the budding yeast *Saccharomyces cerevisiae*.

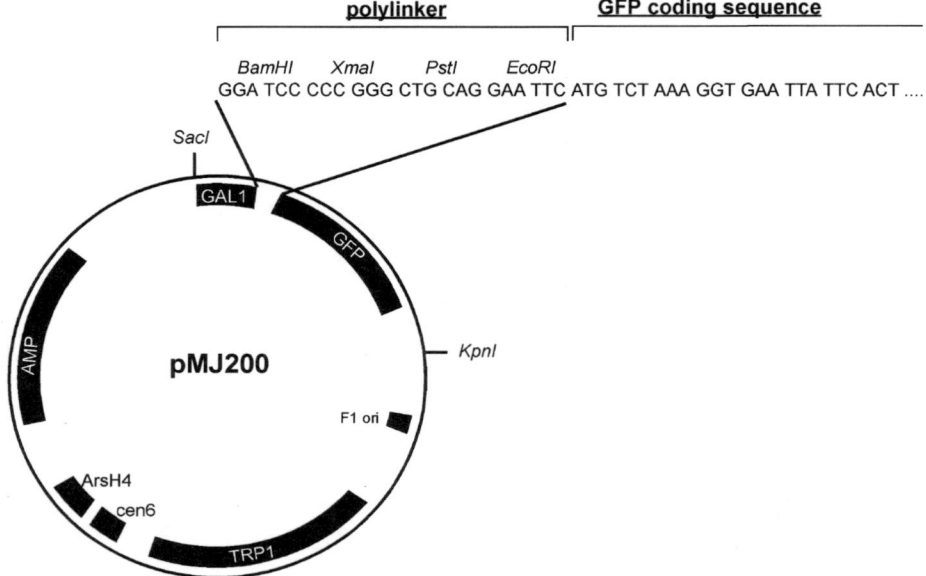

Fig. 1. A schematic drawing of pMJ200 GFP-fusion plasmid *(19)*. Expression of the GFP-fusion protein from this *Saccharomyces cerevisiae* and *Escherichia coli* shuttle plasmid is driven by the inducible *GAL1* promoter. The polylinker facilitates in frame cloning of the sequence of interest, the yeast codon optimized EGFP (*see* **ref. 31**) will be fused to the carboxy-terminus of the protein of interest.

1. The coding sequences of these MAPK, Fus3p and Kss1p (without their STOP codons), were amplified by polymerase chain reaction (PCR) introducing a 5' *Bam*HI restriction site and a 3 *Eco*RI restriction site, the *Eco*RI site was positioned such that there is no frame shift between the MAPK and GFP.
2. The PCR-fragments and vector pMJ200 (*see* **Fig. 1** and **ref. 19**) were digested with *Bam*HI and *Eco*RI and the different fragments isolated from an agarose gel. The two PCR fragments were ligated into the vector.
3. The constructs were transformed into *Escherichia coli* DH5α cells *(16)*. These cells were plated on Luria broth (LB) containing 50 μg/mL ampicillin (amp) and incubated at 37°C for 12 h.
4. Single colonies were selected and grown overnight in LB+amp. The plasmid DNA was then isolated using a standard miniprep method, and presence of the insert was checked using restriction enzymes and sequencing *(16)*.

3.2. Control of Expression and Functionality of the GFP-Fusion Construct

A number of methods can be used to show the expression and functionality of the GFP-fusion construct. Not all the methods described here might be applicable or necessary to confirm expression of the construct of interest.

1. Expression of the GFP construct can easily be detected using a normal fluorescence microscope using an excitation wavelength around 488 nm and an appropriate emission filter to detect fluorescence.
2. Comparing the localization of the GFP-fusion product with the immuno-fluorescence localization of the endogenous protein can rule out the influence of the GFP tag on localization or relative subcellular distribution of the protein.
3. To prevent artifacts caused by cleavage, it is recommended to confirm by western blot that the size of the expressed product is equal to the sum of the sizes of GFP (27 kD) and the protein of interest. α-GFP antibodies are available from different commercial sources; however, most of these antibodies display numerous background bands.
4. Immunoprecipitations can prove the fusion product is still able to interact with essential components or is still incorporated in its physiological complex with a normal stoichiometry.
5. The proof of physiological behavior of the GFP fusion product can be obtained by checking its ability to function similar to the endogenous, untagged protein. In lower eukaryotic systems, this is mostly achieved by determining the ability of the GFP-fusion protein to complement a deletion strain or conditional mutant. This is more difficult in higher eukaryotes, especially if there are no null lines available. An alternative to endogenous complementation is in vitro assays for functionality.

The two proteins used as examples throughout this chapter, Fus3p and Kss1p, are two partially redundant MAPK functioning in the pheromone response pathway in haploid *Saccharomyces cerevisiae* (20,21). A schematic overview of this MAPK cascade is given in **Fig. 2A**. Briefly, binding of pheromone to a seven-transmembrane receptor activates a trimeric G-protein, which leads to the recruitment of the scaffold protein Ste5p as well as the recruitment and activation of a PAK-like kinase (Ste20p) (22) which in turn activates Ste11p (an MEKK) (23). Ste11p activates the MEK Ste7p, which is upstream of the two MAPK.

A yeast strain in which both the *FUS3* and *KSS1* genes have been deleted by homologous recombination is not able to activate the pheromone response pathway, and consequently is not able to mate. Introduction of either Fus3p or

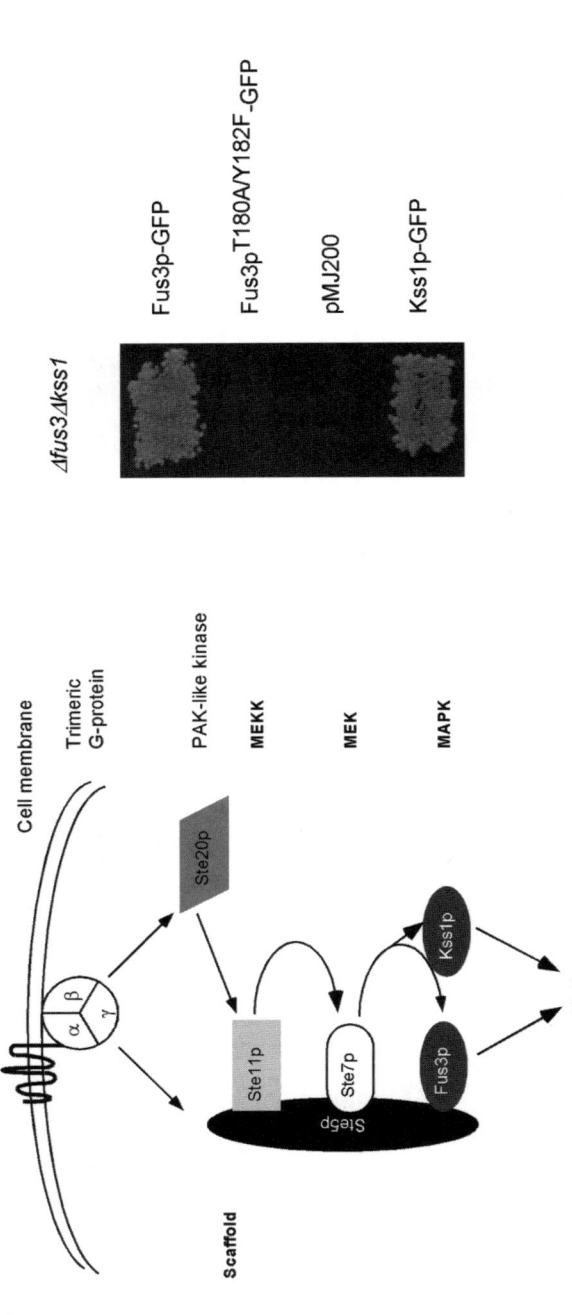

Fig. 2. Yeast pheromone response pathway is functional with GFP-tagged proteins. (**A**) Overview of the *S. cerevisiae* pheromone-response pathway. The budding yeast has two different haploid forms (**a** and α) that are able to conjugate and form diploids (**a**/α). Each haploid form secretes a specific pheromone, which can be recognized by cells of the opposite mating type, inducing the pheromone-response pathway. This MAPK module is activated by peptide pheromones in haploid cells and induces a number of downstream responses preparing cells for mating. (**B**) Mating assay with *Δfus3Δkss1* cells. Haploid cells of the *MAT* **a** mating type deleted for *FUS3* and *KSS1* by homologous recombination bearing the described plasmids were mixed with *MAT* α cells. Both *MAT* **a** cells and *MAT* α cells miss a number of autotrophic markers, in such a fashion that diploids formed by conjugation of both types will be autotrophic for all markers. Hence, when plated on minimal medium, only cells that have successfully mated and formed diploids will be able to grow. *Δfus3Δkss1* cells are not able to mate, unless complemented by expressing Fus3p-GFP or Kss1p-GFP. Cells expressing Fus3p$^{T180A/Y182F}$-GFP, which is a mutant that cannot be activated and is thus not able to activate downstream events, are not able to mate.

Kss1p alone is sufficient to reconstitute the pathway, and cells are no longer sterile. To confirm the functionality of GFP-tagged Fus3p and Kss1p fusion proteins, they were introduced in a *kss1Δfus3Δ* strain and their ability to activate the pheromone response pathway in response to the appropriate stimulus was determined (**Fig. 2B**).

3.3. Optimalization of Signal Detection

At the present time, the only convenient way to perform photobleaching experiments is with the use of a laser-scanning confocal microscope (LCSM). Because the acquisition of images with a confocal microscope is different from that with a standard wide-field microscope, a brief introduction to confocal microscopes is given in this section, followed by several considerations that have to be taken in account when optimizing image acquisition with the LCSM for the live imaging necessary for photobleaching experiments.

3.3.1. Laser-Scanning Confocal Microscopes

A conventional wide field fluorescence microscope illuminates the entire sample for image collection. This facilitates the use of film cameras or CCD cameras to collect the data, or the field can be viewed by eye. Although lenses focus on one horizontal section, signals from above and below this focal plane will be captured during data collection. Although there are (mathematical) means of correcting this, the stray light has a negative influence on image quality. In contrast, in confocal microscopy a beam of laser light is scanned onto the sample. Because the laser beam is focused in one plane, the image produced by scanning the sample represents one optical section *(24,25)*.

LSCMs are generally built onto a conventional light microscope, but a laser is used as a light source and generally sensitive photomultiplier tubes (PMTs) are used for data collection. The key characteristic of confocal microscopy is the use of pinholes to eliminate out-of-focus light *(26)*. As depicted in **Fig. 3**, the beam emitted by the laser passes through a pinhole before being focused onto the sample by an objective lens. The fluorescence emitted by the sample passes through the objective lens and is focused on a second pinhole, which it has to pass through before reaching the PMT. This arrangement with pinholes prevents light emitted above or below the focal plane from being detected by the PMT. The signal detected by the PMT is converted into a gray value for the specific point. In general, mirrors are used to scan the sample in the X and Y planes, whereas either a moveable stage or a moveable lens is responsible for movement in the third dimension.

3.3.2. Image Acquisition

Most modern confocal microscopes are equipped with both a laser source, and a mercury bulb. Although it is possible to look at the slide to find a region

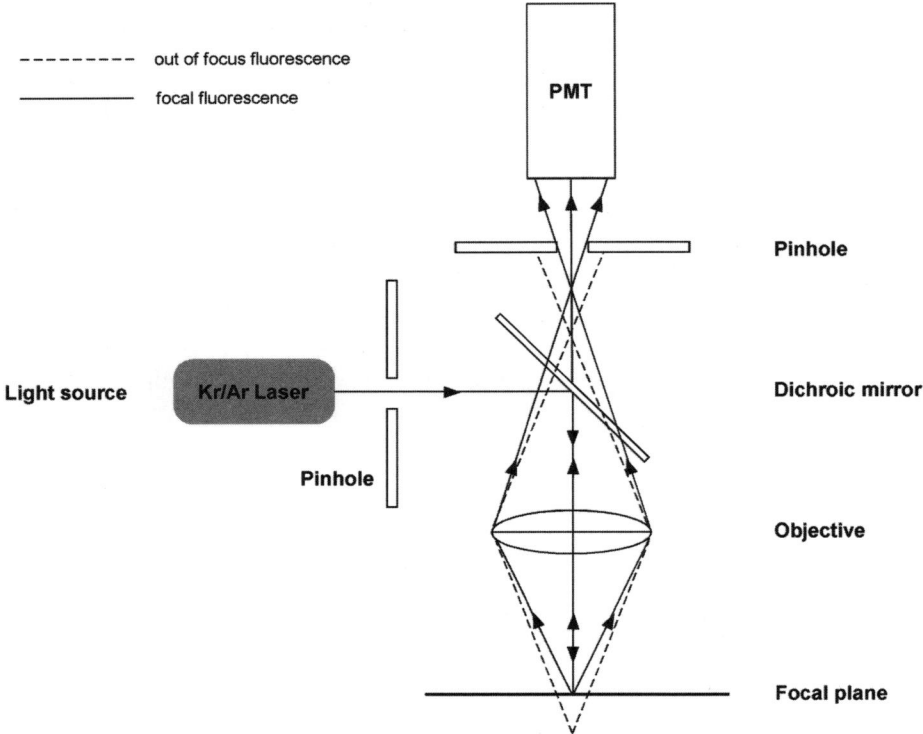

Fig. 3. Overview of the light path in a LSCM. The pinhole prevents fluorescence emitted outsite of the focal plane from reaching the PMT. See text for details.

of interest and focus using the scanning laser, most users find it more convenient to use the eyepieces and conventional illumination to find a region of interest in the sample. However focusing is best achieved using the scanning laser, because minor differences between the optimal focal plane of the eyepieces and the PMTs might occur.

Optical sections are normally scanned more then once, and the signal for each individual point is averaged to improve the signal to noise ratio. Signal intensity is normally represented as an 8- or 16-bit value (256 or 4096 gray values). Although storage of a 16-bit image takes more memory, this is generally recommended for photobleaching experiments, because it represents a greater sensitivity.

Resolution can be adjusted in two different ways on a confocal laser-scanning system. Obviously, the choice of lens (magnification and numerical apperature) influences resolution as in any microscopy system. Another way of improving resolution is through the ability of LSCMs to zoom in on an image without changing the objective lens. The zoom is controlled by the scanning mirrors, which are able to increase or decrease the area that is targeted by the laser. However, an improved resolution directly and linearly correlates with increased photobleaching, whether this improvement is achieved by changing the objective or by zooming in on the sample. A useful advantage of the fact that the movements of the laser are controllable is the fact that users can define one or more regions of interest in the field that can be scanned. This means the time to make one image is reduced, and also protects regions that are not illuminated by the laser from inappropriate bleaching.

Regardless of whether an 8- or a 16-bit mode of gray-tone representation is used, it is recommended to adjust the settings and parameters of the microscopes in such a fashion that images are collected at the full dynamic range. Most LSCMs have a mode that facilitates this by representing areas generating no signal as green pixels and areas saturating the detector as red pixels. While laser intensity is kept as low as feasable, the detector sensitivity, gain, and black levels should be adjusted until there are a few red and a few green pixels in the images, indicating the full dynamic range (4069 gray colors in the 16-bit mode) is used.

3.3.3. Reduction of Intrinsic Photobleaching

Paradoxically, one of the main objectives during photobleaching experiments is to reduce inappropriate bleaching during measurements. In this respect, microscope settings might differ slightly from those used to collect static images. When static images are collected, or even during the collection of a Z-stack, the quality of the final image is important and generally a significant amount of bleaching is tolerated. However, during photobleaching experiments, during a shorter or longer time period, a series of images is collected of the same cells and fluorescence intensities are compared through time. Ideally, the fluorescence intensity should not be influenced by the scans made during these time series. Furthermore, photobleaching experiments are by definition performed on live specimens on which intensive laser light might have adverse effects (mainly owing to heat stress and possibly oxidative damage caused by the excited fluorochromes). Thus, the aim should be to use the least amount of laser power to quickly collect high quality-data.

As stated in **Subheading 3.3.2.**, an optical section is generally scanned more then once, followed by signal averaging to improve the signal to noise ratio. Although a high signal to noise ratio is obviously desirable during photobleaching experiments, reducing the number of scans per image to a maximum of four or less is an effective means to prevent intrinsic photobleaching. Other methods that allow reduction of laser intensity for image acquisition are the increase of detector sensitivity and gain. When bleaching still cannot be sufficiently prevented, or when the signal to noise ratios are too low, it might be advisable to increase expression levels of the GFP-tagged protein, or alternatively fuse more then one GFP tag in tandem to the protein of interest.

3.4. Fluorescence Recovery After Photobleaching

This section is divided into three parts. First, data collection during FRAP experiments is described, including the recommended microscopy settings and procedures. The second part of this section deals with standard data analyses, and, finally, advanced data analysis and a number of possible control experiments are presented.

3.4.1. Data Collection

1. When microscope settings have been optimized for photobleaching experiments, a specimen can be selected for bleaching. To reduce variables as much as possible, try to choose cells that have approximately the same properties (size and orientation) and fluorescence intensity within the experiment. Although the procedures described here are for single cell analysis, FRAP can be performed on more then one cell simultaneously (*see* **Note 5**). An example of a typical FRAP experiment is depicted in **Fig. 4**.
2. To reduce the time needed to perform the scans, and to reduce the size of the file with the collected data, it may be preferable to choose a region of interest around the cell.
3. The subcellular area to be bleached is chosen. Most modern LSCMs have the ability to define a number of regions of interest (ROIs); the region containing the complete cell can be defined as ROI-1 and the one containing the structure that will be bleached as ROI-2 (*see* **Note 6**).
4. Generally, the number of bleach iterations (number of times the laserbeam passes each point in ROI-2) and both the laser output and transmission can be chosen.
5. During image acquisition, the laser output is normally kept as low as possible, both to prevent bleaching and to prolong laser lifetime. Because it has been reported that fast switches in output levels are detrimental to the laser, it is recommended to increase the transmission but not the output. The best solution is to find an optimum between increased transmission and an increased number of iterations (*see* **Note 7**). The fluorescence levels immediately after the bleaching procedure should be between 15% and 30% of the initial fluorescence (**Fig. 4**).
6. Other parameters to be set are the number of initial scans before the bleaching and the time lapse between scans (both before and after bleaching).

a. The initial (prebleach) scans are used to calculate the fluorescence intensity before bleaching, which is arbitrarily set to 100% (**Fig. 4B, D**). Because there are always fluctuations in fluorescence intensity, more than one scan should be made, with the highest value representing 100% fluorescence intensity.

b. The time lapse between scans—that is, between datapoints during the recovery—is dependent mainly on recovery rate and, once again, on intrinsic bleaching. When maximum recovery is reached within a minute, it is optimal to perform one or more scans per second. When recovery is a matter of minutes, larger delays between scans can be chosen (*see* **Note 8**). Ideally the recovery period should be covered by 50–100 datapoints, with the consideration that less points cause less inappropriate bleaching and more points generally gives rise to better quality data.

3.4.2. Standard Data Analysis

1. Generally, the data collected during a FRAP experiment will initially be presented as a stack of images, a subset of an example is shown in **Fig. 4C**. Each point (x,y) represents a gray value (when using a 16-bit mode: 0-4095). The data-analysis programs of most LSCMs (in the case of our example, a Zeiss LSM510) offer the possibility to select areas by drawing an ROI around it, and calculating the average fluorescence intensity (*see* **Note 6**).

2. The values obtained from the procedure described in **step 1** can be shown as a table or a line chart, or they can be exported as a data file to general spreadsheet programs such as Microsoft Excel.

3. In the spreadsheet program, the relative fluorescence values of all points as a percentage can be calculated, relative to the point with the highest absolute fluorescence, and plotted in a chart (**Fig. 4B, D**).

4. To compare recovery rates of different experiments, the time required to reach half the recovery ($\tau_{1/2}$) can be calculated. First, the intensity at $\tau_{1/2}$ is determined according to equation 1:

$$I_{(1/2)} = ((I_{(t)} - I_{(0)}) / 2) + I_{(0)} \qquad (1)$$

where $I_{(1/2)}$ is the intensity at which half the recovery has occured, $I_{(0)}$ represents the fluorescence intensity immediately after bleaching and $I_{(t)}$ is the terminal, or maximum intensity reached after bleaching (**Fig. 4B**). The time between bleaching and the point at which $I_{(1/2)}$ is reached is $\tau_{1/2}$.

Because FRAP experiments are performed on single cells, a relatively high variability between measurements might occur, and it is necessary therefore to carry out a large number of experiments and use statistical analysis to show variation between measurements. Experiments done on

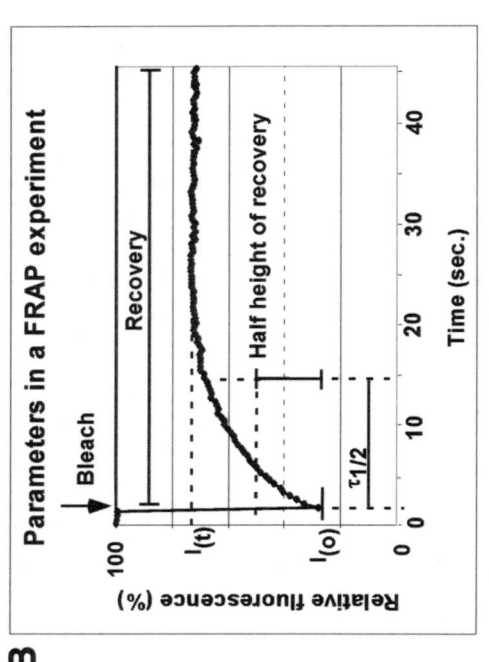

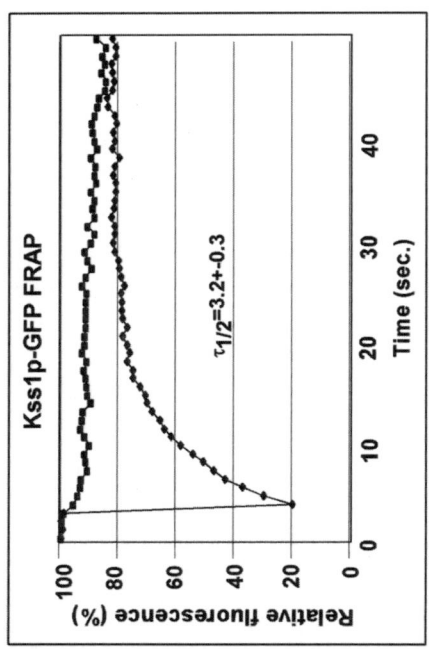

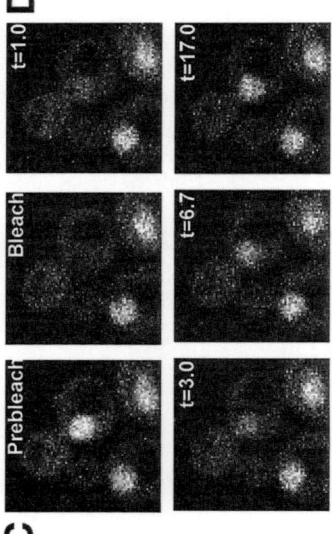

Fus3p-GFP as published in *(9)* and shown in the examples used here were repeated a minimum of 15 times, $\tau_{1/2}$ was calculated for each individual experiment, and the standard deviation was determined.

3.4.3. Advanced Data Analysis

There are a number of "obligatory" control experiments that always should be performed when doing FRAP; furthermore, it is possible to do a number of additional experiments to confirm the data obtained. These experiments are described here.

1. Although in general all possible measures are taken to avoid intrinsic bleaching, fluorescence intensity from a sample that is not bleached, but otherwise has undergone the same treatment as bleached samples, can be shown. Except for the elimination of the bleaching step, all other imaging and data collection procedures obviously should be kept identical. When unavoidable intrinsic bleaching is severe (the final fluorescence intensity being 10–20% lower then the initial intensity), the percentage of bleaching per data point can be calculated for the control, and measured values in the bleached samples can be adjusted accordingly.

2. When recovery time is longer, i.e., several minutes, there is the possibility that new protein synthesis accounts for part of the measured recovery. Inhibition of protein synthesis, for example via addition of cycloheximide, rules this possibility out. However, the time it takes GFP to fold properly, (a prerequisite for fluorescence) is notoriously long (around 30 min to 1 h), and is not inhibited by cycloheximide. Sufficient time should be taken into account between the addition of the protein-synthesis inhibitor and the FRAP experiment.

3. After photobleaching, full recovery is rarely obtained. This is owing to two main reasons. First, there is often an immobile fraction of the GFP-tagged

---

Fig. 4. Nuclear FRAP experiment of Kss1p-GFP. **(A)** Schematic representation of the performed FRAP experiment. The GFP-signal is specifically photobleached in the nucleus (rectangle), and recovery of nuclear fluorescence (circle) is measured in a time-dependent manner. **(B)** Parameters in a standard FRAP experiment. The intensity of nuclear GFP-fluorescence is plotted as a function of time in seconds. The recovery of half-maximum nuclear fluorescence $(\tau_{1/2})$ is determined as indicated. The arrow marks the time of photobleaching. **(C)** The nucleus of a vegetatively growing wild-type cell expressing Kss1p-GFP (prebleach) was photobleached (t = 0; bleach) as depicted in **(A)**. The cell was photographed at the times indicated (in s). **(D)** Recovery of Kss1p-GFP in wild-type cells was measured by nuclear FRAP (diamonds). Nuclear recovery was quantified as described in the text, and shown as $(\tau_{1/2})$ with its standard deviation. As a control, the fluorescence intensity of a nucleus that was not bleached is shown (squares).

protein that is not able to move freely throughout the cell. When the pool of fluorescent molecules in the cell is orders of magnitude larger than the amount of molecules bleached, repetition of the bleaching experiment on the same area should yield a recovery rate of 100% (with $I_{(t)}$ of experiment 2 being equal to $I_{(t)}$ of experiment 1). Secondly, depending on the experiment, the number of molecules bleached can represent a substantial fraction of the total cellular amount. In these cases, full recovery will never be possible. Repetition of bleaching experiments on these cells will give a lower $I_{(t)}$ (compared to $I_{(i)}$ of the first experiment) after each round. Furthermore, when fluorescence intensity of the complete cell is determined, it will show a decrease after the bleaching step. However, in both situations $\tau_{1/2}$ should remain constant when experiments are repeatedly done on the same cell or area.

4. A standard negative control is mildly fixing the cells, or crosslinking the cellular proteins with for example formaldehyde. Because protein movement is prohibited, no recovery should be observed in these cases.
5. Cells are three-dimensional structures, whereas a confocal microscope only illuminates and bleaches a single horizontal plane within a cell. Vertical movement of GFP-tagged proteins in and out of the focal plane is therefore a concern. The use of small, or at least flat cell types can partially alleviate this problem. Collecting a number of z-sections immediately after photobleaching should confirm that the fluorescence intensity of the structure of interest is indeed reduced and that no variations exist within the structure. The collection of z-sections can be repeated a number of times during recovery to make sure no variation in fluorescence intensity occurs.

3.5. Fluorescence Loss in Photobleaching

Whereas FRAP experiments concentrate on recovery of fluorescence in the cellular structure that has been bleached, FLIP experiments look at the relation between pools of fluorescent molecules in different cellular compartments. When there is an exchange between pools, bleaching of pool 1 will eventually lead to a reduction in fluorescence intensity of pool 2 (**Fig. 5A**). Instead of bleaching with one relatively long, intensive laser pulse as applied for FRAP, short pulses are given repeatedly during a FLIP experiment. A number of scans are made to collect data between each of two pulses. First, this will lead to a more gradual reduction of fluorescence intensity at the bleached structure. Second, when there is an exchange of fluorescent molecules with another structure, fluorescence intensity will decrease at this structure as well.

1. During FLIP, an ROI is chosen around one or more cells of interest (ROI-1).
2. Subsequently, the region which is to be bleached is chosen (ROI-2).

A

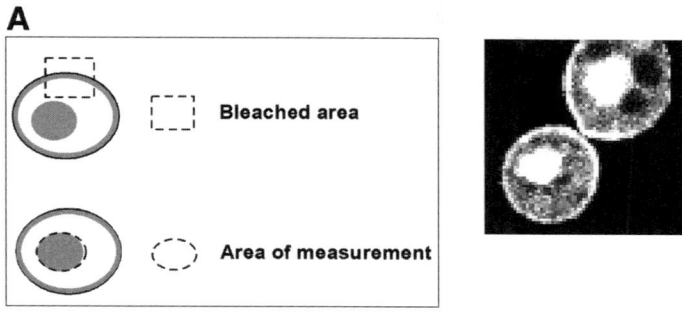

B

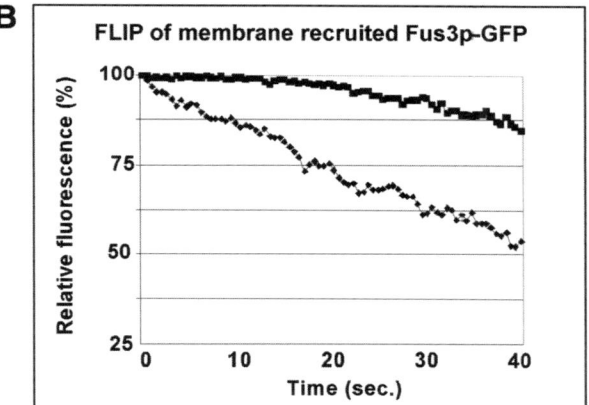

Fig. 5. Nuclear FLIP with membrane recruited Fus3p-GFP. (**A**) Schematic overview of a FLIP experiment. The area that is bleached is shown as a rectangle, the area of measurement is shown as a circle. On the right, an image of Fus3p-GFP that is artificially recruited to the membrane by expressing a CTM-fusion of the scaffold protein Ste5p. (**B**) The decrease of fluorescence intensity in the nucleus, after repeated bleaching Fus3p-GFP at the membrane demonstrates that Fus3p dissociates from Ste5p and translocates to the nucleus (diamonds). As a control, nuclear fluorescence in a cell not undergoing bleaching is shown (squares).

3. A third ROI is chosen at the area where loss of fluorescence will be measured (**Fig. 5A**).
4. A number of scans is made to be able to determine prebleach fluorescence intensities.
5. The bleaching cycle is started. We generally used one short bleach pulse, followed by four scans before the next pulse was given, with a delay of one second between scans or bleach pulses.
6. After data collection, a spreadsheet program is used to calculate relative fluorescence intensities and their decay during time at ROI-3 (**Fig. 5B**).

7. As a control, relative fluorescence intensities at a structure of interest that is not bleached can be determined.

4. Notes

1. Screens done on randomly mutated GFPs, as well as rational introduced mutations in the fluorochrome, have given rise to a number of variants with different excitation and emission spectra. The most common of these are blue fluorescent protein (BFP) (Tyr^{66}→ His, excitation peak 384 nm, emission peak 448 nm) *(27)* cyan fluorescent protein (CFP) (Tyr66→ Trp, excitation peak 434 nm, emission peak 476 nm) *(27)*, and yellow fluorescent protein (YFP) (Ser^{65}→Gly and Thr^{203} → Tyr, excitation peak 514 nm, emission peak 527 nm) *(28)*. Variants fluorescent in the red wavelength range have not been identified, however, red fluorescent proteins (RFPs) have been isolated from other organisms, the best known of which is DsRed (excitation peak 558 nm, emission peak 583 nm) *(29,30)*.

   There are two reasons why EGFP is the fluorochrome of choice. First, owing to the nature of the photobleaching experiments, using the most photostable variant (at present EGFP) will facilitate more accurate measurements. Second, most conventional confocal microscopes will be equipped with an argon laser, with an output wavelength of 488 nm, and filters most suitable for working with GFP. However, under certain circumstances experimental design could require the use of one of the other variants (for example when photobleaching is used as a control during fluorescent resonance energy transfer), which should not cause major problems. In general, the procedures described here can be followed when the change in excitation and emission wave-lengths are taken into account.

2. Although expression levels as close to physiological as possible for the protein of interest are recommended, there are a number of considerations. When expression of the (GFP-fusion) product is toxic, an inducible system should be considered. The main concern, however, is the detectability of the GFP-fusion protein during the FRAP and FLIP experiments. If the product is hard to detect, the laser intensity during measurements will need to be increased, resulting in more intrinsic bleaching and less accurate data collection (*see* also **Subheading 3.3.**); this might lead to the necessity of using a stronger promoter. As an alternative, the protein of interest could be tagged with multiple copies of GFP, increasing the signal accordingly. However, this leads to a significant increase in the size of the protein, which might influence its behavior.

3. Some proteins are not functional when fused to GFP (or any other tag) whether it is an amino-terminal or carboxyl-terminal fusion. Under these

circumstances, it can be considered to place the GFP sequence "internally." When the domain structure of the protein is known or predicted, the GFP sequence can be cloned between different domains that do not need to be proximal in order to function.

4. Expression of fusion products from certain promoters might lead to the transcription of two proteins. Often this is caused by presence of two start sites in the beginning of the sequence, relatively close to the promoter. This could occur when GFP is present 5′ of the protein of interest, or when the protein of interest is relatively small. Deleting or mutating the ATG of the second part of the sequence might prevent the transcription of two different proteins under these circumstances. However, in most cases, deletion of the second start site is not necessary.

5. When recovery times are relatively long compared to the time it takes to scan or bleach the sample, two or more cells within the same ROI can be chosen or more than one ROI can be defined to collect data from more then one cell. Because different bleach ROI also can be defined, different cells in the ROI(s) can be bleached simultaneously. On the other hand, having more then one cell in the field facilitates direct comparison of fluorescence levels in bleached and unbleached cells during data collection (*see* **Subheading 3.4.3.**).

6. When recovery occurs extremely quickly, it is necessary to take a large number of data points in a small amount of time. In the system we used, a Zeiss LSM510, the laserbeam is directed from left to right, and from top to bottom over the sample, and mirrors regulate the on/off status of the laser. We noticed that choosing an irregular ROI-shape (in contrast to a rectangular ROI-shape) increased the time to scan an area significantly. Therefore, in all experiments shown we chose rectangular shapes for all ROIs. Furthermore, we noticed that scanning an area of 500 × 200 pixels took less time than scanning an area of 200 × 500 pixels, apparently because the laser moves from left to right over the X-axis, and is then moved up one pixel on the Y axis and scans again from left to right.

7. As an example: during the experiments with Fus3p our Krypton/Argon laser was consistanly used at 25 mW output. During normal image capture/data collection, it was used at 0.1% transmission with four iterations. For the purpose of bleaching, the transmission was increased to 25% transmission and 20 iterations of the ROI were generally required to reduce fluorescence levels as desired.

8. For samples that reached a maximum recovery within 2 min, but were sensitive to bleaching, we chose a staged time-lapse schedule. This reduces

Table 1 Example of Staged Time-Lapse Schedule

Description	Time	Delay between scans	Number of scans
Initial scans	−4.0 to 1.0 s	1.0 s	4
Bleach	0.0 s	—	—
Recovery phase I	0.0 to 5.0 s	0.5 s	11
Recovery phase II	6.0 to 10.0 s	1.0 s	5
Recovery phase III	10.0 to 30.0 s	2.5 s	8
Recovery phase IV	30.0 to 120.0 s	5.0 s	16

bleaching to a minimum and still ensures sufficient sensitivity, especially during the initial stages of the recovery. An example of such a staged schedule is given in **Table 1**.

Acknowledgments

We wish to thank Malika Jaquenoud for providing plasmid pMJ200. We are grateful to Adrain Smith and Audrey van Drogen for critical reading of the manuscript, to members of the laboratory for discussion, and to the microscopy core facilities at the Swiss Institute for Experimental Cancer Research (ISREC), and The Scripps Research Institute (TSRI) for excellent technical assistance. Matthias Peter was supported by the Swiss National Science Foundation, the Swiss Cancer League, and the ETH Zurich. Frank van Drogen was supported by an EMBO and a HFSP postdoctoral fellowship.

References

1. Morise, H., Shimomura, O., Johnson, F. H., and Winant, J. (1974) Intermolecular energy transfer in the bioluminescent system of Aequorea. *Biochemistry* **13,** 2656–2662.
2. Chalfie, M., Tu, Y., Euskirchen, G., et al. (1994) Green fluorescent protein as a marker for gene expression. *Science* **263,** 802–805.
3. Lippincott-Schwartz, J., Snapp, E., and Kenworthy, A. (2001) Studying protein dynamics in living cells. *Nat. Rev. Mol. Cell Biol.* **2,** 444–456.
4. Lippincott-Schwartz, J. and Patterson, G. H. (2003) Development and use of fluorescent protein markers in living cells. *Science* **300,** 87–91.
5. Stephens, D. J. and Allan, V. J. (2003) Light microscopy techniques for live cell imaging. *Science* **300,** 82–86.
6. Ellenberg, J. and Lippincott-Schwartz, J. (1998) Fluorescence photobleaching techniques, in *Cells: A Laboratory Manual* (Spector, D., Goldman, R., and Leinwand,

L., eds.), Cold Spring Harbor Laboratory Press, Cold Spring Harbor, NY, pp. 79.1–79.23.

7. Phair, R. D. and Misteli, T. (2001) Kinetic modelling approaches to in vivo imaging. *Nat. Rev. Mol. Cell Biol.* **2**, 898–907.

8. Nehls, S., Snapp, E. L., Cole, N. B., et al. (2000) Dynamics and retention of misfolded proteins in native ER membranes. *Nat. Cell Biol.* **2**, 288–295.

9. van Drogen, F., Stucke, V. M., Jorritsma, G., and Peter, M. (2001) MAP kinase dynamics in response to pheromones in budding yeast. *Nat. Cell Biol.* **3**, 1051–1059.

10. van Drogen, F. and Peter, M. (2001) MAP kinase dynamics in yeast. *Biol. Cell* **93**, 63–70.

11. van Drogen, F. and Peter, M. (2002) Spa2p functions as a scaffold-like protein to recruit the Mpk1p MAP kinase module to sites of polarized growth. *Curr. Biol.* **12**, 1698–1703.

12. Leidel, S. and Gonczy, P. (2003) SAS-4 is essential for centrosome duplication in C elegans and is recruited to daughter centrioles once per cell cycle. *Dev. Cell* **4**, 431–439.

13. Stenoien, D. L., Sen, S., Mancini, M. A., and Brinkley, B. R. (2003) Dynamic association of a tumor amplified kinase, Aurora-A, with the centrosome and mitotic spindle. *Cell Motil. Cytoskel.* **55**, 134–46.

14. Carballido-Lopez, R. and Errington, J. (2003) The bacterial cytoskeleton: in vivo dynamics of the actin-like protein Mbl of Bacillus subtilis. *Dev. Cell* **4**, 19–28.

15. Gerlich, D., Beaudouin, J., Kalbfuss, B., et al. (2003) Global chromosome positions are transmitted through mitosis in mammalian cells. *Cell* **112**, 751–764.

16. Sambrook, J., Fritsch, E. F., and Maniatis, T. (1989) *Molecular Cloning: A Laboratory Manual.* Cold Spring Harbor Laboratory Press, Cold Spring Harbor, NY.

17. Cormack, B. P., Valdivia, R. H., and Falkow, S. (1996) FACS-optimized mutants of the green fluorescent protein (GFP). *Gene* **173**, 33–38.

18. Heim, R., Cubitt, A. B., and Tsien, R. Y. (1995) Improved green fluorescence. *Nature* **373**, 663–664.

19. Jaquenoud, M., Gulli, M. P., Peter, K., and Peter, M. (1998) The Cdc42p effector Gic2p is targeted for ubiquitin-dependent degradation by the SCF_{Grr1} complex. *EMBO J.* **17**, 5360–5373.

20. Breitkreutz, A. and Tyers, M. (2002) MAPK signaling specificity: It takes two to tango. *Trends Cell Biol.* **12**, 254–257.

21. Elion, E. A. (2000) Pheromone response, mating and cell biology. *Curr. Opin. Microbiol.* **3**, 573–581.

22. Dohlman, H. G. (2002) G proteins and pheromone signaling. *Annu. Rev. Physiol.* **64**, 129–152.

23. van Drogen, F., O'Rourke, S. M., Stucke, V. M., et al. (2000) Phosphorylation of the MEKK Ste11p by the PAK-like kinase Ste20p is required for MAP kinase signaling in vivo. *Curr. Biol.* **10**, 630–639.

24. Paddock, S. W. (1999) Confocal laser scanning microscopy. *Biotechniques* **27**, 992–996, 998–1002, 1004.

25. Paddock, S. W. (1999) An introduction to confocal imaging. *Methods Mol. Biol.* **122,** 1–34.
26. Pawley, J. B. (1995) *Handbook of Biological Confocal Microscopy.* Plenum Press, New York, NY.
27. Heim, R., Prasher, D. C., and Tsien, R. Y. (1994) Wavelength mutations and post-translational autoxidation of green fluorescent protein. *Proc. Natl. Acad. Sci. USA* **91,** 12,501–12,504.
28. Ormo, M., Cubitt, A. B., Kallio, K., et al. (1996) Crystal structure of the Aequorea victoria green fluorescent protein. *Science* **273,** 1392–1395.
29. Matz, M. V., Fradkov, A. F., Labas, Y. A., et al. (1999) Fluorescent proteins from nonbioluminescent Anthozoa species. *Nat. Biotechnol.* **17,** 969–973.
30. Bevis, B. J. and Glick, B. S. (2002) Rapidly maturing variants of the Discosoma red fluorescent protein (DsRed). *Nat. Biotechnol.* **20,** 83–7.
31. Cormack, B. P., Bertram, G., Egerton, M., et al. (1997) Yeast-enhanced green fluorescent protein (yEGFP)a reporter of gene expression in Candida albicans. *Microbiology* **143 (Pt 2),** 303–311.

21

Assaying Cytochrome c Translocation During Apoptosis

Nigel J. Waterhouse, Rohan Steel, Ruth Kluck, and Joseph A. Trapani

Summary

Translocation of proteins from the mitochondrial intermembrane space to the cytoplasm is a critical event during apoptosis. There are several methods for assaying this event cited in the literature. In this chapter, we highlight separation of cytosolic and mitochondrial fractions of cultured cells using digitonin as the method for measuring cytochrome c release that, in our hands has been the simplest and most reproducible.

Key Words: Cytochrome c; apoptosis; fractionation; outer membrane permeabilization; mitochondria; digitonin.

1. Introduction

It has been known for many years that Bcl-2 family members regulate apoptotic cell death (*1*). It is now known that this is done at least in part by regulating specific permeabilization of the mitochondrial outer membrane (*2–4*). Proapoptotic Bcl-2 family members induce mitochondrial outer membrane permeabilization (MOMP) resulting in the release of several proteins from the mitochondrial intermembrane space (MIS). Anti-apoptotic Bcl-2 family members antagonize the pro-apoptotic family members to block the permeabilization event.

Of the proteins released, cytochrome c, second mitochondrial activator of caspases (SMAC)/Diablo, and HtrA$_2$/Omi have been shown to participate in the activation of caspases; the proteases that orchestrate apoptotic cell death. On

From: *Methods in Molecular Biology, vol. 284:*
Signal Transduction Protocols
Edited by: R. C. Dickson © Humana Press Inc., Totowa, NJ

entering the cytosol, cytochrome c forms a protein complex with apoptotic protease activating factor-1 (APAF-1) and the zymogen form of caspase-9 *(5)*. Formation of this complex, "the apoptosome," brings several pro-caspase-9 molecules within sufficient proximity to allow auto-activation *(6)*. SMAC/Diablo and htRA2/Omi deregulate caspases by displacing endogenous caspase inhibitors, the inhibitor of apoptosis proteins (IAPs) *(7–11)* allowing unfettered caspase activation. MOMP is therefore a critical event in apoptosis and as such the measurement of this event is a component of many studies on cell death.

MOMP can be assayed in several ways. In many instances, measuring loss of the mitochondrial transmembrane potential ($\Delta\Psi$m) provides an indicator of MOMP as the electrochemical potential across the mitochondrial inner membrane is lost as a consequence of cytochrome c release *(12)*. However, this loss of $\Delta\Psi$m is only transient in the absence of caspase activation, such that in cells with caspase-9 or APAF-1 deficiency, measuring loss of $\Delta\Psi$m may not yield an accurate indication of MOMP *(12)*. By far the most common method to detect MOMP is by following the translocation of proteins from the MIS to the cytosol. Although there is debate as to whether all MIS proteins are co-released, cytochrome c appears to be the most abundant and easily detectable of the MIS proteins.

Detection of mitochondrial cytochrome c release is accomplished by Western blot of cellular fractions, by immunocytochemistry or by following cytochrome c fused to green fluorescent protein (GFP) *(13–15)*. Immunocytochemistry gives an accurate representation of the number of cells with translocated cytochrome c, although in suspension cells or in cells with a low cytoplasmic to nucleus ratio, it can be difficult to distinguish between cytosolic and mitochondrial cytochrome c. Measurement of translocated cytochrome c-GFP obviously requires transfection of the fusion protein into cells, which is not always possible. Analysis of cytochrome c release by Western blot of cellular fractions is relatively simple and can be achieved in many primary and transformed populations of cells. This technique therefore remains one of the most widely applicable ways to assay for cytochrome c release. This chapter describes the digitonin method of isolating mitochondrial and cytosolic fractions from cultured cells for analysis of cytochrome c release by Western blot or flow cytometry.

2. Materials

1. Apoptotic stimulus (e.g., granzyme B/Perforin *(16,17)*, heat shock *(18)*, irradiation *(19)*, cytotoxic drugs *(12)*.
2. Plasma membrane permeabilization buffer: 200 µg/mL digitonin, 80 mM KCl in PBS (*see* **Notes 1** and **2**).
3. Total cell lysis buffer: 50 mM Tris-HCl, pH 7.4, 150 mM NaCl, 2 mM EGTA, 2 mM EDTA, 0.2% Triton X-100, 0.3% NP-40, 1X Complete™ Protease Inhibitor (Bohringer Mannheim).

4. Phosphate-buffered saline (PBS): 8 g NaCl, 0.2 g KCl, 1.15 g $Na_2HPO_4 \cdot 7H_2O$, 0.2 g KH_2PO_4/L, pH approx 7.3.
5. Paraformaldehyde (4 % in PBS).
6. Blocking buffer: 3% Bovine serum albumin (BSA), 0.05% saponin in PBS (make fresh).
7. V-bottomed 96-well plate.
8. Anti-cytochrome c (clone 6H2.B4, BD Pharmingen, San Diego, CA).
9. Anti-cytochrome c antibody (clone 7H8.2C12, BD Pharmingen, San Diego, CA).
10. Electrophoresis apparatus for sodium dodecyl sulfate polyacrylamide gel electrophoresis (SDS-PAGE).

3. Method

The methods below outline: (1) efficient isolation of cytosol from mitochondrial fractions, and (2) detection of cytochrome c by Western blot or fluorescence-activated cell sorting (FACS).

3.1. Isolation of Cytosolic and Mitochondrial Fractions

Early studies used physical disruption (homogenization) of the plasma membrane followed by differential centrifugation to separate cytosolic and mitochondrial fractions.

However, using homogenization, it is difficult to disrupt a large percentage of the cells while ensuring that all mitochondria remain intact, generally resulting in relatively few cells being analyzed. More recently, it has been shown that lytic molecules such as digitonin or streptolysin *O* disrupt plasma membrane at lower concentrations than are required to lyse mitochondrial outer membrane. Treatment of cells with plasma membrane-permeabilization buffer (containing digitonin) allows diffusion of cytosolic proteins out of cells. Separation of the cytosol from the mitochondria can then be obtained by a single centrifugation step.

1. Resuspend 1×10^6 cells in 100 µL of ice-cold plasma-membrane-permeabilization buffer and incubate on ice for 5 min (*see* **Notes 3–6**).
2. Centrifuge lysates (800 *g* for 5 min at 4°C).
3. Store the supernatant (cytosolic fraction) at $-70°C$ (for Western blot, *see* **Subheading 3.2.**). *Do not* discard the pellet.
4. Continue to **Subheading 3.2.** for Western blot or **Subheading 3.3.** for FACS.

3.2. Detection of Cytochrome c by Western Blot

With this procedure about 80% of the cytochrome c is released from HeLa cells treated with 1 µ*M* Actinomycin D for 12 h, with ultraviolet (UV) light (80 mJ/m²) and incubated for 6 h, with 230 ng/mL Trail for 2 h or with Jurkat cells treated with 40 μM Etoposide for 8 h, 500 nM Actinomycin D for 8 h or 500 nM Staurosporine for 8 h.

1. Resuspend the pellet from **Subheading 3.1.**, **Step 2** in 100 µL of ice-cold total cell lysis buffer and rock gently at 4°C for 10 min.
2. Centrifuge the lysate (10,000 g for 10 min at 4°C).
3. Store the supernatant (mitochondria/nuclear/membrane fraction) at −70°C.
4. Load pellet fractions (**Subheading 3.2.**, **Step 3**) and cytosolic fractions (**Subheading 3.1.**, **Step 3**), normalized for protein content, on a 12 or 15% SDS-PAGE gel and transfer the proteins to nitrocellulose (*see* **Notes 7–9**).
5. Detect cytochrome c (12.3 kDa) with anticytochrome c clone 7H8.2C12 using standard immunoblotting protocol.

3.3. Detection of Cytochrome c by FACS Analysis

1. Resuspend the pellet from **Subheading 3.1.**, **Step 2** in 100 µL of paraformaldehyde and incubate for 20 min at room temperature.
2. Place the suspension in individual wells of a 96-well plate and wash the pellet three times in PBS (*see* **Note 10**).
3. Incubate the pellet in blocking buffer for 1 h at room temperature.
4. Resuspend the pellet in anticytochrome c clone 6H2.B4 diluted 1:200 in blocking buffer and incubate overnight at 4°C.
5. Wash the pellet three times in blocking buffer.
6. Resuspend the pellet in phycoerythrin (PE) labeled secondary antibody diluted 1:200 in blocking buffer and incubate for 1 h at room temperature.
7. Wash the pellet and analyze by flow cytometry detecting PE fluorescence in the FL-2 channel of a flow cytometer (*see* **Notes 11** and **12**). Cells that have undergone cytochrome c release will have low PE fluorescence.

4. Notes

1. This buffer should be made fresh. We use digitonin from Sigma. We make a 20 mg/mL stock of digitonin by adding PBS just before it is needed and heat the mixture at 95°C until dissolved.
2. At least 80 mM KCl is required in the buffer to allow cytochrome c that has been released from mitochondria to dissociate from membranes. KCl is present in cytoplasm at approx 137 mM.
3. The amount of permeabilization buffer should be titrated for each cell line such that at least 95% of cells are permeabilized. This can be monitored by staining small aliquots of the permeabilized cells with trypan blue.
4. Ensure that the concentration of digitonin is not lysing mitochondria. Untreated, intact cells should exhibit only background cytochrome c release,

e.g., reduced fluorescence by FACS or a faint band in the cytosolic fractions assayed by Western blot.

5. The number of cells assayed can be varied as long as the digitonin concentration is titrated according to **Notes 3** and **4.**

6. In permeabilized cells, Ca^{2+} will access mitochondria and induce permeability transition and swelling. EDTA should therefore be added to the permeabilization buffer if the cells were treated in media containing high $Ca2+$.

7. The abundance of cytochrome c in many cells makes it easy to detect by Western blot, however this also means that small changes may appear significant. It is therefore essential to assay both the cytosolic and mitochondrial fractions.

8. Because Western blotting shows an averaged result from a population of cells, it is not possible to determine whether all the cytochrome c is cytoplasmic in a small percentage of cells or all cells have partially redistributed their cytochrome c. Immunocytochemistry/FACS-based analysis (**Subheading 3.3.**) is recommended to quantitate the percentage of cells having undergone cytochrome c release.

9. Western blotting may not be possible in cells that have very low levels of cytochrome c (e.g., primary T cells *[20]*). In the event of low yields of cytochrome c in the cytosolic fractions, we have successfully concentrated the proteins from this fraction by acetone precipitation. High levels of digitonin, however, can interfere with precipitation.

10. We have found that V-shaped bottoms in 96-well plates facilitates pellet washing without losing samples.

11. Using the immunocytochemistry/FACS-based analysis, cells can be visualized by fluorescence microscopy for confirmation.

12. The FACS-based assay will only be useful in cells that have sufficient levels of cytochrome c such that the fluorescence of stained vs unstained cells is easily distinguishable.

Acknowledgments

NJW is a Peter Doherty Fellow (REG Key 165405), and JAT is a Principal Research Fellow of the National Health and Medical Research Council Australia.

References

1. Vaux, D. L., Cory, S., and Adams, J. M. (1988) Bcl-2 gene promotes haemopoietic cell survival and cooperates with c-myc to immortalize pre-B cells. *Nature* **335,** 440–442.

2. Yang, J., Liu, X., Bhalla, K., et al. (1997) Prevention of apoptosis by Bcl-2: release of cytochrome c from mitochondria blocked. *Science* **275,** 1129–1132.

3. Waterhouse, N. J., Ricci, J. E., and Green, D. R . (2002) And all of a sudden it's over: mitochondrial outer-membrane permeabilization in apoptosis. *Biochimie* **84,** 113–121.

4. Kluck, R. M., Bossy-Wetzel, E., Green, D. R., and Newmeyer, D. D. (1997) The release of cytochrome c from mitochondria: a primary site for Bcl-2 regulation of apoptosis. *Science* **275,** 1132–1136.

5. Zou, H., Li, Y., Liu, X., and Wang, X. (1999) An APAF-1.cytochrome c multimeric complex is a functional apoptosome that activates procaspase-9. *J. Biol. Chem.* **274,** 11,549–11,556.

6. Salvesen, G. S. and Dixit, V. M. (1999) Caspase activation: the induced-proximity model. *Proc. Natl. Acad. Sci. USA* **96,** 10,964–10,967.

7. Verhagen, A. M., Silke, J., Ekert, P. G., et al. (2002) HtrA2 promotes cell death through its serine protease activity and its ability to antagonize inhibitor of apoptosis proteins. *J. Biol. Chem.* **277,** 445–454.

8. Verhagen, A. M., Ekert, P. G., Pakusch, M., et al. (2000) Identification of DIABLO, a mammalian protein that promotes apoptosis by binding to and antagonizing IAP proteins. *Cell* **102,** 43–53.

9. van Loo, G., van Gurp, M., Depuydt, B., et al. (2002) The serine protease Omi/HtrA2 is released from mitochondria during apoptosis. Omi interacts with caspase-inhibitor XIAP and induces enhanced caspase activity. *Cell Death Differ.* **9,** 20–26.

10. Suzuki, Y., Imai, Y., Nakayama, H., et al. (2001) A serine protease, HtrA2, is released from the mitochondria and interacts with XIAP, inducing cell death. *Mol. Cell* **8,** 613–621.

11. Du, C., Fang, M., Li, Y., Li, L., and Wang, X. (2000) Smac, a mitochondrial protein that promotes cytochrome c-dependent caspase activation by eliminating IAP inhibition. *Cell* **102,** 33–42.

12. Waterhouse, N. J., Goldstein, J. C., von Ahsen, O., et al. (2001) Cytochrome c maintains mitochondrial transmembrane potential and ATP generation after outer mitochondrial membrane permeabilization during the apoptotic process. *J. Cell Biol.* **153,** 319–328.

13. Goldstein, J. C., Waterhouse, N. J., Juin, P., et al. (2000) The coordinate release of cytochrome c during apoptosis is rapid, complete and kinetically invariant. *Nat. Cell Biol.* **2,** 156–162.

14. Waterhouse, N. J., Goldstein, J. C., Kluck, R. M., et al. (2001) The (Holey) study of mitochondria in apoptosis. *Methods Cell Biol.* **66,** 365–391.

15. Finucane, D. M., Bossy-Wetzel, E., Waterhouse, N. J., et al. (1999) Bax-induced caspase activation and apoptosis via cytochrome c release from mitochondria is inhibitable by Bcl-xL. *J. Biol. Chem.* **274,** 2225–2233.

16. Sutton, V. R., Davis, J. E., Cancilla, M., et al. (2000) Initiation of apoptosis by granzyme B requires direct cleavage of bid, but not direct granzyme B-mediated caspase activation. *J. Exp. Med.* **192,** 1403–1414.

17. Pinkoski, M. J., Waterhouse, N. J., Heibein, J. A., et al. (2001) Granzyme B-mediated apoptosis proceeds predominantly through a Bcl-2-inhibitable mitochondrial pathway. *J. Biol. Chem.* **276,** 12,060–12,067.

18. Buzzard, K. A., Giaccia, A. J., Killender, M., and Anderson, R. L. (1998) Heat shock protein 72 modulates pathways of stress-induced apoptosis. *J. Biol. Chem.* **273,** 17,147–17,153.

19. Waterhouse, N., Kumar, S., Song, Q., et al. (1996) Heteronuclear ribonucleoproteins C1 and C2, components of the spliceosome, are specific targets of interleukin 1beta-converting enzyme-like proteases in apoptosis. *J. Biol. Chem.* **271,** 29,335–29,341.

20. Murphy, B. M., O'Neill, A. J., Adrain, C., et al. (2003) The apoptosome pathway to caspase activation in primary human neutrophils exhibits dramatically reduced requirements for cytochrome c. *J. Exp. Med.* **197,** 625–632.

Index